Alexander A. Boichuk, Anatolii M. Samoilenko
Generalized Inverse Operators

Inverse and Ill-Posed Problems Series

Edited by
Sergey I. Kabanikhin, Novosibirsk,
Russia; Almaty, Kazakhstan

Volume 59

Alexander A. Boichuk, Anatolii M. Samoilenko

Generalized Inverse Operators

And Fredholm Boundary-Value Problems

2nd Edition

DE GRUYTER

Mathematics Subject Classification 2010
Primary: 47A53, 15A09, 34B15; Secondary: 34A37, 34B40, 34K26.

Authors
Prof. Dr. Alexander Andreevych Boichuk
National Academy of Sciences of Ukraine
Institute of Mathematics
Tereshchenkovskaya st.,3
KIEV-4
01601
UKRAINE
boichuk.aa@gmail.com

Prof. Dr. Anatoly Samoilenko
National Academy of Sciences of Ukraine
Institute of Mathematics
Tereshchenkovskaya st.,3
KIEV-4
01601
UKRAINE
sam@imath.kiev.ua

Translated from the Russian by Dr. Peter V. Malyshev
e-mail: malysh@voliacable.com

ISBN 978-3-11-037839-9
e-ISBN (PDF) 978-3-11-037844-3
e-ISBN (EPUB) 978-3-11-038735-3
Set-ISBN 978-3-11-037845-0
ISSN 1381-4524

Library of Congress Cataloging-in-Publication Data
A CIP catalog record for this book has been applied for at the Library of Congress.

Bibliographic information published by the Deutsche Nationalbibliothek
The Deutsche Nationalbibliothek lists this publication in the Deutsche Nationalbibliografie; detailed bibliographic data are available on the Internet at http://dnb.dnb.de.

Printing and binding CPI books GmbH, Leck
♾ Printed on acid-free paper
Printed in Germany

MIX
Papier aus verantwortungsvollen Quellen
FSC
www.fsc.org
FSC® C083411

www.degruyter.com

Contents

PREFACE

The problems of development of constructive methods for the analysis of linear and weakly nonlinear boundary-value problems for a broad class of functional differential equations, including systems of ordinary differential and difference equations, systems of differential equations with delay, systems with pulse action, and integro-differential systems, traditionally occupy one of the central places in the qualitative theory of differential equations [5, 51, 94, 104, 118, 148]. This is explained, first of all, by the practical significance of the theory of boundary-value problems for various applications — theory of nonlinear oscillations [2, 7, 14, 15, 70, 89, 97, 101, 139], theory of stability of motion [50, 51, 97, 98, 108], control theory [128, 158], and numerous problems in radioengineering, mechanics, biology, etc. [81, 89, 103, 121, 152].

As a specific feature of the analyzed boundary-value problems, we can mention the fact that their linear part is, in most cases, an operator without inverse. This fact makes it impossible to use traditional methods based on the fixed-point principle for the investigation of boundary-value problems of this kind. The uninvertibility of the linear part of the operator is a consequence of the fact that the number m of boundary conditions does not coincide with the number n of unknown variables in the operator system. It is said that problems of this kind for systems of functional differential equations are problems of Fredholm type (or with Fredholm linear parts). They include extremely complicated and insufficiently studied (both underdetermined and overdetermined) critical and noncritical boundary value-problems. The applicability of the well-known Schmidt lemma [148] to the investigation of boundary-value problems regarded as operator equations with bounded operators in the linear part with an aim to construct a generalized inverse operator resolving the original boundary-value problem is restricted by the requirement that the corresponding boundary-value problem must be of Fredholm type with index zero, i.e., that $m = n$. Therefore, a major part of the works dealing with problems of this sort were carried out under the assumption that these problems are of Fredholm type with index zero (Azbelev, Maksimov, and Rakhmatullina [8], Vejvoda [149], Wexler [153], Grebenikov, Lika, and Ryabov [65, 93], Malkin [101], Mitropol'skii and Martynyuk [107], Samoilenko, Perestyuk, and Ronto [139, 140]). Moreover, a significant part of these results was, in fact, obtained under the assumption that the operator in the linear part of the original boundary-value problem has the inverse operator (noncritical case). We do not use this assumption.

It is known (Atkinson [6], Vainberg and Trenogin [148], Pyt'yev [123], Turbin [147], and Nashed [112]) that the classical Schmidt procedure [141] is applicable to the construction of generalized inverse operators only in the case of Fredholm operators of index zero. Thus, for boundary-value problems regarded as operator systems in abstract spaces [153], we suggest some methods for the construction of the generalized inverse (or pseudo-inverse) operators for the original linear Fredholm operators in Banach (or Hilbert) spaces. As a result of systematic application and development of the theory of generalized inverse operators [112, 123] and matrices [91, 109, 117, 147], new criteria of solvability were obtained and the structure of solutions was determined for linear Fredholm boundary-value problems for various classes of systems of functional differential operators. The methods used for the construction of the generalized Green's operators (and generalized matrices playing the role of kernels of their integral representations) for semihomogeneous boundary-value problems for systems of this sort are presented from the common viewpoint. We also study basic properties of the generalized Green's operator. In particular, it is shown how, using this operator, one can construct the generalized inverse of the operator of the original boundary-value problem.

New efficient methods of perturbation theory were developed in analyzing weakly nonlinear boundary-value problems. These methods, including the Lyapunov–Poincaré method of small parameter [97, 120], asymptotic methods of nonlinear mechanics developed by Krylov, Bogolyubov, Mitropol'skii, and Samoilenko [14, 15, 90], some methods proposed by Tikhonov [144, 145] and the Vishik–Lyusternik method [150], are extensively used for the solution of various problems encountered in different fields of science and engineering, such as radioengineering [101, 121], shipbuilding [89], celestial mechanics [65, 81], biology [103, 152], etc. These methods were developed and used in numerous works (Vainberg and Trenogin [148], Vejvoda [149], Grebenikov and Ryabov [65], Kato [80], Malkin [101], Mishchenko and Rozov [106], and Hayashi [74]). However, the application of the methods of perturbation theory to the analysis of weakly nonlinear boundary-value problems for various classes of differential systems was, for the most part, restricted to the case of ordinary periodic boundary-value problems in the theory of nonlinear oscillations (Grebenikov and Ryabov [65], Hale [70], Malkin [101], Proskuryakov [122], and Yakubovich and Starzhinskii [154] for systems of ordinary differential equations, Mitropol'skii and Martynyuk [107] and Shimanov [143] for systems with delay, and Samoilenko and Perestyuk [139] and Bainov and Simeonov [10] for systems with pulse action).

We show that the principal results in the theory of weakly nonlinear periodic oscillations remain valid (with necessary refinements, changes, and supplements) for general weakly perturbed (with Fredholm-type linear parts) boundary-value problems for systems of functional differential equations.

The boundary-value problems are specified by linear or weakly nonlinear vector functionals such that the number of their components does not coincide with the dimension of the operator system. In our monograph, we develop a general theory of boundary-value problems of this kind, give a natural classification of critical and noncritical cases,[1] establish efficient conditions for the coefficients guaranteeing the existence of solutions, and develop iterative algorithms for the construction of solutions of these problems. Numerous results presented in the monograph were originally obtained and approved in analyzing boundary-value problems for systems of ordinary differential equations (Boichuk [19]). Later, it was discovered that the proposed procedures of investigation and algorithms are applicable to the analysis of much more general objects, including boundary-value problems for ordinary systems with lumped delay [32, 37, 157], systems with pulse action [29, 30, 135], autonomous differential systems [25, 26, 36], and operator equations in functional spaces whose linear part is a normally resolvable operator but they are not everywhere solvable [33–35].

In the first chapters, to make our presentation more general, we give some results from the theory of generalized inversion of bounded linear operators in abstract spaces, which are then used for the investigation of boundary-value problems for systems of functional differential equations. Some of these results are of independent interest for the theory of linear operators, although our main aim was just to develop the tools required for the analysis of boundary-value problems for systems of functional differential equations. The methods used for the construction of generalized inverse operators in Banach and Hilbert spaces are presented separately because these spaces are characterized by absolutely different geometries. The construction of the generalized inverse operator for a linear Fredholm operator acting in Banach spaces is based on the Atkinson theorem [6] obtained as a generalization of the Nikol'skii theorem [113], which states that any bounded Fredholm operator can be represented in the form of a unilaterally invertible and completely continuous (finite-dimensional) operator. By using this fact, we arrive at the construction of the generalized inverse of a Fredholm operator similar to the well-known Schmidt procedure [148] applicable only in the case of generalized inversion of Fredholm operators of index zero in Banach spaces.

The theory of generalized inversion and pseudoinversion of linear Fredholm operators in Banach and Hilbert spaces enabled us to develop a unified procedure for the investigation of Fredholm boundary-value problems for operator equations solvable either everywhere or not everywhere (Chapter 4).

The proposed approach is then improved for the analysis of boundary-value problems for standard operator systems, including systems of ordinary differential equations and equations with delay (Chapter 5) and systems with

[1] In the literature, these cases are sometimes called resonance and nonresonance.

pulse action (Chapter 6). We obtain necessary and sufficient conditions for the existence of solutions of linear and nonlinear differential and difference systems bounded on the entire axis (Chapter 7).

This enables us to take into account specific features of each analyzed differential system and present necessary examples. The readers interested primarily in the theory of boundary-value problems for specific differential systems may focus their attention on the corresponding chapters and omit the chapters containing preliminary information.

The authors do not even try to present the complete bibliography on the subject, which is quite extensive, and mention only the works required for the completeness of presentation.

In conclusion, the authors wish to express their deep gratitude to all participants of numerous seminars and conferences on the theory of differential equations and nonlinear oscillations, where all principal results included in the book were reported and discussed.

The authors,
Kiev, May 2004

Preface to the second edition

For more than 11 years that have passed since the appearance of the first edition, numerous new publications of the authors have appeared. Thus, it becomes necessary to supplement the book with the new accumulated material and introduce necessary corrections to the previous edition, which remains to be of considerable interest for the researchers working in this field, quite actual, and well-cited.

We give more correct definitions of differences in the classification of resonance and nonresonance boundary-value problems in the investigation of general Fredholm problems (of nonzero index) and special (periodic) cases of these problems. The classical definition of the equation for generating amplitudes that gives necessary condition for the branching of solutions of nonlinear periodic problems is generalized to the case of general Fredholm boundary-value problems. Thus, the well-known results obtained in the works by A. M. Lyapunov, I. G. Malkin, Yu. A. Ryabov, V. A. Yakubovich, and V. M. Starzhinskii are generalized.

We add new sections dealing the theory of differential-algebraic systems with singular matrix of derivatives, impulsive, and boundary-value problems for these systems of equations extensively studied for the last ten years. The results presented in these sections illustrate the possibility of application of the methods of investigations proposed in our monograph to a broad class of operator equations and represent the latest actual achievements of the authors in this direction. We also add new results obtained for delay differential systems. In our opinion, they illustrate the previous sections of the monograph devoted to the boundary-value problems for these systems and show the efficiency of the proposed algorithm in the investigation of delay systems.

New references are added, some necessary corrections are made, and misprints found in the first edition are corrected. We hope that the second supplemented and revised edition of the monograph will be more interesting for the readers.

The authors,
Kiev, May 2016

NOTATION

B	Banach space of vector functions $z\colon [a,b] \to R^n$
$C[a,b]$	space of vector functions $x\colon [a,b] \to R^n$ continuous on an interval $[a,b]$
$C^1[a,b]$	space of vector functions $x\colon [a,b] \to R^n$ continuously differentiable on an interval $[a,b]$
$D_p^n[a,b]$	Banach space of vector functions $z\colon [a,b] \to R^n$ absolutely continuous on an interval $[a,b]$ with a norm $\|\|z\|\|_{D_p^n} = \|\|\dot{z}\|\|_{L_p^n} + \|\|z(a)\|\|_{R^n}$
$\dim \operatorname{im} L$	dimension of the image of an operator L
$\dim \ker L$	dimension of the kernel of an operator L
$(G\cdot)(t)$	generalized Green operator for a semihomogeneous boundary-value problem
$G(t,\tau)$	generalized Green matrix
H	Hilbert space of vector functions $z\colon [a,b] \to R^n$
$\operatorname{im} L\ (R(L))$	image of a linear operator L
$\ker L\ (N(L))$	kernel (null space) of a linear operator L
$K(t,\tau)$	Cauchy matrix
l	linear vector functional
$lX(\cdot)$	$m \times n$ constant matrix that is the result of the action of an m-dimensional linear vector functional l on the columns of an $n \times n$ matrix $X(t)$
L^-, L^+	operator generalized inverse or pseudoinverse to an operator L
$L^+_{l(r)}$	left (right) pseudoinverse of an operator L
L^{-1}	inverse of an operator L
$L^{-1}_{l(r)}$	left (right) inverse of an operator L
$L_p^n[a,b]$	Banach space of vector functions $z\colon [a,b] \to R^n$ integrable to the pth power $(1 < p < \infty)$ equipped with a norm $\|\|z\|\|_{L_p^n} = \left(\int\limits_a^b \sum\limits_{i=1}^n \|z_i(t)\|^p \, dt \right)^{1/p}$

$\mathcal{L}(B_1, B_2)$	space of bounded linear operators L acting from a Banach space B_1 into a Banach space B_2
$\mathcal{P}_L$	linear operator (projector) projecting a Banach space B onto the null space of an operator L
$\mathcal{P}_{Y_L}$	linear operator (projector) projecting a Banach space B onto a subspace $Y \subset B$ isomorphic to the null space $N(L^*)$ of the operator adjoint to an operator L
$P_L = P_{N(L)}$	linear operator (orthoprojector) projecting a Hilbert space H onto the null space of an operator L
R^n	Euclidean vector space of constant vectors
$\operatorname{sign} t$	sign function
δ_{ij}	Kronecker symbol
$\chi_{[a,b]}(t)$	characteristic function of an interval $[a, b]$

Chapter 1

PRELIMINARY INFORMATION

In Chapter 1, we present some well-known definitions and results from functional analysis, the theory of linear operators in Hilbert and Banach spaces, and matrix theory, required for our subsequent presentation. The readers who are not familiar with the theory of linear operators in function spaces can find here an elementary presentation of the facts essentially used in what follows. The other readers can use this material for references. The theorems are presented without proofs, but the reader is referred to the sources for further details. In this chapter, we also introduce necessary notation.

1.1 Metric and Normed Spaces

Definition 1.1. A metric space is defined as a set X equipped with a metric $\rho(\cdot,\cdot)$, i.e., a real function defined in the set X and such that

(1) $\rho(x, y) \geq 0 \quad (\rho(x, y) = 0$ iff $x = y)$;

(2) $\rho(x, y) = \rho(y, x)$;

(3) $\rho(x, z) \leq \rho(x, y) + \rho(y, z)$ (triangle inequality).

Thus, an arbitrary set equipped with a metric is a metric space.

Example 1. A set X whose points are collections of n-dimensional real vectors $x = (x_1, \dots, x_n)$ turns into a metric space if we set

$$\rho(x, y) = \left[\sum_{i=1}^{n} (x_i - y_i)^2\right]^{1/2}.$$

The same set X can be also equipped with other metrics, e.g.,

$$\rho_1(x, y) = \max_{1 \leq i \leq n} |x_i - y_i|,$$

$$\rho_2(x, y) = \sum_{i=1}^{n} |x_i - y_i|.$$

As a result, it turns into different metric spaces.

Example 2. Let Y be the set of continuous functions defined on a segment $[a, b]$. If we introduce a metric by setting

$$\rho(x, y) = \max_{a \le t \le b} |x(t) - y(t)| \quad \text{for} \quad x, y \in Y.$$

In this case, Y turns into the well-known metric space $C[a, b]$. The set of continuous functions can be transformed into other metric spaces by introducing different functions, e.g., as follows:

$$\rho_1(x, y) = \left[\int_a^b |x(t) - y(t)|^p \, dt \right]^{1/p}, \quad p > 1.$$

The set Z of n $(n \ge 1)$ times continuously differentiable functions defined on the segment $[a, b]$ turns into a metric space $C^m[a, b]$ if we use the following metric:

$$\rho(x, y) = \max_{0 \le i \le n} \max_{a \le t \le b} |x^{(i)}(t) - y^{(i)}(t)|,$$

$$x^{(0)}(t) \equiv x(t), \qquad y^{(0)}(t) \equiv y(t), \quad \forall x, y \in Z.$$

Example 3. Consider a set whose points are ordered systems of real numbers $x = (x_1, x_2, \ldots, x_n, \ldots)$ and $y = (y_1, y_2, \ldots, y_n, \ldots)$ such that

$$\sum_{i=1}^{\infty} |x_i|^p < \infty \quad \text{and} \quad \sum_{i=1}^{\infty} |y_i|^p < \infty, \qquad p \ge 1.$$

If the distance is introduced according to the formula

$$\rho(x, y) = \left[\sum_{i=1}^{\infty} |x_i - y_i|^p \right]^{1/p},$$

then we get a metric space denoted by l_p, $p \ge 1$.

Definition 1.2. A sequence $\{x_n\}_{n=1}^{\infty}$ of elements of a metric space X is called convergent to an element $x \in X$ if $\rho(x, x_n) \to 0$ as $n \to \infty$. The element x is called the limit point of the set X.

Definition 1.3. A set $M \subset X$ is called closed if it contains all its limit points. The empty set is always regarded as closed.

Definition 1.4. Let $M \subset X$ and let M' be a set of limit points of M. The set $\overline{M} = M \cup M'$ is called the closure of the set M.

Definition 1.5. A sequence $\{x_n\}_{n=1}^{\infty}$ of elements of a metric space X is called a fundamental (Cauchy) sequence if, for any real $\varepsilon > 0$, there exists N such that $\rho(x_n, x_m) < \varepsilon$ whenever $n, m > N$.

In any convergent sequence, one can select a fundamental subsequence. The example presented below shows that the converse statement is not true.

Example 4. Let Q be a set of rational numbers with metric $\rho(x, y) = |x - y|$ and let x_0 be an irrational number, i.e., $x_0 \in R \setminus Q$, where R is the set of real numbers. We construct a sequence of rational numbers x_n that converges to x_0 in R. Then x_n is a Cauchy sequence in Q but it does not converge in Q to any rational number y (indeed, if $x_n \to y$ in Q, then $x_n \to y$ also in R, and we set $y = x_0$).

Definition 1.6. A metric space in which any fundamental sequence is convergent is called complete.

Thus, the metric space R is complete but the set Q is not a complete space.

In order to transform an incomplete metric space X into a complete space, it is necessary to expand it by adding the limits of all possible fundamental sequences. The original space X is, in this case, dense in the enveloping space $\overline{X}$ in the following sense:

Definition 1.7. A set M is called dense in a metric space X if any element $x \in X$ is the limit of a sequence of elements from M.

As follows from the definition of the operation of closure of a set (Definition 1.4), the statement that a set M is dense in X, $M \subset X$, $M \neq X$, means that $\overline{M} = X$. In other words, the closure of the set M coincides with X.

Definition 1.8. A normed linear space is defined as a vector space X over the field R equipped with the norm $\|\cdot\|$, i.e., a nonnegative function that maps X into R and satisfies the following conditions:

(1) $\|x\| \geq 0$ for all $x \in X$ and $\|x\| = 0$ iff $x = 0$;

(2) $\|\alpha x\| = |\alpha| \, \|x\|$ for all $x \in X$ and $\alpha \in R$;

(3) $\|x + y\| \leq \|x\| + \|y\|$ for all $x, y \in X$.

By using a norm, one can easily introduce a metric in a normed space by setting $\rho(x, y) = \|x - y\|$, and, hence every normed space is a metric space. It is often possible to introduce more than one norm in the same vector space. The normed spaces obtained as a result are regarded as different.

If a normed space X is equipped with two norms $\|\cdot\|_1$ and $\|\cdot\|_2$, then any sequence convergent in the norm $\|\cdot\|_1$ converges in the norm $\|\cdot\|_2$ if and only if there exists a positive constant C_1 such that

$$\|x\|_2 \leq C_1 \|x\|_1 \quad \forall x \in X.$$

Two norms $\|\cdot\|_1$ and $\|\cdot\|_2$ in a linear space X are called topologically equivalent if convergence in one of these norms implies convergence in the second norm, and vice versa.

The norms are topologically equivalent if and only if there exist positive constants C_1 and C_2 such that

$$C_2 \le \frac{\|x\|_2}{\|x\|_1} \le C_1 \qquad \forall x \in X, \quad x \neq 0.$$

Example 5. The set of all n-dimensional vectors x with operations of addition and multiplication by numbers is called the n-dimensional vector space R^n. The number $|x| = \sqrt{\sum_{j=1}^{n} |x_j|^2}$ is called the length (modulus) of a vector x. The vector space R^n turns into a normed space if we introduce a norm in this space by the formula

$$\|x\| = \sqrt{\sum_{j=1}^{n} |x_j|^2},$$

The space R^n with this norm is called the n-dimensional Euclidean space.

Since the length $|x|$ of a vector x coincides with its norm in the Euclidean space R^n, we say that the Euclidean norm of a vector is compatible with its length.

Example 6. $C[a,b]$ is a normed space of functions continuous on $[a,b]$ with the norm

$$\|x\| = \max_{a \le t \le b} |x(t)|.$$

Example 7. L_p is a normed space of functions integrable to the pth power $(1 < p < +\infty)$ with the following norm:

$$\|f\| = \left(\int_a^b |f(x)|^p \, dx \right)^{1/p}.$$

This space is formed by measurable n-dimensional vector functions f defined on a measurable set $[a,b]$ and such that $|f|^p$ is integrable.

The equality $\|f\| = 0$ holds if and only if $f(x) = 0$ almost everywhere in $[a,b]$.

Example 8. D_p is a normed space of absolutely continuous vector functions $f(x)$ defined on a measurable set $[a,b]$ and such that $\dot{f}(x) \in L_p$. The norm in this space is introduced as follows:

$$\|f\|_{D_p} = \|\dot{f}\|_{L_p} + |f(a)|_{R^n}.$$

Example 9. l_p is a normed space with the following norm:

$$\|x\| = \left(\sum_{i=1}^{\infty} |\, x_i \,|^p\right)^{1/p}, \quad 1 < p < +\infty.$$

Since any norm specifies a metric in a normed space, the notions of convergence, continuity, completeness, etc., connected with the distance are naturally defined in this space.

Definition 1.9. A mapping L of a linear space X into a linear space Y, $L\colon X \to Y$, is called a linear operator if the following axioms are satisfied:

(1) $L(x + y) = Lx + Ly$ for any x and y in X;

(2) $L(\alpha x) = \alpha Lx$ for any $x \in X$ and $\alpha \in R$ or C.

Definition 1.10. A linear operator L is called *bounded* if there exists a number $C \geq 0$ such that, for any $x \in X$,

$$\|Lx\| \leq C\,\|x\|.$$

The smallest of these numbers is denoted by $\|L\|$ and is called the norm of the operator L.

Definition 1.11. A linear operator $l_i\colon X \to R$ is called a linear functional and an operator $l = [l_1, \ldots, l_m]\colon X \to R^m$ defined by the equality $lx = \operatorname{col}\{l_1 x, \ldots, l_m x\}$ for all $x \in X$ is called a linear vector functional.

Thus, a linear functional maps a linear space into a field of constant elements, and a vector functional maps a linear space into a field of vectors with constant components.

Definition 1.12. The set of all linear bounded operators mapping a linear space X into a space Y equipped with properly introduced operations of addition and multiplication by constant numbers is a linear space called the *space of linear operators* and denoted by $\mathcal{L}(X, Y)$.

1.2 Hilbert Spaces

Hilbert spaces are fairly simple infinite-dimensional normed spaces. Their relative simplicity is explained by the fact that they possess an additional structure created by the operation of inner product (a generalization of the ordinary inner product in vector algebras). The presence of an additional algebraic structure substantially enriches the geometric properties of Hilbert spaces. It is worth noting that the analyzed structure allows one to introduce the notion of perpendicular vectors, which makes the geometry of Hilbert spaces much closer to the geometry of Euclidean spaces.

Definition 1.13. A Hilbert space is defined as a set H of elements $f, g, h, \ldots$ with the following properties:

(1) H is a linear space, i.e., the operations of addition and multiplication by real or complex numbers are defined in H (depending on the field of constants, the space H is either real or complex).

(2) H is equipped with the operation of inner product, i.e., with a real function (f, g) of two arguments satisfying the following axioms:

(a) $(\alpha f, g) = \alpha(f, g)$ for any number α;

(b) $(f + h, g) = (f, g) + (h, g)$;

(c) $(f, g) = \overline{(g, f)}$, where the overbar denotes the operation of complex conjugation;

(d) $(f, f) > 0$ if $f \neq 0$, and $(f, f) = 0$ whenever $f = 0$.

In what follows, we consider only real Hilbert spaces for which axiom (2c) takes the form $(f, g) = (g, f)$.

(3) H is a complete space with respect to the metric $\rho(f, g) = \|f - g\|$, where the norm of an arbitrary element $h \in H$ is defined as $\|h\| = (h, h)^{1/2}$.

Example 1. Let R^n be the n-dimensional Euclidean space. The inner product of two elements $x = (x_1, \ldots, x_n)$ and $y = (y_1, \ldots, y_n)$ of the space R^n is defined as follows:

$$(x, y) = \sum_{i=1}^{n} x_i y_i.$$

Thus, R^n is a Hilbert space.

Example 2. The space L_2 (a special case of the space L_p with $p = 2$) is a Hilbert space if the inner product of two functions $f\colon [a, b] \to R^n$ and $g\colon [a, b] \to R^n$ is defined as

$$(f, g) = \int_a^b \sum_{i=1}^{n} f_i(x) g_i(x)\, dx.$$

The norm of an element in this space is given by the formula

$$\|f\| = \sqrt{(f, f)} = \left(\int_a^b \sum_{i=1}^{n} |f_i(x)|^2 dx \right)^{1/2}.$$

Example 3. If we consider the space l_p with $p = 2$, then the inner product can be defined as follows:

$$(x, y) = \sum_{i=1}^{\infty} x_i y_i,$$

where x_i and y_i, $i = 1, 2, \ldots$, are real numbers and the norm of an element is given by the formula

$$\|x\| = \left(\sum_{i=1}^{\infty} |x_i|^2 \right)^{1/2}.$$

The space l_2 obtained as a result can be regarded as the simplest example of an infinite-dimensional Hilbert space.

Example 4. The space D_2 is a Hilbert space if the inner product is defined as follows:

$$(x, y)_{D_2} = (\dot{x}, \dot{y})_{L_2} + (x(a), y(a))_{R^n}.$$

We now introduce some geometric notions valid in all spaces with inner product.

Definition 1.14. Two vectors x and y in a space H are called orthogonal if $(x, y) = 0$.

A collection of vectors $\{x_i\}$ in a space H is called orthonormal if $(x_i, x_i) = 1$ for all i and $(x_i, x_j) = 0$ whenever $i \neq j$.

Definition 1.15. Let H_1 and H_2 be Hilbert spaces. Then the set $\{x + y\}$, where $x \in H_1$ and $y \in H_2$, is a Hilbert space called the *direct sum of the spaces* H_1 *and* H_2 and denoted by $H_1 \oplus H_2$.

Let M be a closed subspace of a given Hilbert space H. Then M is a Hilbert space with natural inner product inherited from H. Let $M^\perp$ be the set of vectors from H orthogonal to M. The set $M^\perp$ is called the orthogonal complement of M. The linearity and continuity of the inner product imply that $M^\perp$ is a closed linear subspace of the space H and, hence, $M^\perp$ is a Hilbert space. The only common element of the subspaces M and $M^\perp$ is zero [129]. The theorem presented below states that, for any closed proper subspace, one can find vectors orthogonal to it. In fact, we have

$$H = M \oplus M^\perp.$$

This geometric property is one of the most important factors due to which Hilbert spaces are much easier for analysis than Banach spaces (see Section 1.3).

Lemma 1.1 ([1]). *Let H be a Hilbert space, let M be its closed subspace, and let $x \in H$. Then there exists a unique element $z \in M$ nearest to x.*

Theorem 1.1 ((theorem on projection) [1]). *Let H be a Hilbert space and let M be its closed subspace. Then any element $x \in H$ admits a unique representation of the form*

$$x = z + h, \qquad z \in M, \quad h \in M^{\perp}.$$

Definition 1.16. Two Hilbert spaces H_1 and H_2 are called *isomorphic* if there exists a continuous linear operator U mapping from H_1 onto H_2 and such that

$$(Ux, Uy)_{H_2} = (x, y)_{H_1}$$

for all $x, y \in H_1$. The operator U is called unitary and the mapping realized by this operator is called an isomorphism. An isomorphism is called isometric if it preserves the norm.

For all g from a real Hilbert space H, we define an operator $g^*\colon H \to R$ by the requirement that the equality $g^*(f) = (f, g)$ must hold for all $f \in H$. By virtue of the Schwarz inequality [129], we have $\|g^*\| \leq \|g\|$ (for the norms in $\mathcal{L}(H, R)$ and H, respectively). Therefore, g^* is a continuous linear functional on H, i.e., it is an element of the space $\mathcal{L}(H, R) = H^*$. It remains to clarify whether an arbitrary element of the dual space H^* can be determined in a similar way using an element of H. The Riesz theorem gives a positive answer to this question:

Theorem 1.2 ([129]). *Any element g^* of the space H^* dual to a Hilbert space H is associated with a single element $g \in H$ such that $g^*(f) = (f, g)$ for all $f \in H$ and, moreover, $\|g^*\|_{H^*} = \|g\|_H$.*

Thus, the space dual to a Hilbert space coincides with this space to within an isomorphism.

1.3 Banach Spaces

Various concepts of finite-dimensional analysis admit generalizations to infinite-dimensional normed vector spaces. At the same time, the required analysis cannot be developed in all spaces of this sort. Serious difficulties are encountered in analyzing the problem of convergence of sequences, which is of crucial importance for analysis. Indeed, a sequence expected to be convergent may appear to be divergent. In this case, it is necessary to restrict the class of spaces under consideration. To do this, one can, e.g., impose the condition of completeness on the norm. The assumption of completeness substantially simplifies abstract analysis and, at the same time, a broad class of normed vector spaces satisfies this assumption [130].

Definition 1.17. A complete normed linear space B is called a Banach space.

In the metric spaces R^n, $C[a,b]$, $C^n[a,b]$ $(n \geq 1)$, l_p, and L_p the norm of an element x is defined as its distance from the zero element, i.e., $\|x\| = \rho(x,0)$, and, hence, these are Banach spaces.

In Hilbert spaces, the norm is introduced with the help of the operation of inner product. In Banach spaces, it satisfies a weaker condition, namely, every element $f \in B$ is associated with a number $\|f\| \geq 0$ called its norm and such that

$$\|f+g\| \leq \|f\| + \|g\| \quad \text{and} \quad \|af\| = |\alpha|\,\|f\|.$$

In general, this norm is not induced by the inner product and, hence, Banach spaces do not possess numerous important properties of the Hilbert spaces.

Example 1 [73]**.** Let us show that the geometry of Banach spaces may exhibit some unexpected features.

Let $B = R^2$. The norm of an element $f = (f_1, f_2)$ is defined by setting $\|f\| = |f_1| + |f_2|$.

A square with vertices at the points $(1,0)$, $(0,-1)$, $(-1,0)$, and $(0,1)$ lying on the Ox- and Oy-axes plays the role of a unit ball in this space. We draw a line passing through the origin of coordinates at an angle of 45° to the Ox-axis and consider this line as a subspace B_0 of the space B. Any vector $g \in B_0$ has coordinates (g_1, g_1), i.e., $g = (g_1, g_1)$. Therefore, if we consider the vector $f(1,0)$ and write $\|f - g\| = |1 - g_1| + |g_1|$, then the norm of the difference between the vectors f and g attains its minimum equal to one for any g_1, $0 \leq g_1 \leq 1$.

Thus, the minimum distance from the vector f to the subspace B_0 is attained on an infinite set of vectors $g \in B_0$.

It is clear that, in the plane R^2 with ordinary Euclidean distance, the indicated minimum is attained on a single vector.

In some normed spaces, for infinite-dimensional subspaces B_0 this minimum may not be attained at all.

In Section 1.1, we have introduced the notion of bounded linear operators and denoted the set of all operators of this sort from X into Y by $\mathcal{L}(X,Y)$. If Y is complete, then $\mathcal{L}(X,Y)$ is a Banach space. In the case where Y consists of real numbers $y \in R$, the space $\mathcal{L}(X,R)$ is denoted by X^* and is called the space dual to X. The elements of X^* are bounded linear functionals defined in the space X. The dual space is a complete normed space.

For some Banach spaces, the dual space can be described explicitly. This description is given by theorems on the general form of linear bounded functionals in these spaces.

Example 2. Consider the space R^n. Any bounded linear functional f in the space R^n has the form $f(x) = \sum_{i=1}^{n} x_i y_i$, where $y_1, \ldots, y_n$ are given constants. Thus, we obtain an isomorphism between $(R^n)^*$ and R^n.

Example 3. Consider a Hilbert space. According to Theorem 1.2, any bounded linear functional admits a unique representation in the form $f(x) = (x, u)$ with $\|f\| = \|u\|$.

Example 4. For $1 < p < +\infty$, the space $(l_p)^*$ is isomorphic to the space l_q with $1/p + 1/q = 1$. The isomorphism is specified by the formula

$$f(x) = \sum_{i=1}^{\infty} x_i y_i,$$

where

$$x = (x_1, \ldots, x_n, \ldots) \in l_p, \qquad y = (y_1, \ldots, y_n, \ldots) \in l_q, \quad f \in (l_p)^*.$$

Example 5. For $1 < p < +\infty$, the space $(L_p)^*$ is isomorphic to the space L_q with $1/p + 1/q = 1$. The corresponding isomorphism is established by the formula

$$f(x) = \int_a^b x(t)g(t)dt, \qquad x(t) \in L_p, \quad g(t) \in L_q, \quad f \in (L_p)^*.$$

Example 6. The space $(C[a,b])^*$ is isomorphic to the space $V[a,b]$ of left-continuous functions with bounded variation on $[a,b]$ satisfying the condition $g(a) = 0$. The norm in this space is introduced as follows: $\|g\| = \overset{b}{\underset{a}{V}} g$. The required isomorphism is established by the formula

$$f(x) = \int_a^b x(t)dg(t),$$

where

$$g(t) \in V[a,b], \qquad x(t) \in C[a,b], \quad \text{and} \quad f \in (C[a,b])^*.$$

In working with Banach spaces, it is often necessary to solve the problems of construction of linear functionals with given properties. As a rule, this program is realized in two steps: first, the functional is defined on a subspace of a Banach space where the required properties can easily be verified, and then a general theorem is proved according to which any functional of this sort can be extended to the entire space with preservation of the required properties.

Theorem 1.3 ((Hahn–Banach) [79]). *Let M be a linear subspace of a Banach space B and let f^* be a continuous linear functional defined on M. Then f^* can be extended to a continuous linear functional on B with the same norm.*

Corollary 1.1. *Let M be a linear subspace of B and let g be an arbitrary element of B lying outside $\overline{M}$. Then there exists a continuous linear functional f^**

on B such that

(a) $f^*(f) = 0$ *for* $f \in M$;

(b) $f^*(g) = 1$;

(c) $\|f^*\| = \operatorname{dist}^{-1}(g, M)$, *where* $\operatorname{dist}(g, M)$ *is the distance between the point* g *and the set* M.

Corollary 1.2. *For any nonzero* $f \in B$, *one can find* $f^* \in B^*$ *such that* $\|f^*\| = 1$ *and* $f^*(f) = \|f\|$.

Corollary 1.3. *If* $f^*(f) = 0$ *for all* $f^* \in B$, *then* $f = 0$.

Corollary 1.4. *For any* $f \in B$,

$$\|f\| = \sup_{\|f^*\|=1} |f^*(f)|.$$

Definition 1.18. Let B_1 and B_2 be two subspaces of a space B intersecting only at the origin. The set of all vectors of the form $x_1 + x_2$, where $x_1 \in B_1$ and $x_2 \in B_2$, is called the direct (algebraic) sum of the subspaces B_1 and B_2 and is denoted as follows:

$$B = B_1 \dot{+} B_2.$$

Moreover, if B_1 and B_2 are closed subspaces of the space B, then we say that the space B decomposes into a direct (topological) sum of closed subspaces. In this case, we write

$$B = B_1 \oplus B_2.$$

Definition 1.19. If $B_1 \subset B$ is a closed subspace and there exists a subspace $B_2 \subset B$ such that $B = B_1 \oplus B_2$, then B_1 is called complemented in B, and B_2 is called a direct complement of B_1 with respect to B.

Clearly, a direct complement is not unique.

The direct sum is a linear manifold but, generally speaking, it is not a subspace (because it can be nonclosed). The following theorem gives necessary and sufficient conditions for the direct sum to be closed:

Theorem 1.4 ([63]). *Let* B *be a Banach space and let* B_1 *and* B_2 *be two subspaces of* B *intersecting only at zero. In order that the direct sum* $B_1 \oplus B_2$ *be a subspace, it is necessary and sufficient that there exist a constant* $k \geq 0$ *such that*

$$\|x_1 + x_2\| \geq k(\|x_1\| + \|x_2\|), \qquad x_1 \in B_1, \quad x_2 \in B_2.$$

Corollary 1.5. *If one of the subspaces* B_1 *and* B_2 *is finite-dimensional, then the direct sum* $B_1 \oplus B_2$ *is closed, i.e., any finite-dimensional subspace of a Banach space can always be complemented.*

If $B = H$ is a Hilbert space, then any subspace $H_1 \subset H$ possesses a direct complement, e.g., the orthogonal complement of H_1.

For Banach spaces, one can introduce the following analog of orthogonality based on the duality of the spaces B and B^*:

Definition 1.20 ([88]). For any given sets $M \subset B$ and $M^* \subset B^*$, the sets

$$M^{\perp} = \{f^* \in B^*\colon f^*(f) = 0, \ \ f \in M\},$$

$$M^{*\perp} = \{f \in B\colon f(f^*) = 0, \ \ f^* \in M^*\}$$

are called the orthogonal complements of the sets M and M^*, respectively.

In this case, the analogy with Hilbert spaces is not so close as desired because $M^{\perp}$ is a subspace of the dual space B^* but not of the original space B. This is an additional manifestation of the complex geometric structure of Banach spaces.

Generally speaking, direct complements cannot be constructed for some subspaces of a Banach space. However, as already indicated, finite-dimensional subspaces can always be complemented.

Theorem 1.5 ((on complements) [4]). *1. If B_1 is an n-dimensional subspace of a Banach space B, then there exists a closed complement of B_1, which can be specified by n linearly independent functionals.*

2. If B_2 is a closed subspace of a Banach space B defined by a finite set of n linearly independent functionals $B_2 = \{x\colon f_i(x) = 0, \ i = 1, \dots, n\}$, then B_2 possesses an n-dimensional complement with respect to the space B.

1.4 Linear Operators

Let B_1 and B_2 be Banach spaces and let $\mathcal{L}(B_1, B_2)$ be the Banach space of bounded linear operators acting from B_1 into B_2.

Definition 1.21. The set $D(L) \subseteq B_1$ where an operator $L \in \mathcal{L}(B_1, B_2)$ acts is called the domain of this operator.

Definition 1.22. The set of all solutions of the homogeneous equation $Lx = 0$ forms a subspace $\ker L \subseteq B_1$ called the kernel of the operator L. The kernel of the operator L is often denoted by $N(L)$ and is called the null space of the operator L.

Definition 1.23. The set of all values of an operator $L \in \mathcal{L}(B_1, B_2)$ forms a linear manifold $\operatorname{Im} L \subseteq B_2$ called the image of the operator L. The image of the operator L is also denoted by $R(L)$.

In connection with these definitions, we want to make the following remark: In the definition of an operator L acting from B_1 into B_2, it is possible that $D(L)$ is a subset of B_1. However, in what follows, the notation $L\colon B_1 \to B_2$ always means that $D(L) = B_1$.

Example 1. We now illustrate the notions introduced above by using the space $C[a,b]$ as an example. First, we not necessarily consider the operation of ordinary differentiation. Since continuous functions are differentiable, we first restrict ourselves to the case of smooth functions, say, from $C^\infty[a,b]$. For $x \in C^\infty[a,b]$, we set $(Lx)(t) = x'(t)$. This relation specifies an operator L mapping the space $C[a,b]$ into itself with the domain $C^\infty[a,b]$. Thus, we can write $L\colon C^\infty[a,b] \to C[a,b]$.

However, the operator of differentiation is also meaningful in a broader space, e.g., in $C^1[a,b]$, and we can define another operator $\tilde{L}$ by setting $(\tilde{L}x)(t) = x'(t)$ for $x \in C^1[a,b]$. Despite the fact that $\tilde{L}x = Lx$ for $x \in D(L)$, the operators L and $\tilde{L}$ are regarded as different because $D(L) \neq D(\tilde{L})$. Note that the operators can be regarded as identical only in the case where both their images and domains are identical. The operator $\tilde{L}$ is called a continuation (extension) of the operator L.

Definition 1.24. Two operators $L, \tilde{L} \in \mathcal{L}(B_1, B_2)$ are called equal if $D(L) = D(\tilde{L})$ and $Lx = \tilde{L}x$ for all $x \in D(L)$. The operator $\tilde{L}$ is called an extension (continuation) of the operator L and the operator L is called a restriction of $\tilde{L}$ if $D(\tilde{L}) \supset D(L)$ and $Lx = \tilde{L}x$ for all $x \in D(L)$. This relationship between the operators is denoted as follows: $L \subset \tilde{L}$.

Definition 1.25. The dimensionality of an operator $L \in \mathcal{L}(B_1, B_2)$ is defined as the dimensionality of its image, i.e.,

$$\dim L = \dim \operatorname{Im} L.$$

An operator is called finite-dimensional if $\dim L < \infty$.

Definition 1.26. An operator $L \in \mathcal{L}(B_1, B_2)$ is called bounded if there exists a constant $C \geq 0$ independent of x and such that $\|Lx\| \leq C\,\|x\|$ for $x \in D(L)$.

The smallest of these constants is called the norm of the operator L and is denoted by $\|L\|$:

$$\|L\| = \sup_{x \in B_1} \frac{\|Lx\|_{B_2}}{\|x\|_{B_1}}.$$

In the collection of all linear operators acting in Banach spaces, we can select a class of operators (generally speaking, unbounded) whose properties are quite close to the properties of bounded operators. This is the class of closed operators.

Definition 1.27. A linear operator L is called closed if the facts that $x_n \to x$ and $Lx_n \to y$ imply that $x \in D(L)$ and $Lx = y$.

Any bounded operator defined in the entire space B is closed but a closed operator can be unbounded.

If a closed operator L has a nontrivial null space $N(L)$, then this space is closed.

Theorem 1.6 ([84]). *A closed operator L is bounded if and only if its domain $D(L)$ is closed.*

Thus, a closed operator L defined in the entire space B_1 is bounded.

Definition 1.28. An operator $L \in \mathcal{L}(B_1, B_2)$ is called densely defined in the space B_1 if its domain $D(L)$ is dense in B_1.

An operator defined in the entire space B_1 is densely defined. The operator $L^*\colon B_2^* \to B_1^*$ adjoint to a densely defined operator $L\colon B_1 \to B_2$, is uniquely determined:

$$(L^*g)(x) = g(Lx) \quad \forall x \in B_1, \quad \forall g \in B_2^*. \tag{1.1}$$

In the case of Hilbert spaces, the adjoint operator $L^*\colon H_2^* \to H_1^* = H_1$ is defined in terms of the inner product as follows:

$$(L^*y, x)_{H_1} = (y, Lx)_{H_2}. \tag{1.2}$$

The adjoint operator is always closed. Indeed, if $g_n \to g$ and $L^*g_n \to f$, then, for $x \in D(L)$, we have $g_n(Lx) \to g(Lx)$ by virtue of the first relation and $g_n(Lx) \to f(x)$ by virtue of the second relation, i.e., $g(Lx) = f(x)$ for all $x \in (D(L))$ and $f = L^*g$. The operator L^* adjoint to a bounded linear operator is also a bounded linear operator.

Definition 1.29. The kernel $\ker L^*$ of the operator L^* is called the cokernel of the operator L and is denoted by $\operatorname{Coker} L$ or $N(L^*)$ (the null space of the operator L^*).

If B_1 and B_2 are reflexive spaces, then $L^{**} = (L^*)^* = L$, whence

$$\ker L = \operatorname{Coker} L^*.$$

Example 2. Let $H_1 = R^n$, let $H_2 = R^m$, and let a linear operator $L\colon H_1 \to H_2$ be specified by an $n \times m$ matrix A with entries a_{ij} according to the formula

$$(Lx)_j = \sum_{i=1}^{n} a_{ij}x_i, \quad j = 1, \ldots, m.$$

The space dual to R^m is isomorphic to R^m, and the correspondence between these spaces is established by the formula

$$f(x) = \sum_{j=1}^{m} x_j \xi_j = (\xi, x).$$

It follows from relation (1.2) that

$$(L^* f)(x) = f(Lx) = \sum_{j=1}^{m} \sum_{i=1}^{n} a_{ij} x_i \xi_j = (\xi, Lx).$$

The functional thus constructed can be represented in the form

$$(L^* \xi, x) = \sum_{j=1}^{n} \left(\sum_{i=1}^{m} a_{ij} \xi_i \right) x_j = \sum_{j=1}^{n} x_j \tau_j,$$

where $\tau_j = \sum_{i=1}^{m} a_{ij} \xi_i$.

Thus, the action of the adjoint operator is determined by the $m \times n$ transpose $A^T = (a_{ji})$ of the matrix $A = (a_{ij})$.

Example 3. In the space $L_p[0,1]$, we consider an integral operator

$$(Lx)(t) = \int_0^1 K(t,s) x(s)\, ds,$$

where $K(t,s)$ is a bounded function measurable in the square $[0,1] \times [0,1]$.

Any functional $f \in (L_p[0,1])^*$ can be represented in the form

$$f(x) = \int_0^1 x(t) g(t)\, dt,$$

where $g \in L_q[0,1]$ (see Section 1.3). We fix $g \in L[0,1]$ and study, by using relation (1.1), the action of the operator L^* on the corresponding functional:

$$(L^* g)(x) = g(Lx) = \int_0^1 \Big(\int_0^1 K(t,s) x(s)\, ds \Big) g(t)\, dt.$$

Changing the order of integration (by virtue of the Fubini theorem [84], this is possible), we obtain

$$(L^* g)(x) = \int_0^1 x(s) \Big[\int_0^1 K(t,s) g(t)\, dt \Big]\, ds = \int_0^1 x(s) v(s)\, ds,$$

where

$$v(s) = \int_0^1 K(t,s)g(t)dt.$$

Thus, the adjoint operator L^* transforms the functional corresponding to the function g into the functional corresponding to the function v, i.e.,

$$(L^*g)(s) = \int_0^1 K(s,t)g(t)\,dt.$$

Hence, the operator adjoint to the integral operator with kernel $K(t,s)$ is an integral operator with kernel $K(s,t)$.

Example 4. In the space $D_2[0,2]$ of absolutely continuous functions whose derivatives belong to the space $L_2[0,2]$, we consider the operator

$$(Lx)(t) = \dot{x}(t) - x(1) + x(0) \equiv \dot{x}(t) - \int_0^1 \dot{x}(s)\,ds, \quad t \in [0,2],$$

acting into the space $L_2[0,2]$.

As in the previous example, we choose $g \in L_2[0,2]$ and consider the action of the operator L^* on the corresponding functional:

$$\begin{aligned}(L^*g)(x) = g(Lx) &= \int_0^2 \Big(\dot{x}(t) - \int_0^1 \dot{x}(s)ds\Big)g(t)\,dt \\ &= \int_0^2 \Big(\dot{x}(t) - \int_0^2 \chi_{[0,1]}(s)\dot{x}(s)\,ds\Big)g(t)\,dt \\ &= \int_0^2 \dot{x}(t)g(t)\,dt - \int_0^2 \Big(\int_0^2 \chi_{[0,1]}(s)\dot{x}(s)\,ds\Big)g(t)\,dt,\end{aligned}$$

where $\chi_{[0,1]}(t)$ is the characteristic function of the interval $[0,1]$.

Changing the order of integration in the second integral, we get

$$(L^*g)(x) = \int_0^2 \Big(g(s) - \chi_{[0,1]}(s)\int_0^2 g(t)dt\Big)\dot{x}(s)\,ds.$$

Thus, the adjoint operator has the form

$$(L^*g)(t) = g(t) - \chi_{[0,1]}(t)\int_0^2 g(s)\,ds.$$

Example 5. Consider an operator $L\colon L_p[0,1] \to L_p[0,1]$, $1 < p < \infty$, of the form $(Lx)(t) = x(\alpha(t))$, where $\alpha\colon [0,1] \to [0,1]$ is a continuously differentiable invertible mapping of the interval $[0,1]$ onto itself with $\alpha'(t) \neq 0$ [64]. The adjoint operator L^* is constructed as follows:

$$g(Lx) = \int_0^1 g(t)x(\alpha(t))\,dt$$

$$= \begin{bmatrix} \tau = \alpha(t) & dt = \beta'(\tau)d\tau \\ t = \beta(\tau) & \end{bmatrix} = \int_0^1 g(\beta(\tau))x(\tau)\beta'(\tau)d\tau$$

$$= \int_0^1 \beta'(\tau)g(\beta(\tau))x(\tau)\,d\tau = (L^*g)(x),$$

where

$$\tau = \alpha(t), \quad t = \beta(\tau), \quad \text{and} \quad dt = \beta'(\tau)d\tau.$$

Thus, the adjoint operator L^* acts in the space $L_q[0,1]$ according to the formula $(L^*g)(t) = \beta'(t)g(\beta(t))$, where $\beta = \alpha^{-1}$.

In the set of closed operators, we distinguish the class of normally resolvable operators.

Definition 1.30. A densely defined operator L acting from B_1 into B_2 is called normally resolvable if its image is closed, i.e., $R(L) = \overline{R(L)}$.

Thus, an operator $L\colon B_1 \to B_2$ for which $R(L) = B_2$ is a simple example of a normally resolvable operator.

We now consider the case of strict inclusion: $R(L) \subset B_2$. Let us show that if L is a normally resolvable operator and $R(L)$ does not coincide with B_2, then the null space $N(L^*)$ of the adjoint operator L^* is nontrivial.

Let $f \in B_2$ but $f \overline{\in} R(L)$. Since, according to our assumption, $R(L)$ is a closed linear manifold, we can use the Hahn–Banach theorem for the construction of a linear functional $\varphi \in B_2^*$ such that $\varphi(f) = 1$ and $\varphi(h) = 0$ for all $h \in R(L)$. The last equality means that $0 = \varphi(h) = \varphi(Lx) = (L^*\varphi)(x)$, where x is a vector from $D(L)$. Since $D(L)$ is dense in B_1, we have $L^*\varphi = 0$ and, hence, $\varphi \in N(L^*)$. By construction, the functional φ is nonzero, which proves that $N(L^*)$ is nontrivial.

Theorem 1.7 ([85]). *A closed densely defined operator $L \in \mathcal{L}(B_1, B_2)$ with $R(L) \neq B_2$ is normally resolvable if and only if one of the following conditions is satisfied:*

(1) $N(L^*)^{\perp} = R(L)$, (1.3)

(2) the equation $Lx = y$ is solvable only for $y \in B_2$ satisfying the condition

$$\varphi(y) = 0,$$

where φ is an arbitrary solution of the homogeneous conjugate equation

$$L^*\varphi = 0.$$

Assume that the functionals $\varphi_1, \varphi_2, \dots, \varphi_m$ form a basis in $N(L^*)$. Then the requirement $\varphi(y) = 0$ is equivalent to the following conditions:

$$\varphi_i(y) = 0, \quad i = 1, \dots, m, \tag{1.4}$$

necessary and sufficient for the solvability of the equation $Lx = y$.

We now present another sufficient condition for an operator to be normally resolvable.

Lemma 1.2 ([85]). *If $L \in \mathcal{L}(B_1, B_2)$ is a closed operator and there exists a closed linear subspace M of the space B_2 such that*

$$B_2 = M \oplus R(L),$$

then L is a normally resolvable operator.

If an operator L maps a space B into itself, then we can introduce the following definition:

Definition 1.31 ([85]). A closed densely defined operator $L\colon B \to B$ is called reducibly invertible if

$$B = N(L) \oplus R(L).$$

By virtue of Lemma 1.2, $R(L)$ is a closed space, i.e., the reducibly invertible operator is normally resolvable, and the equation $Lx = y$, $y \in N(L)$, $y \neq 0$, is unsolvable in B.

In what follows, an important role is played by the dimensions

$$n = n(L) = \dim\ker L,$$

$$m = m(L) = \dim \operatorname{Coker} L.$$

Definition 1.32. The index of an operator L is defined as follows:

$$\operatorname{ind} L = n(L) - m(L).$$

It follows from the arguments presented above that $\operatorname{ind} L^* = -\operatorname{ind} L$. The notion of index is also applicable in the case where one of the numbers $\big(n(L)$ or $m(L)\big)$ is infinite. In this case, the index is equal to $\pm\infty$.

Definition 1.33. A normally resolvable closed operator L with finite $n(L)$ $(m(L))$ is called n-normal (d-normal).

In the literature, n- and d-normal operators are also known as semi-Fredholm operators (including the classes of Φ_+-operators ($n(L) < \infty$) and Φ_--operators ($m(L) < \infty$)) [88, 112].

For a densely defined closed operator L, the properties of n- and d-normality are symmetric with respect to the operation of conjugation.

We now consider another class of normally resolvable linear operators. Numerous problems in the theory of functional differential equations [8] are based on the analysis of so-called Fredholm operators.[1]

Definition 1.34. A bounded linear operator $L \in \mathcal{L}(B_1, B_2)$ is called a Fredholm operator if its image is closed, i.e., $R(L) = \overline{R(L)}$, and the dimensions of its kernel and cokernel are finite: $n(L) < \infty$ and $m(L) < \infty$.

Definition 1.35. If $\operatorname{ind} L = 0$ $(n(L) = m(L))$, then the operator L is called a Fredholm operator of index zero.

Example 6. A rectangular $m \times n$ matrix L with constant entries is a simple example of Fredholm operator. Let

$$\operatorname{rank} L = n_1 \leq \min(m, n).$$

Since the dimension of the null space $N(L)$ is equal to the defect of the matrix L and $\operatorname{rank} L = n_1$, we have

$$\dim N(L) = n - \operatorname{rank} L = n - n_1 = r.$$

Taking into account that $\operatorname{rank} L = \operatorname{rank} L^*$ [151], we obtain

$$\dim N(L^*) = m - n_1 = d$$

and, therefore,

$$\operatorname{ind} L = \dim \ker L - \dim \ker L^* = r - d = n - m \neq 0$$

whenever $n \neq m$.

If $m = n$ $(r = d)$, then $\operatorname{ind} L = 0$. A square $n \times n$ matrix L is a simple example of a Fredholm operator of index zero.

Example 7. The integrodifferential operator $L\colon D_2[0,2] \to L_2[0,2]$ defined by the formula

$$(Lx)(t) = \dot{x}(t) - \int_0^1 \dot{x}(s)ds$$

is a Fredholm operator [8].

---

[1] Sometimes, especially in the Russian-language literature (see [38, 88, 100]), Fredholm operators are called Noetherian operators. The Noetherian operators are operators for which all conditions of Definition 1.34 are satisfied but $n(L)$ may differ from $m(L)$. This class of operators is named after F. Noether who studied a class of singular integral equations with operators of this sort for the first time as early as in 1921 [114].

Indeed, the homogeneous operator equation $(Lx)(t)=0$ has two linearly independent solutions $x_1(t) = 1$ and $x_2(t) = t$, whereas the equation $(L^*g)(t) = 0$ possesses a single solution $g_1(t) = \chi_{[0,1]}(t)$. Therefore, L is a Fredholm operator of index one.

Example 8. Consider the operator $L\colon l_2 \to l_2$ defined by the formula $Lx = (x_2, x_3, \ldots)$ for all $x = (x_1, x_2, x_3, \ldots)$ [64].

Let us show that L and L^* are Fredholm operators. The operator L is linear and continuous and, therefore, it possesses the adjoint operator $L^*\colon l_2 \to l_2$. The identity

$$(Lx, y) = (x, L^*y),$$

which is valid for all $x, y \in l_2$, implies that $L^*y = (0, y_1, y_2, \ldots)$ for $y = (y_1, y_2, \ldots) \in l_2$.

The kernel of the operator L consists of all vectors of the form $(c, 0, 0, \ldots)$, where c is an arbitrary number. For the kernel of the operator L^*, we have $\ker L^* = \{0\}$ and, hence,

$$\dim\ker L = 1 \quad \text{and} \quad \dim\ker L^* = 0.$$

Example 9. Let us show that the operators L^k and $(L^*)^k$, where $Lx = (x_2, x_3, \ldots)$ for all $x = (x_1, x_2, x_3, \ldots)$, are also Fredholm operators [64].

The operator L^k transforms an element $x = (x_1, x_2, x_3, \ldots) \in l_2$ according to the formula $L^k x = (x_{k+1}, x_{k+2}, \ldots)$. The adjoint operator $(L^*)^k$ is defined by the equality

$$(L^*)^k y = (0, \ldots, 0, y_1, y_2, \ldots) \quad \text{for all} \quad y = (y_1, y_2, \ldots) \in l_2.$$

Thus, $\ker L^k$ consists of all elements of the form $x = (\alpha_1, \alpha_2, \ldots, \alpha_k, 0, 0, \ldots)$, where $\alpha_1, \ldots, \alpha_k$ are arbitrary numbers. Then $\dim\ker L^k = k$. The kernel of the operator $(L^*)^k$ is formed by the zero element, i.e., $\dim\ker (L^*)^k = \{0\}$. Hence,

$$\operatorname{ind} L^k = k \quad \text{and} \quad \operatorname{ind} (L^*)^k = -k.$$

Therefore, L^k and $(L^*)^k$ are Fredholm operators.

Let $L \in \mathcal{L}(B_1, B_2)$ be a bounded linear Fredholm operator defined in the entire space B_1. Since the null spaces $N(L)$ and $N(L^*)$ are finite-dimensional, they have direct complements (by virtue of Theorem 1.5).

We now study the structure of the null space $N(L^*)$ of the operator L^*. If $g \in N(L^*) \subseteq B_2^*$, then

$$0 = (L^*g)x = g(Lx)$$

for $x \in B_1$, i.e., the functional g is orthogonal to $R(L)$. If $g \in R(L)^\perp$, then $g(Lx) = 0$ for all $x \in B_1$ and, hence, $g \in B_2^*$ and $L^*g = 0$.

Thus, the null space $N(L^*)$ of the adjoint operator L^* is an orthogonal complement of the image $R(L)$ of the original operator L and

$$B_2 = Y_0 \oplus R(L), \tag{1.5}$$

where the space Y_0 is isomorphic to the null space $N(L^*)$.

We now consider the relationship between $N(L)$ and $R(L^*)$. If $x \in N(L)$ and $f \in R(L^*)$, then

$$f(x) = (L^*g)(x) = g(Lx) = 0,$$

i.e., $N(L)$ and $R(L^*)$ are orthogonal. However, the sets $N(L)$ and $R(L^*)$ are not necessarily the orthogonal complements to each other. In fact, the following inclusion is true: $R(L^*) \subset N(L)^{\perp}$.

However, if the operator L is bounded and defined everywhere in B_1, then the operator L^* is defined everywhere in B_2, and we have $N(L) = R(L^*)^{\perp}$ [88], whence

$$B_1 = N(L) \oplus X_0. \tag{1.6}$$

1.5 Unilateral Inverse, Generalized Inverse, and Pseudoinverse Operators

Definition 1.36. An operator $L \in \mathcal{L}(B_1, B_2)$ is called left invertible if there exists an operator $X \in \mathcal{L}(B_2, B_1)$ such that $XL = I_{B_1}$. The operator X is called left inverse to L and is denoted by L_l^{-1}.

Definition 1.37. An operator $L \in \mathcal{L}(B_1, B_2)$ is called right invertible if there exists an operator $Y \in \mathcal{L}(B_2, B_1)$ such that $LY = I_{B_2}$. The operator Y is called right inverse to L and is denoted by L_r^{-1}.

Here, I_{B_1} and I_{B_2} are the identity operators in the spaces B_1 and B_2, respectively.

Necessary and sufficient conditions for the existence of the left (L_l^{-1}) and right (L_r^{-1}) inverse operators are established by the following theorems:

Theorem 1.8 ([63]). *In order that an operator $L \in \mathcal{L}(B_1, B_2)$ be right invertible, it is necessary and sufficient that the following conditions be satisfied:*

(i) $\operatorname{Im} L = B_2$;

(ii) $\ker L$ *possesses a direct complement in* B_1.

Relation (1.1) implies that the condition $\operatorname{Im} L = B_2$ is equivalent to the equality $\ker L^* = \{0\}$. By virtue of Corollary 1.5, the validity of the second condition is guaranteed if $\ker L$ is finite-dimensional.

Theorem 1.9 ([63]). *In order that an operator $L \in \mathcal{L}(B_1, B_2)$ be left invertible, it is necessary and sufficient that the following conditions be satisfied:*

(i) $\operatorname{Im} L$ *is a subspace with direct complement in* B_2;

(ii) $\ker L = \{0\}$.

If $\ker L^*$ is finite-dimensional, then the first condition is satisfied by virtue of relation (1.1) and Corollary 1.5.

Definition 1.38. An operator $L \in \mathcal{L}(B_1, B_2)$ is called invertible in the generalized sense if there exists an operator $X \in \mathcal{L}(B_2, B_1)$ such that

$$LXL = L.$$

The operator X is called the generalized inverse of the operator L and is denoted by L^-.

If an operator $L \in \mathcal{L}(B_1, B_2)$ is left and right invertible, then it is invertible. Unilaterally invertible bounded linear operators are invertible in the generalized sense.

Theorem 1.10 ([63]). *An operator* $L \in \mathcal{L}(B_1, B_2)$ *is invertible in the generalized sense if and only if*

(i) L *is a normally resolvable operator;*

(ii) the subspace $\ker L$ *possesses a direct complement in* B_1;

(iii) the subspace $\operatorname{Im} L$ *possesses a direct complement in* B_2.

Corollary 1.6. *If* B_1 *and* B_2 *are Hilbert spaces, then the operator* L *is invertible in the generalized sense if and only if it is normally resolvable.*

Corollary 1.7. *Any finite-dimensional operator from* $\mathcal{L}(B_1, B_2)$ *is invertible in the generalized sense.*

If an operator $L \in \mathcal{L}(B_1, B_2)$ is invertible in the generalized sense, then the operator $L^- \in \mathcal{L}(B_2, B_1)$ can be chosen to guarantee the validity of the equalities

$$\text{(a)}\ \ LL^-L = L \qquad \text{and} \qquad \text{(b)}\ \ L^-LL^- = L^-. \tag{1.7}$$

Assume that an operator $L \in \mathcal{L}(H_1, H_2)$ acts from a Hilbert space H_1 into a Hilbert space H_2. Then the set of generalized inverse operators $L^- \in \mathcal{L}(H_2, H_1)$ contains a unique operator satisfying the following conditions:

$$\begin{aligned} &\text{(a)}\ \ LL^-L = L, &&\text{(b)}\ \ L^-LL^- = L^-,\\ &\text{(c)}\ \ (LL^-)^* = LL^-, &&\text{(d)}\ \ (L^-L)^* = L^-L. \end{aligned} \tag{1.8}$$

Definition 1.39. An operator L^- satisfying conditions (1.8) is called a Moore–Penrose pseudoinverse operator [109, 112, 117] and is denoted by L^+.

Chapter 2

GENERALIZED INVERSE OPERATORS IN BANACH SPACES

In Chapter 2, we study the problems encountered in the construction of generalized inverse operators for linear bounded Fredholm operators acting in Banach spaces. Note that the well-known Schmidt lemma [146, 148] and its subsequent generalization (i.e., the Nikol'skii theorem [113] on the general form of Fredholm operators of index zero) used for the construction of the generalized inverse operator L^- are not true in the case of Fredholm operators.

The Atkinson theorem [6] generalizes the Nikol'skii theorem to the case of Fredholm operators and states that any normally resolvable Fredholm operator can be represented in the form of a sum of unilateral inverse and finite-dimensional operators. The Atkinson theorem enables us to generalize the Schmidt lemma and thus construct the generalized inverse operators for Fredholm operators of any finite index in Banach spaces. This construction is realized by complementing the original operator L to an operator $\overline{L}$ of complete rank. The procedure of complementing is carried out by constructing special finite-dimensional operators added to the original Fredholm operator. We study the properties of these finite-dimensional operators and their relationship with the projectors to the kernel and cokernel of the Fredholm operator.

2.1 Finite-Dimensional Operators

In this section, we consider a class of linear operators of special kind encountered in the general theory of linear operators and required, in particular, for the construction of generalized inverse operators, namely, the class of projectors.

Definition 2.1. A linear map $\mathcal{P}\colon B \to B$ is called a projector in the space B if

$$\mathcal{P}^2 = \mathcal{P}, \tag{2.1}$$

i.e., $\mathcal{P}(\mathcal{P}x) = \mathcal{P}x$ for any $x \in B$.

Example 1. As a trivial example of projectors in any Banach space, we can mention the identity operator I and the operator 0. If $\mathcal{P}$ is a projector, then $I - \mathcal{P}$ is also a projector. Indeed, since $\mathcal{P}^2 = \mathcal{P}$, we can write

$$(I - \mathcal{P})^2 = (I - \mathcal{P})(I - \mathcal{P}) = I - \mathcal{P} + \mathcal{P}^2 - \mathcal{P} = I - \mathcal{P}.$$

Example 2. In any space of sequences $x = (x_1, x_2, x_3, \ldots, x_n, \ldots)$, every fixed combination of the numbers of coordinates $i_1 < i_2 < \ldots < i_k < \ldots$

specifies a projector $\mathcal{P}$ such that $u = \mathcal{P}x$, where $u = (0, \ldots, 0, x_{i_1}, 0, \ldots, 0, x_{i_2}, 0, \ldots, 0, x_{i_k}, 0, \ldots)$.

Lemma 2.1 ([125]). *Let $\mathcal{P}$ be a projector in the space B with kernel $N(\mathcal{P})$ and image $R(\mathcal{P})$. Then the following statements are true:*

(a) $R(\mathcal{P}) = N(I - \mathcal{P}) = \{x \in B \colon \mathcal{P}x = x\}$;

(b) $N(\mathcal{P}) = R(I - \mathcal{P})$; (2.2)

(c) $R(\mathcal{P}) \cap N(\mathcal{P}) = \{0\}$ *and* $B = R(\mathcal{P}) \dot{+} N(\mathcal{P})$;

(d) if B_1 and B_2 are subspaces of B such that $B_1 \cap B_2 = \{0\}$ and $B = B_1 \dot{+} B_2$, then the space B contains a unique projector such that $B_1 = R(\mathcal{P})$ and $B_2 = N(\mathcal{P})$.

Proof. Since $(I - \mathcal{P})\mathcal{P} = 0$, we have $R(\mathcal{P}) \subset N(I - \mathcal{P})$, and if $x \in N(I - \mathcal{P})$, then $x - \mathcal{P}x = 0$. Therefore, $x = \mathcal{P}x \in R(\mathcal{P})$, which proves relation (a). The second relation is obtained by applying (a) to the projector $I - \mathcal{P}$. If $x \in R(\mathcal{P}) \cap N(\mathcal{P})$, then $x = \mathcal{P}x = 0$ and any vector $x \in B$ admits a representation $x = \mathcal{P}x + (x - \mathcal{P}x)$ with $x - \mathcal{P}x \in N(\mathcal{P})$. This proves assertion (c). If the subspaces B_1 and B_2 satisfy the conditions in (d), then every vector $x \in B$ admits a unique representation in the form $x = x_1 + x_2$, where $x_1 \in B_1$ and $x_2 \in B_2$. We set $\mathcal{P}x = x_1 \in R(\mathcal{P})$. Then $x_2 = x - x_1 = x - \mathcal{P}x \in N(\mathcal{P})$ because $\mathcal{P}x_2 = \mathcal{P}x - \mathcal{P}^2 x = 0$. Note that, in this case, B_1 and B_2 are not necessarily closed subspaces of the space B. At the same time, if $\mathcal{P}$ is a continuous projector, then the subspaces are definitely closed.

Theorem 2.1 ([125]). *If $\mathcal{P}$ is a continuous projector in a Banach space B, then*

$$B = R(\mathcal{P}) \oplus N(\mathcal{P}), \tag{2.3}$$

where $R(\mathcal{P})$ and $N(\mathcal{P})$ are closed subspaces of the space B.

Proof. Since the operators $\mathcal{P}$ and $I - \mathcal{P}$ are continuous, the subspaces $N(\mathcal{P})$ and $R(\mathcal{P}) = N(I - \mathcal{P})$ (see assertion (a)) are closed and the statement of the theorem follows from (c).

Let $L \in \mathcal{L}(B_1, B_2)$ be a bounded linear Fredholm operator ($\operatorname{ind} L = \dim\ker L - \dim\ker L^* = r - d < \infty$) defined in the entire Banach space B_1 and mapping this space into a Banach space B_2.

Let $\{f_i\}_{i=1}^r$ and $\{\varphi_s\}_{s=1}^d$ be bases in the null spaces $N(L)$ and $N(L^*)$ of the operators L and L^*, respectively. According to Corollary 1.1 of Theorem 1.3 (Hahn–Banach theorem), there exist systems of continuous linearly independent functionals $\{\gamma_j\}_{j=1}^r \subset B_1^*$ and elements $\{\psi_k\}_{k=1}^d \subset B_2^*$ biorthogonal to these bases, namely,

$$\gamma_j(f_i) = \delta_{ji}, \quad i, j = 1, \ldots, r, \qquad \text{and} \qquad \varphi_s(\psi_k) = \delta_{sk}, \quad s, k = 1, \ldots, d.$$

We introduce finite-dimensional operators

$$\mathcal{P}_L x = \sum_{i=1}^{r} \gamma_i(x)\, f_i \qquad \text{and} \qquad \mathcal{P}_{L^*} y = \sum_{s=1}^{d} \varphi_s(y)\, \psi_s \tag{2.4}$$

and study the relationship between the subspaces into which the spaces B_1 and B_2 are decomposed by the operators $\mathcal{P}_L$ and $\mathcal{P}_{L^*}$, respectively.

Lemma 2.2. *The operators $\mathcal{P}_L$ and $\mathcal{P}_{L^*}$ are continuous projectors acting in Banach spaces B_1 and B_2 and partitioning these spaces into direct sums of closed subspaces, namely,*

$$B_1 = N(L) \oplus X_0 \quad \textit{and} \quad B_2 = Y_0 \oplus R(L). \tag{2.5}$$

Proof. We first prove that the operators $\mathcal{P}_L$ and $\mathcal{P}_{L^*}$ are projectors, i.e., that they satisfy condition (2.1):

$$\begin{aligned} \mathcal{P}_L^2 x = \mathcal{P}_L(\mathcal{P}_L x) &= \sum_{i=1}^{r} \gamma_i\Big(\sum_{j=1}^{r} \gamma_j(x) f_j\Big)\, f_i = \sum_{i=1}^{r}\sum_{j=1}^{r} \gamma_i(f_j)\,\gamma_j(x)\, f_i \\ &= \sum_{i=1}^{r}\sum_{j=1}^{r} \delta_{ij}\gamma_j(x)\, f_i = \sum_{i=1}^{r} \gamma_i(x)\, f_i = \mathcal{P}_L x, \\ \mathcal{P}_{L^*}^2 y = \mathcal{P}_{L^*}(\mathcal{P}_{L^*} y) &= \sum_{s=1}^{d} \varphi_s\Big(\sum_{k=1}^{d} \varphi_k(y)\psi_k\Big)\, \psi_s = \sum_{s=1}^{d}\sum_{k=1}^{d} \varphi_s(\psi_k)\varphi_k(y)\, \psi_s \\ &= \sum_{s=1}^{d}\sum_{k=1}^{d} \delta_{sk}\varphi_k(y)\, \psi_s = \sum_{s=1}^{d} \varphi_s(y)\, \psi_s = \mathcal{P}_{L^*} y. \end{aligned}$$

The continuity of the projectors $\mathcal{P}_L$ and $\mathcal{P}_{L^*}$ follows from the continuity of the functionals γ_j, $j = 1, \dots, r$, and φ_s, $s = 1, \dots, d$, and the fact that the sums used to define the operators $\mathcal{P}_L$ and $\mathcal{P}_{L^*}$ are finite-dimensional.

Thus, by Theorem 2.1, the projectors $\mathcal{P}_L$ and $\mathcal{P}_{L^*}$ split the spaces B_1 and B_2 into direct topological sums of closed subspaces, namely,

$$B_1 = N(\mathcal{P}_L) \oplus R(\mathcal{P}_L) \quad \text{and} \quad B_2 = N(\mathcal{P}_{L^*}) \oplus R(\mathcal{P}_{L^*}).$$

To prove relations (2.5), it is necessary and sufficient to show that

$$\begin{aligned} &\text{(a)}\ \ N(L) = R(\mathcal{P}_L), && \text{(b)}\ \ R(L) = N(\mathcal{P}_{L^*}), \\ &\text{(c)}\ \ Y_0 = R(\mathcal{P}_{L^*}), && \text{(d)}\ \ X_0 = N(\mathcal{P}_L). \end{aligned} \tag{2.6}$$

Since

$$L\mathcal{P}_L x = L\Big(\sum_{i=1}^{r} \gamma_i(x)\, f_i\Big) = \sum_{i=1}^{r} \gamma_i(x)\, L f_i = 0,$$

we have $R(\mathcal{P}_L) \subset N(L)$. If $x \in N(L)$, then $x = \sum_{i=1}^{r} c_i f_i$. By applying the functionals γ_j, $j = 1, \dots, r$, to the last equality, we find $c_i = \gamma_i(x)$, i.e., $x = \sum_{i=1}^{r} \gamma_i(x) f_i$. Hence, $x = \mathcal{P}_L x$ and it follows from assertion (a) in Lemma 2.1 that $x \in R(\mathcal{P}_L)$. Consequently, $N(L) \subset R(\mathcal{P}_L)$, which completes the proof of equality (a) in relation (2.6).

Since

$$\mathcal{P}_{L^*} L x = \sum_{s=1}^{d} \varphi_s(Lx)\, \psi_s = \sum_{s=1}^{d} (L^* \varphi_s)\,(x)\, \psi_s = 0,$$

where φ_s are basis vectors of the null space of the operator L^*, we conclude that $R(L) \subset N(\mathcal{P}_{L^*})$. On the other hand, if $y \in N(\mathcal{P}_{L^*})$, then

$$\mathcal{P}_{L^*} y = \sum_{i=1}^{d} \varphi_s\,(y)\, \psi_s = 0,$$

i.e., $\varphi_s(y) = 0,\ s = 1, \ldots, d$. By virtue of the normal resolvability of the operator L, this means that $y \in R(L)$. Hence, $N(\mathcal{P}_{L^*}) \subset R(L)$ and equality (b) in (2.6) is proved.

Equalities (c) and (d) are proved similarly.

Thus, relations (2.3) and (2.6) imply that the projectors $\mathcal{P}_L$ and $\mathcal{P}_{L^*}$ split the spaces B_1 and B_2 into direct sums of closed subspaces, which completes the proof of Lemma 2.2.

We now construct finite-dimensional operators mentioned in the introduction to this chapter and study some of their properties.

Let $p = \min(r, d)$. Following [6], we introduce two operators

$$\overline{\mathcal{P}}_L x = \sum_{i=1}^{p} \gamma_i(x)\, \psi_i, \quad \overline{\mathcal{P}}_L \colon X \to \begin{cases} N_1(L^*) \subseteq N(L^*) & \text{for } r \leq d, \\ N(L^*) & \text{for } r \geq d, \end{cases}$$

$$\overline{\mathcal{P}}_{L^*} y = \sum_{s=1}^{p} \varphi_s(y)\, f_s, \quad \overline{\mathcal{P}}_{L^*} \colon Y \to \begin{cases} N(L) & \text{for } r \leq d, \\ N_1(L) \subseteq N(L) & \text{for } r \geq d. \end{cases} \tag{2.7}$$

Our aim is to show that the images of the operators $\overline{\mathcal{P}}_L$ and $\overline{\mathcal{P}}_{L^*}$ indeed belong to subspaces of the null spaces $N(L^*)$ and $N(L)$, respectively. As follows from equalities (a) in relations (2.6) and (2.2), the subspaces $N(L)$ and $N(L^*)$ are formed of elements satisfying the relations $\mathcal{P}_L x_0 = x_0$ for all $x_0 \in B_1$ and $\mathcal{P}_{L^*} y_0 = y_0$ for all $y_0 \in B_2$, respectively.

Let $x_0 = \overline{\mathcal{P}}_{L^*} y$ and $y_0 = \overline{\mathcal{P}}_L x$. Then

$$\begin{aligned} \mathcal{P}_L x_0 = \mathcal{P}_L \overline{\mathcal{P}}_{L^*} y &= \sum_{i=1}^{r} \gamma_i \Big(\sum_{s=1}^{p} \varphi_s(y)\, f_s \Big)\, f_i \\ &= \sum_{i=1}^{r} \sum_{s=1}^{p} \gamma_i(f_s)\, \varphi_s(y)\, f_i = \sum_{s=1}^{p} \varphi_s(y)\, f_s = x_0 \end{aligned}$$

because

$$\gamma_i(f_s) = \begin{cases} \delta_{is} & \text{for } i, s = 1, \dots, p, \\ 0 & \text{for } i > p. \end{cases}$$

Similarly,

$$\begin{aligned} \mathcal{P}_{L^*} y_0 = \mathcal{P}_{L^*} \overline{\mathcal{P}}_L x &= \sum_{s=1}^{d} \varphi_s \Big(\sum_{i=1}^{p} \gamma_i(x)\, \psi_i \Big)\, \psi_s \\ &= \sum_{s=1}^{d} \sum_{i=1}^{p} \varphi_s(\psi_i)\, \gamma_i(x)\, \psi_s = \sum_{i=1}^{p} \gamma_i(x)\, \psi_s = y_0 \end{aligned}$$

because

$$\varphi_s(\psi_i) = \begin{cases} \delta_{si} & \text{for } s, i = 1, \dots, p, \\ 0 & \text{for } s > p. \end{cases}$$

Thus, the images of the operators $\overline{\mathcal{P}}_L$ and $\overline{\mathcal{P}}_{L^*}$ are subspaces of the spaces $N(L^*)$ and $N(L)$, respectively.

The next lemma establishes relations between projectors (2.4) and operators (2.7).

Lemma 2.3. *The operators $\mathcal{P}_L$, $\mathcal{P}_{L^*}$, $\overline{\mathcal{P}}_L$, and $\overline{\mathcal{P}}_{L^*}$ satisfy the following relations:*

$$\begin{aligned} &(a)\ \mathcal{P}_{L^*}\overline{\mathcal{P}}_L = \overline{\mathcal{P}}_L \mathcal{P}_L = \overline{\mathcal{P}}_L, && (b)\ \mathcal{P}_L \overline{\mathcal{P}}_{L^*} = \overline{\mathcal{P}}_{L^*} \mathcal{P}_{L^*} = \overline{\mathcal{P}}_{L^*}, \\ &(c)\ \overline{\mathcal{P}}_L \overline{\mathcal{P}}_{L^*} = \sum_{s=1}^{p} \varphi_s(y)\, \psi_s, && (d)\ \overline{\mathcal{P}}_{L^*} \overline{\mathcal{P}}_L x = \sum_{i=1}^{p} \gamma_i(x)\, f_i. \end{aligned} \tag{2.8}$$

Lemma 2.3 is proved by the immediate verification of relations (2.8). We check, e.g., relation (a).

The equality $\mathcal{P}_{L^*}\overline{\mathcal{P}}_L = \overline{\mathcal{P}}_L$ has already been proved. Therefore,

$$\begin{aligned} \overline{\mathcal{P}}_L \mathcal{P}_L x &= \sum_{j=1}^{p} \gamma_j \Big(\sum_{i=1}^{r} \gamma_i(x)\, f_i \Big)\, \psi_j \\ &= \sum_{j=1}^{p} \sum_{i=1}^{r} \gamma_j(f_i)\, \gamma_i(x)\, \psi_j = \sum_{j=1}^{p} \gamma_j(x)\, \psi_j = \overline{\mathcal{P}}_L x \end{aligned}$$

because

$$\gamma_j(f_i) = \begin{cases} \delta_{ji} & \text{for } j, i = 1, \dots, p, \\ 0 & \text{for } i > p. \end{cases}$$

The remaining relations are proved by analogy.

2.2 An Analog of the Schmidt Lemma for Fredholm Operators

As already indicated, the Schmidt lemma is not true for Fredholm operators. By using the Atkinson theorem [6] on the representation of Fredholm operators in the form of a sum of unilaterally invertible and finite-dimensional operators, we arrive at the following analog of the Schmidt lemma for Fredholm operators:

Lemma 2.4. *Let $L\colon B_1 \to B_2$ be a bounded linear Fredholm operator of any index. Then the operator $\overline{L} = L + \overline{\mathcal{P}}_L$ has the bounded inverse*

$$\overline{L}_{l,r}^{-1} = \begin{cases} (L + \overline{\mathcal{P}}_L)_l^{-1} & \textit{left for } r \le d, \\ (L + \overline{\mathcal{P}}_L)_r^{-1} & \textit{right for } r \ge d. \end{cases}$$

Proof. Let $r \le d$. It follows from Theorem 1.9 that the operator $\overline{L}_l^{-1}$ (left inverse to $\overline{L}$) exists if and only if (a) $\ker \overline{L} = \{0\}$ and (b) $\operatorname{Im} \overline{L}$ is a subspace with direct complement in B_2.

We first show that $\ker \overline{L} = \{0\}$. Assume that there exists $x_0 \ne 0$, $x_0 \in B_1$, such that $(L + \overline{\mathcal{P}}_L)\, x_0 = 0$. Then

$$Lx_0 = -\sum_{i=1}^{p} \gamma_i\,(x_0)\,\psi_i.$$

By applying the functionals φ_s, $s = 1, \ldots, d$, to both sides of the last equality, we arrive at the system of equations

$$0 = (L^*\varphi_s)(x_0) = \varphi_s\,(Lx_0) = -\sum_{i=1}^{p} \gamma_i\,(x_0)\,\varphi_s\,(\psi_i) = -\gamma_s\,(x_0),$$
$$s = 1, \ldots, p = r.$$

Since the functionals γ_s, $s = 1, \ldots, p = r$, are linearly independent, by virtue of Corollary 1.3 of Theorem 1.3, we conclude that $x_0 = 0$, which contradicts the assumption. Thus, $\ker \overline{L} = \{0\}$.

Let us show that $\operatorname{Im} \overline{L}$ possesses a direct complement in B_2. To do this, we deduce an expression for $\overline{\mathcal{P}}_L^*$. Let $x \in B_1$ and $g \in B_2^*$. Then

$$g\,(\overline{\mathcal{P}}_L x) = g\,\Big(\sum_{i=1}^{p} \gamma_i\,(x)\,\psi_i\Big) = \sum_{i=1}^{p} \gamma_i\,(x)\,g\,(\psi_i) = \sum_{i=1}^{p} g\,(\psi_i)\,\gamma_i\,(x),$$

whence

$$\overline{\mathcal{P}}_L^* g = \sum_{i=1}^{p} g\,(\psi_i)\,\gamma_i.$$

We now determine the general form of functionals $g \in B_2^*$ satisfying the equation $(L+\overline{\mathcal{P}}_L)^* g = 0$. By using this equation, we get

$$L^* g = -\sum_{i=1}^{p} g(\psi_i)\gamma_i. \tag{2.9}$$

Applying the functionals $L^* g \in B_1^*$ and $\sum_{i=1}^{p} g(\psi_i)\gamma_i \in B_1^*$ to the elements f_k and using relation (2.9), we obtain the following system of equations:

$$0 = (L^* g) f_k = g(Lf_k) = -\sum_{i=1}^{p} g(\psi_i)\gamma_i(f_k) = -g(\psi_k), \quad k = 1,\dots,r.$$

Therefore, relation (2.9) takes the form $L^* g = 0$ and, hence, $g = \sum_{j=1}^{d} c_j \varphi_j$, where φ_j are basis vectors of the kernel $N(L^*)$. It has already been shown that $g(\psi_k) = 0$, $k = 1,\dots,p = r$. Thus, we can write

$$0 = g(\psi_k) = \sum_{j=1}^{d} c_j\varphi_j(\psi_k) = \sum_{j=1}^{p} c_j\varphi_j(\psi_k) + \sum_{j=p+1}^{d} c_j\varphi_j(\psi_k).$$

Since

$$\varphi_j(\psi_k) = \begin{cases} \delta_{jk} & \text{for } j,k = 1,\dots,p, \\ 0 & \text{for } j > p, \end{cases}$$

we conclude that c_j are arbitrary for $j = r+1,\dots,d$ and $c_j = 0$ for $j = 1,\dots,p = r$. Therefore,

$$g = \sum_{i=1}^{d-r} c_i\varphi_i \in N(L^*) \qquad \text{and} \qquad \dim\ker \overline{L}^* = d - r < \infty.$$

Thus, we have proved the existence of the left inverse $\overline{L}_l^{-1}$ for the operator $\overline{L}$.

Now let $r \geq d$. As follows from Theorem 1.8, for the existence of the right inverse operator $\overline{L}_r^{-1}$ it is necessary and sufficient that the following conditions be satisfied: (a) $\operatorname{im}\overline{L} = B_2$ and (b) the subspace $\ker\overline{L}$ has a direct complement in B_1.

We prove that $\ker\overline{L}^* = \{0\}$. Assume that there exists $g_0 \neq 0$, $g_0 \in B_2^*$, such that $(L+\overline{\mathcal{P}}_L)^* g_0 = 0$ and, hence,

$$L^* g_0 = -\sum_{i=1}^{p} g_0(\psi_i)\gamma_i. \tag{2.10}$$

Applying the functionals $L^* g_0 \in B_1^*$ and $\sum_{i=1}^{p} g_0(\psi_i)\gamma_i \in B_1^*$ to the elements f_j, $j = 1,\dots,r$, and using relation (2.10), we obtain the following system of equations:

$$0 = g_0(Lf_j) = (L^* g_0)\, f_j = -\sum_{i=1}^{p} g_0(\psi_i)\gamma_i(f_j) = -g_0(\psi_i), \quad i = 1, \ldots, p = d.$$

Since the elements ψ_i, $i = 1, \ldots, d$, are linearly independent, the equality $g_0\,(\psi_i) = 0$ is possible only for $g_0 \equiv 0$. The contradiction thus obtained proves that $\ker \overline{L}^* = \{0\}$.

Further, we establish the general form of elements $x \in B_1$ satisfying the equation $(L + \overline{\mathcal{P}}_L)\, x = 0$. This equation implies that

$$Lx = -\sum_{i=1}^{p} \gamma_i\,(x)\,\psi_i. \tag{2.11}$$

By applying thc functionals φ_s, $s = 1, \ldots, d$, to both sides of relation (2.11), we get

$$0 = (L^* \varphi_s)\,(x) = \varphi_s\,(Lx) = -\sum_{i=1}^{p} \gamma_i\,(x)\,\varphi_s\,(\psi_i) = -\gamma_i\,(x).$$

As a result, relation (2.11) takes the form $Lx = 0$, whence $x = \sum_{j=1}^{r} c_j f_j$, where f_j are basis vectors of the kernel $N(L)$. Since $\gamma_i(x) = 0$ for $i = 1, \ldots,$ $p = d \leq r$, we obtain

$$0 = \gamma_i(x) = \gamma_i \left(\sum_{j=1}^{r} c_j f_j \right) = \sum_{j=1}^{p} c_j \gamma_i(f_j) + \sum_{j=p+1}^{r} c_j \gamma_i(f_j).$$

In view of the relations

$$\gamma_i\,(f_j) = \begin{cases} \delta_{ij} & \text{for } \; i, j = 1, \ldots, p, \\ 0 & \text{for } \; j > p, \end{cases}$$

we conclude that c_j are arbitrary constants for $j = d + 1, \ldots, r$, and $c_j = 0$ for $j = 1, \ldots, p = d$. Therefore,

$$x = \sum_{i=1}^{r-d} c_i f_i \in N(\overline{L})$$

and $\operatorname{Im} \overline{L} = B_2$, which proves the existence of the right inverse $\overline{L}_r^{-1}$ of the operator $\overline{L}$.

Since the bounded linear operator L is a Fredholm (normally resolvable) operator [6], its image $R(L)$ is closed. The fact that the operator $(L + \overline{\mathcal{P}}_L)_{l,r}^{-1}$ is closed and, hence, bounded follows from the facts that $R(L)$ is closed and $R(\overline{\mathcal{P}}_L) = N_1(L^*)$ is finite-dimensional.

Remark 2.1. For $r = d = p$, L is a Fredholm operator of index zero, and the operator $\overline{L}$ possesses both left $(\overline{L}_l^{-1})$ and right $(\overline{L}_r^{-1})$ inverse operators. This means [146] that the inverse operator $\overline{L}^{-1}$ exists and, hence, Lemma 2.4 turns into the well-known Schmidt lemma [148].

We now establish some properties of the operator $\overline{L}_{l,r}^{-1} \in \mathcal{L}(B_2, B_1)$.

Lemma 2.5. *The operator* $\overline{L}_{l,r}^{-1} \in \mathcal{L}(B_2, B_1)$ *satisfies the relations*

$$\begin{aligned} &(a)\ \mathcal{P}_L \overline{L}_{l,r}^{-1} = \overline{\mathcal{P}}_{L^*}, && (b)\ L\overline{L}_{l,r}^{-1} = I_{B_2} - \mathcal{P}_{L^*}, \\ &(c)\ \overline{L}_{l,r}^{-1} \mathcal{P}_{L^*} = \overline{\mathcal{P}}_{L^*}, && (d)\ \overline{L}_{l,r}^{-1} L = I_{B_1} - \mathcal{P}_L, \end{aligned} \tag{2.12}$$

where I_{B_1} *and* I_{B_2} *are the identity operators in the spaces* B_1 *and* B_2*, respectively.*

Proof. Let $r \leq d$. We have

$$\overline{\mathcal{P}}_{L^*} L = 0 \qquad \text{and} \qquad \overline{\mathcal{P}}_{L^*} \overline{\mathcal{P}}_L = \mathcal{P}_L \qquad (p = \min(r, d) = r).$$

By applying the operator $L + \overline{\mathcal{P}}_L$ to both sides of relation (a) in (2.12) from the right, we arrive at the identity

$$\mathcal{P}_L = \overline{\mathcal{P}}_{L^*}(L + \overline{\mathcal{P}}_L) = \overline{\mathcal{P}}_{L^*} L + \overline{\mathcal{P}}_{L^*} \overline{\mathcal{P}}_L = \mathcal{P}_L,$$

which proves the required relation (a).

It follows from Lemma 2.3 that

$$\mathcal{P}_{L^*} \overline{\mathcal{P}}_L = \overline{\mathcal{P}}_L, \quad \mathcal{P}_{L^*}(Lx) = \sum_{s=1}^{d} \varphi_s(Lx)\, \psi_s = \sum_{s=1}^{d} (L^*\varphi_s)(x) \psi_s = 0.$$

Thus, by applying the operator $L + \overline{\mathcal{P}}_L$ to both sides of relation (b) in (2.12) from the right, we get

$$L = (I_{B_2} - \mathcal{P}_{L^*})(L + \overline{\mathcal{P}}_L) = L + \overline{\mathcal{P}}_L - \mathcal{P}_{L^*} L - \mathcal{P}_{L^*} \overline{\mathcal{P}}_L = L + \overline{\mathcal{P}}_L - \overline{\mathcal{P}}_L = L,$$

which proves the required relation. Since

$$\mathcal{P}_{L^*}^2 = \mathcal{P}_{L^*}, \quad L(\overline{\mathcal{P}}_{L^*} y) = L\Big(\sum_{s=1}^{p} \varphi_s(y)\, f_s\Big) = \sum_{s=1}^{p} \varphi_s(y)\, L f_s = 0, \tag{2.13}$$

by applying the operator L to both sides of relation (c) in (2.12) from the left and using relation (b) we obtain the identity

$$0 = (I_{B_2} - \mathcal{P}_{L^*}) \mathcal{P}_{L^*} = L\overline{L}_{l,r}^{-1} \mathcal{P}_{L^*} = L\overline{\mathcal{P}}_{L^*} = 0,$$

which proves relation (c).

Finally, applying the operator $\overline{L}_{l,r}^{-1}$ to both sides of relation (d) from the right and using the other properties in (2.12), we arrive at the identity

$$\overline{L}_{l,r}^{-1} I_{B_2} - \overline{L}_{l,r}^{-1} \mathcal{P}_{L^*} = \overline{L}_{l,r}^{-1} L \overline{L}_{l,r}^{-1} = (I_{B_1} - \mathcal{P}_L)\, \overline{L}_{l,r}^{-1} = I_{B_1} \overline{L}_{l,r}^{-1} - \mathcal{P}_L \overline{L}_{l,r}^{-1},$$

which proves the required relation. For $r \geq d$, the proof of the lemma is similar.

2.3 Generalized Inverse Operators for Bounded Linear Fredholm Operators

Various problems for ordinary differential, functional differential, integral, partial differential, and other equations are often represented in the form of a linear operator equation $Lx = y$, where $L\colon B_1 \to B_2$ is a Fredholm operator. This representation enables us to forget about specific difficulties encountered in every special case and focus our attention on more general regularities. Thus, if the operator L possesses the generalized inverse L^-, then a partial solution of the analyzed equation can be represented in explicit form as $x = L^- y$. Hence, the problem of finding conditions under which the equation $Lx = y$ is solvable and the development of procedures for the construction of generalized inverse operators seems to be quite urgent. Lemmas 2.4 and 2.5 proved in Section 2.2 enable us to suggest the following structure of the generalized inverse operator $L^-\colon B_2 \to B_1$ for a bounded linear Fredholm operator $L\colon B_1 \to B_2$.

Theorem 2.2. *The bounded generalized inverse of a bounded linear Fredholm operator L has the form*

$$L^- = \overline{L}_{l,r}^{-1} - \overline{\mathcal{P}}_{L^*}. \tag{2.14}$$

Proof. To prove the theorem, it is necessary and sufficient to show that the operator L^- satisfies conditions (1.7) specifying generalized inverse operators. We first prove that

$$LL^- = I_{B_2} - \mathcal{P}_{L^*} \quad \text{and} \quad L^- L = I_{B_1} - \mathcal{P}_L. \tag{2.15}$$

By using relation (b) in (2.12) and relation (2.13), we can write

$$LL^- = L\,(\overline{L}_{l,r}^{-1} - \overline{\mathcal{P}}_{L^*}) = L\overline{L}_{l,r}^{-1} - L\overline{\mathcal{P}}_{L^*} = I_{B_2} - \mathcal{P}_{L^*}.$$

In view of the relation

$$\overline{\mathcal{P}}_{L^*}\,(Lx) = \sum_{s=1}^{p} \varphi_s\,(Lx)\,f_s = \sum_{s=1}^{p} (L^*\varphi_s)\,(x)\,f_s = 0$$

and relation (d) in (2.12), we get

$$L^- L = (\overline{L}_{l,r}^{-1} - \overline{\mathcal{P}}_{L^*})\,L = \overline{L}_{l,r}^{-1}\,L - \overline{\mathcal{P}}_{L^*}\,L = I_{B_1} - \mathcal{P}_L.$$

We now use relation (2.15) to check the validity of (1.7). As a result, we obtain

$$LL^- L = L\,(I_{B_1} - \mathcal{P}_L) = L - L\mathcal{P}_L = L,$$

$$L^- LL^- = (I_{B_1} - \mathcal{P}_L)\,L^- = L^- - \mathcal{P}_L L^- = L^-$$

because $\mathcal{P}_L L^- = \mathcal{P}_L \overline{L}_{l,r}^{-1} - \mathcal{P}_L \overline{\mathcal{P}}_{L^*} = \overline{\mathcal{P}}_{L^*} - \overline{\mathcal{P}}_{L^*} = 0$ by virtue of Lemmas 2.4 and 2.5. This completes the proof of Theorem 2.2.

With the help of the proposed expression for the generalized inverse operator L^-, one can easily represent the general solution of the linear operator equation in the explicit form:

$$Lx = y, \tag{2.16}$$

where $L\colon B_1 \to B_2$ is a bounded linear Fredholm operator. The general solution of equation (2.16) can be represented in the form of the direct sum of the general solution of the homogeneous equation $Lx = 0$ and a special solution $L^- y$ of the inhomogeneous equation (2.16). The general solution of the homogeneous equation has the form of a linear combination of basis vectors f_i, $i = 1, \ldots, r$, i.e.,

$$x = \sum_{i=1}^{r} c_i f_i, \quad c_i \in R^1,$$

which can be rewritten as

$$x = (f_1, \ldots, f_r)\, c,$$

where $c = \operatorname{col}(c_1, \ldots, c_i, \ldots, c_r)$ is an arbitrary r-dimensional column vector of constants, $c \in R^r$, $f_i \in \ker L$, $i = 1, \ldots, r$.

Since any linear operator equation with Fredholm operator L is normally resolvable [6], Theorem 1.7 implies that the analyzed equation is solvable if and only if

$$\varphi_s(y) = 0, \quad s = 1, \ldots, d, \tag{2.17}$$

where φ_s are the solutions of the homogeneous equation $L^* g = 0$. If condition (2.17) is satisfied, then y necessarily belongs to the image $R(L)$ of the operator L. It follows from relation (b) in (2.6) that $R(L) = N(\mathcal{P}_{L^*})$. Therefore, relation (2.17) is equivalent to the following condition:

$$\mathcal{P}_{L^*} y = 0. \tag{2.18}$$

The reasoning presented above enables us to formulate the following theorem:

Theorem 2.3. *The Fredholm operator equation (2.16) is solvable if and only if its right-hand side $y \in B_2$ satisfies (2.18). Moreover, under the indicated condition, equation (2.16) possesses an r-parameter ($r = \dim\ker L$) family of solutions, which can be represented in the form of a direct sum*

$$x = (f_1, \ldots, f_r)\, c + L^- y, \tag{2.19}$$

the first term of which is the general solution of the corresponding homogeneous equation, and the second term is a special solution of the operator equation (2.16).

Proof. Substituting relation (2.19) in the original equation (2.16) and using relations (2.15) and (2.18), we get

$$Lx = L(f_1, \ldots, f_r)c + LL^- y = LL^- y$$
$$= (I_{B_2} - \mathcal{P}_{L^*})y = I_{B_2}y - \mathcal{P}_{L^*}y = I_{B_2}y = y.$$

Theorem 2.3 also covers the important "extreme" cases where either $\ker L = \{0\}$ or $\ker L^* = \{0\}$. In these cases, we say that L is an operator of full rank.

Corollary 2.1. *Suppose that* $\ker L = \{0\}$. *Then the operator equation (2.16) is solvable if and only if its right-hand side* $y \in B_2$ *satisfies condition (2.18). Moreover, under the indicated condition, equation (2.16) possesses the unique solution*

$$x = L_l^{-1} y. \tag{2.20}$$

Indeed, since $\ker L = \{0\}$, we have $r = 0$, $\overline{\mathcal{P}}_L = 0$, and $\overline{\mathcal{P}}_{L^*} = 0$. Therefore, relation (2.19) takes the form (2.20) and $L^- = \overline{L}_{l,r}^{-1} = L_l^{-1}$.

Corollary 2.2. *Assume that* $\ker L^* = \{0\}$. *Then the operator equation (2.16) is solvable for all* $y \in B_2$ *and possesses an* r*-parameter family of solutions*

$$x = (f_1, \ldots, f_r)c + L_r^{-1} y.$$

Indeed, since $\ker L^* = \{0\}$, we have $\mathcal{P}_{L^*} = 0$, and condition (2.18) is satisfied for all $y \in B_2$. Furthermore, $\overline{\mathcal{P}}_L = 0$, $\overline{\mathcal{P}}_{L^*} = 0$, and $L^- = L_{l,r}^{-1} = L_r^{-1}$.

In the first case, we say that the operator equation (2.16) satisfies the uniqueness theorem. In the second case, this equation satisfies the existence theorem.

2.4 Generalized Inverse Matrices

In the present section, the results obtained in the previous sections are used for the construction of generalized inverse matrices for rectangular real matrices, which can be regarded as a simple example of Fredholm operators.

Let A be an $m \times n$ matrix with constant elements and let $\operatorname{rank} A \leq \min(m, n)$. The basis vectors of the null spaces $N(A)$ and $N(A^*) = N(A^T)$ of the matrices A and $A^T = A^*$,[1] respectively, are denoted by $\{f_i\}_{i=1}^r$ and $\{\varphi_s\}_{s=1}^d$, i.e.,

$$Af_i = 0, \quad i = 1, \ldots, r, \qquad \text{and} \qquad A^T \varphi_s = 0, \quad s = 1, \ldots, d.$$

We choose systems $\{\gamma_j\}_{j=1}^r$ and $\{\psi_k\}_{k=1}^d$ biorthogonal to the basis systems:

$$(\gamma_j, f_i) = \delta_{ji}, \quad j, i = 1, \ldots, r, \qquad \text{and} \qquad (\varphi_s, \psi_k) = \delta_{sk}, \quad s, k = 1, \ldots, d,$$

where $(\cdot, \cdot)$ denotes an inner product in the Euclidean space.

[1] Since we consider only real spaces, for the operation T of transposition of matrices, we also use the symbol $*$ (here in what follows).

The projectors $\mathcal{P}_A$ and $\mathcal{P}_{A^*}$ are introduced by relations (2.4), i.e.,

$$\mathcal{P}_A x = \sum_{i=1}^{r} (\gamma_i, x)\, f_i, \qquad \mathcal{P}_{A^*} y = \sum_{s=1}^{d} (\varphi_s, y)\, \psi_s. \tag{2.21}$$

By using the definition of inner product in the Euclidean space, we can rewrite these expressions as follows:

$$\mathcal{P}_A x = \Big(\sum_{i=1}^{r} f_i \gamma_i^T\Big) x \quad \text{and} \quad \mathcal{P}_{A^*} y = \Big(\sum_{s=1}^{d} \psi_s \varphi_s^T\Big) y.$$

Thus, the projectors

$$\mathcal{P}_A = \sum_{i=1}^{r} f_i \gamma_i^T \quad \text{and} \quad \mathcal{P}_{A^*} = \sum_{s=1}^{d} \psi_s \varphi_s^T \tag{2.22}$$

are $n \times n$ and $m \times m$ matrices with constant entries.

Similarly, the finite-dimensional operators $\overline{\mathcal{P}}_A$ and $\overline{\mathcal{P}}_{A^*}$ defined by relations (2.7) are $m \times n$ and $n \times m$ matrices, respectively:

$$\overline{\mathcal{P}}_A = \sum_{i=1}^{p} \psi_i \gamma_i^T \quad \text{and} \quad \overline{\mathcal{P}}_{A^*} = \sum_{s=1}^{p} \gamma_s \varphi_s^T, \qquad p = \operatorname{min}(m, n).$$

Then $\overline{A} = A + \overline{\mathcal{P}}_A$ is an $m \times n$ matrix of full rank: $\operatorname{rank}(A + \overline{\mathcal{P}}_A) = \operatorname{min}(m, n)$. By virtue of Lemma 2.4, the matrix $\overline{A}$ possesses the left (right) inverse whenever $r \leq d$ $(r \geq d)$.

If we now determine the left (right) inverse matrix $\overline{A}_{l,r}^{-1}$, then, by using relation (2.14), we can find a matrix A^- generalized inverse to the matrix A.

Example 1. Let us find a matrix A^- generalized inverse to the matrix

$$A = \begin{bmatrix} 1 & 0 & 0 \\ 1 & 0 & 0 \\ 0 & 0 & 1 \\ 1 & 0 & 1 \end{bmatrix}.$$

The matrix conjugate to this matrix has the form

$$A^* = A^T = \begin{bmatrix} 1 & 1 & 0 & 1 \\ 0 & 0 & 0 & 0 \\ 0 & 0 & 1 & 1 \end{bmatrix}.$$

The vector

$$f_1 = \operatorname{col}(0,\ 1,\ 0),$$

is a basis vector of the null space $N(A)$, and the vectors

$$\varphi_1 = \text{col}\,(1,0,1,-1) \qquad \text{and} \qquad \varphi_2 = \text{col}\,(0,1,1,-1)$$

are basis vectors of the null space $N(A^*)$. The systems of vectors biorthogonal to the indicated basis vectors can be represented, e.g., in the form

$$\gamma_1 = \text{col}\,(0,1,0) \quad \text{and} \quad \psi_1 = \text{col}\,(1,0,0,0), \quad \psi_2 = \text{col}\,(0,1,0,0),$$

respectively. Then

$$\mathcal{P}_A = f_1\gamma_1^T = \begin{bmatrix} 0 \\ 1 \\ 0 \end{bmatrix} \begin{bmatrix} 0 & 1 & 0 \end{bmatrix} = \begin{bmatrix} 0 & 0 & 0 \\ 0 & 1 & 0 \\ 0 & 0 & 0 \end{bmatrix},$$

$$\begin{aligned} \mathcal{P}_{A^*} &= \psi_1\varphi_1^T + \psi_2\varphi_2^T \\ &= \begin{bmatrix} 1 \\ 0 \\ 0 \\ 0 \end{bmatrix} \begin{bmatrix} 1 & 0 & 1 & -1 \end{bmatrix} + \begin{bmatrix} 0 \\ 1 \\ 0 \\ 0 \end{bmatrix} \begin{bmatrix} 0 & 1 & 1 & -1 \end{bmatrix} \\ &= \begin{bmatrix} 1 & 0 & 1 & -1 \\ 0 & 1 & 1 & -1 \\ 0 & 0 & 0 & 0 \\ 0 & 0 & 0 & 0 \end{bmatrix}. \end{aligned}$$

Since $p = \min\,(r,\,d) = 1$, we obtain

$$\overline{\mathcal{P}}_A = \psi_1\gamma_1^T = \begin{bmatrix} 1 \\ 0 \\ 0 \\ 0 \end{bmatrix} \begin{bmatrix} 0 & 1 & 0 \end{bmatrix} = \begin{bmatrix} 0 & 1 & 0 \\ 0 & 0 & 0 \\ 0 & 0 & 0 \\ 0 & 0 & 0 \end{bmatrix},$$

$$\overline{\mathcal{P}}_{A^*} = f_1\varphi_1^T = \begin{bmatrix} 0 \\ 1 \\ 0 \end{bmatrix} \begin{bmatrix} 1 & 0 & 1 & -1 \end{bmatrix} = \begin{bmatrix} 0 & 0 & 0 & 0 \\ 1 & 0 & 1 & -1 \\ 0 & 0 & 0 & 0 \end{bmatrix}.$$

The matrix

$$\overline{A} = A + \overline{\mathcal{P}}_A = \begin{bmatrix} 1 & 1 & 0 \\ 1 & 0 & 0 \\ 0 & 0 & 1 \\ 1 & 0 & 1 \end{bmatrix}$$

has $\mathrm{rank}\,\overline{A} = 3$ and, thus, its left inverse exists and takes the form

$$\overline{A}_l^{-1} = \begin{bmatrix} 0 & 0 & -1 & 1 \\ 1 & 0 & 1 & -1 \\ 0 & 0 & 1 & 0 \end{bmatrix}.$$

Therefore, by virtue of Theorem 2.2, the matrix

$$\begin{aligned} A^- &= \overline{A}_l^{-1} - \overline{\mathcal{P}}_{A^*} \\ &= \begin{bmatrix} 0 & 0 & -1 & 1 \\ 1 & 0 & 1 & -1 \\ 0 & 0 & 1 & 0 \end{bmatrix} - \begin{bmatrix} 0 & 0 & 0 & 0 \\ 1 & 0 & 1 & -1 \\ 0 & 0 & 0 & 0 \end{bmatrix} \\ &= \begin{bmatrix} 0 & 0 & -1 & 1 \\ 0 & 0 & 0 & 0 \\ 0 & 0 & 1 & 0 \end{bmatrix} \end{aligned}$$

is the generalized inverse of the matrix A. This fact is confirmed by the validity of relations (1.7).

Chapter 3

PSEUDOINVERSE OPERATORS IN HILBERT SPACES

In the present chapter, we study the methods used for the construction of generalized inverse operators in Hilbert spaces. The geometry of Hilbert spaces formed by the presence of the operation of scalar multiplication (and, hence, the possibility of unique decomposition into direct sums of orthogonal subspaces and the fact that the dual spaces are isomorphic) makes it possible to establish more sophisticated results concerning inverse operators in Hilbert spaces. Indeed, in the set of generalized inverse operators, it is possible to select a single pseudoinverse operator (see [109, 112, 117]) with numerous remarkable properties. Thus, in particular, it minimizes the norm of the residual for the inhomogeneous operator equation $Lx = y$. This minimum can be attained due to the existence of the operators of orthogonal projection P_L and P_L^* with unit norm in Hilbert spaces. These operators are constructed in Section 3.1. In the same section, we also introduce finite-dimensional operators $\bar{P}_L$ and $\bar{P}_{L^*}$ and study their properties and relation to the orthoprojectors P_L and P_{L^*}. An analog of relation (2.14) for the construction of an operator pseudoinverse to a bounded linear Fredholm operator L is established.

3.1 Orthoprojectors, Their Properties, and Relation to Finite-Dimensional Operators

Let M_1 and M_2 be subspaces of a Hilbert space H such that

$$H = M_1 \oplus M_2.$$

Hence, every vector $h \in H$ admits a unique representation in the form

$$h = h_1 + h_2,$$

where $h_1 \in M_1$ and $h_2 \in M_2$. The vector h_1 is called a projection of the vector h onto M_1.

Definition 3.1. An operator defined in the entire Hilbert space H and mapping every element of this space $h \in H$ into its projection onto the subspace M_1 is called an operator of projection onto M_1 (or an orthoprojector) and is denoted by P or P_{M_1}, i.e.,

$$h_1 = P_{M_1} h = Ph.$$

It is clear that projection operators P are linear. Moreover, they are bounded and have the unit norm. Indeed, since $\|h\|^2 = \|h_1\|^2 + \|h_2\|^2$, we have $\|h_1\| \leq \|h\|$, i.e., $\|P_{M_1}h\| \leq \|h\|$. At the same time, if $h \in M_1$, then $h_1 = h$ and $\|P_{M_1}h\| = \|h\|$. Therefore, $\|P_{M_1}\| = 1$.

The orthoprojector P has the following properties:

$$\begin{aligned} &\text{(a)} \quad P^2 = P, \\ &\text{(b)} \quad P^* = P. \end{aligned} \tag{3.1}$$

In what follows, we present several simple assertions concerning the operations of multiplication, addition, and subtraction of orthoprojectors. The reader can find the proofs, e.g., in [1].

Theorem 3.1. *The product of two orthoprojectors P_{M_1} and P_{M_2} is an orthoprojector if and only if these orthoprojectors are commutative, i.e.,*

$$P_{M_1}P_{M_2} = P_{M_2}P_{M_1}. \tag{3.2}$$

Moreover, in this case,

$$P_{M_1}P_{M_2} = P_M,$$

where $M = M_1 \cap M_2$.

Corollary 3.1. *Any two subspaces M_1 and M_2 are orthogonal if and only if*

$$P_{M_1}P_{M_2} = 0.$$

Theorem 3.2. *The sum of orthoprojectors*

$$P_{M_1} + P_{M_2} + \ldots + P_{M_n} = P, \quad n < \infty$$

is an orthoprojector if and only if

$$P_{M_j}P_{M_k} = 0, \quad j \neq k,$$

i.e., if and only if the subspaces M_i, $i = 1, \ldots, n$, are mutually orthogonal. In this case, $P = P_M$, where $M = M_1 \oplus M_2 \oplus \ldots \oplus M_n$.

Theorem 3.3. *The difference of two orthoprojectors*

$$P_{M_1} - P_{M_2} = P_M$$

is an orthoprojector if and only if $M_2 \subset M_1$. In this case, P_M is an orthoprojector onto the subspace $M = M_1 \ominus M_2$.

We now study the properties of orthoprojectors related to the construction of the operator L^+ pseudoinverse to a bounded linear Fredholm operator L mapping a real Hilbert space H_1 into a real Hilbert space H_2.

As earlier, by $\{f_i\}_{i=1}^r$ and $\{\varphi_s\}_{s=1}^d$ we denote the bases of the null spaces $N(L)$ and $N(L^*)$ of the operator L and its adjoint operator L^* given by relation (1.2), respectively. Also let P_L and P_{L^*} be operators projecting the spaces H_1 and H_2 onto the corresponding null spaces, i.e.,

$$P_L\colon H_1 \to \ker L \qquad \text{and} \qquad P_{L^*}\colon H_2 \to \ker L^*.$$

The projectors P_L and P_{L^*} are defined by setting

$$\gamma_i(x) = \sum_{j=1}^{r} a_{ij}^{(-1)}(f_j, x), \quad \varphi_s(y) = \sum_{k=1}^{d} \beta_{sk}^{(-1)}(\varphi_k, y), \quad \psi_k = \varphi_k \tag{3.3}$$

in relation (2.4); here, $\alpha_{ij}^{(-1)}$ and $\beta_{sk}^{(-1)}$ are the entries of the matrices α^{-1} and β^{-1} inverse to the $(r \times r)$- and $(d \times d)$-dimensional symmetric Gram matrices

$$\alpha = [(f_i, f_j)] \qquad \text{and} \qquad \beta = [(\varphi_s, \varphi_k)],$$

respectively.

Thus, we have

$$P_L x = \sum_{i,j=1}^{r} \alpha_{ij}^{(-1)}(f_j, x) f_i \qquad \text{and} \qquad P_{L^*} y = \sum_{s,k=1}^{d} \beta_{sk}^{(-1)}(\varphi_k, y)\varphi_s, \tag{3.4}$$

where $(\cdot,\cdot)$ is the inner product in the corresponding Hilbert space.

Lemma 3.1. *The operators P_L and P_{L^*} given by relations (3.4) are orthoprojectors, i.e., they satisfy conditions (3.1).*

Proof. We present the proof for the operator P_L. We have

$$\begin{aligned}
P_L^2 x = P_L(P_L x) &= \sum_{i,j=1}^{r} \alpha_{ij}^{(-1)}\Big(f_j, \sum_{q,l=1}^{r} \alpha_{ql}^{(-1)}(f_l, x) f_q\Big) f_i \\
&= \sum_{i,j=1}^{r} \alpha_{ij}^{(-1)} \sum_{q,l=1}^{r} \alpha_{ql}^{(-1)}(f_l, x)(f_j, f_q) f_i \\
&= \sum_{i,j=1}^{r} \alpha_{ij}^{(-1)} \sum_{q,l=1}^{r} \alpha_{ql}^{(-1)} \alpha_{jq}(f_l, x) f_i \\
&= \sum_{i,j=1}^{r} \alpha_{ij}^{(-1)} \sum_{l=1}^{r} \delta_{lj}(f_l, x) f_i \\
&= \sum_{i,j=1}^{r} \alpha_{ij}^{(-1)}(f_j, x) f_i = P_L x
\end{aligned}$$

for any $x \in H_1$.

In view of the equality $(f_i, z) = (z, f_i)$ and the fact that the matrix α^{-1} is symmetric, we obtain

$$(P_L x, z) = \Big(\sum_{i,j=1}^{r} \alpha_{ij}^{(-1)} (f_j, x) f_i, z\Big)$$
$$= \sum_{i,j=1}^{r} \alpha_{ij}^{(-1)} (f_i, z)(x, f_j)$$
$$= \Big(x, \sum_{i,j=1}^{r} \alpha_{ij}^{(-1)} (f_i, z) f_j\Big) = (x, P_L z)$$

for any $x, z \in H_1$.

Let P_L and P_{L^*} be the orthoprojectors given by relations (3.4) and let $\mathcal{P}_L$ and $\mathcal{P}_{L^*}$ be the skew projectors onto the null spaces $N(L)$ and $N(L^*)$ of the operators L and L^*, respectively. By using the operator of scalar multiplication, we specify the operators $\mathcal{P}_L$ and $\mathcal{P}_{L^*}$ according to relations (2.4) as follows:

$$\mathcal{P}_L x = \sum_{i=1}^{r} (\gamma_i, x) f_i, \quad \text{where} \quad (\gamma_i, f_j) = \delta_{ij},$$
$$\mathcal{P}_{L^*} y = \sum_{s=1}^{d} (\varphi_s, y) \psi_k, \quad \text{where} \quad (\varphi_s, \psi_k) = \delta_{sk}.$$

Lemma 3.2. *The operators P_L, $P_{L^*}, \mathcal{P}_L$, and $\mathcal{P}_{L^*}$ satisfy the following relations:*

(a) $P_L \mathcal{P}_L = \mathcal{P}_L$;

(b) $\mathcal{P}_L P_L = P_L$;

(c) $P_{L^*} \mathcal{P}_{L^*} = P_{L^*}$;

(d) $\mathcal{P}_{L^*} P_{L^*} = \mathcal{P}_{L^*}$.

Proof. Since

$$\sum_{j=1}^{r} \alpha_{ij}^{(-1)} \alpha_{js} = \delta_{is}, \quad i, s = 1, \ldots, r,$$

we have

$$P_L \mathcal{P}_L x = \sum_{i,j=1}^{r} \alpha_{ij}^{(-1)} \Big(f_j, \sum_{s=1}^{r} (\gamma_s, x) f_s\Big) f_i$$
$$= \sum_{i,j=1}^{r} \sum_{s=1}^{r} \alpha_{ij}^{(-1)} \alpha_{js} (\gamma_s, x) f_i = \sum_{s=1}^{r} (\gamma_s, x) f_s = \mathcal{P}_L x.$$

In view of the fact that $(\gamma_s, f_i) = \delta_{si}$, we get

$$\mathcal{P}_L P_L x = \sum_{s=1}^{r}\Big(\gamma_s, \sum_{i,j=1}^{r} \alpha_{ij}^{(-1)}(f_j, x) f_i\Big) f_s$$
$$= \sum_{s=1}^{r}\sum_{i,j=1}^{r} \delta_{si}\alpha_{ij}^{(-1)}(f_j, x) f_s = \sum_{i,j=1}^{r} \alpha_{ij}^{(-1)}(f_j, x) f_i = P_L x.$$

Since $(\varphi_k, \psi_j) = \delta_{kj}$, we obtain

$$P_{L^*}\mathcal{P}_{L^*} y = \sum_{s,k=1}^{d} \beta_{sk}^{(-1)}\Big(\varphi_k, \sum_{j=1}^{d}(\varphi_j, y)\psi_j\Big)\varphi_s$$
$$= \sum_{s,k=1}^{d}\sum_{j=1}^{d} \beta_{sk}^{(-1)}\delta_{kj}(\varphi_j, y)\varphi_s$$
$$= \sum_{s,k=1}^{d} \beta_{sk}^{(-1)}(\varphi_k, y)\varphi_s = P_{L^*} y.$$

By using the equality $\sum_{s=1}^{d} \beta_{sk}^{(-1)}\beta_{js} = \delta_{jk}$, we get

$$\mathcal{P}_{L^*} P_{L^*} y = \sum_{j=1}^{d}\Big(\varphi_j, \sum_{s,k=1}^{d} \beta_{sk}^{(-1)}(\varphi_k, y)\varphi_s\Big)\psi_j$$
$$= \sum_{j=1}^{d}\sum_{s,k=1}^{d} \beta_{sk}^{(-1)}\beta_{js}(\varphi_k, y)\psi_j = \sum_{j=1}^{d}(\varphi_j, y)\psi_j = \mathcal{P}_{L^*} y.$$

To construct finite-dimensional operators $\bar{P}_L$ and $\bar{P}_{L^*}$, in relations (2.7), we set

$$\gamma_i(x) = \sum_{j=1}^{p} \bar{\alpha}_{ij}^{(-1)}(f_j, x), \quad \varphi_s(y) = \sum_{k=1}^{p} \bar{\beta}_{sk}^{(-1)}(\varphi_k, y), \quad \psi_s = \varphi_s \tag{3.5}$$

for $i, s = 1, \ldots, p = \min(r, d)$; here, $\bar{\alpha}_{ij}^{(-1)}$ and $\bar{\beta}_{sk}^{(-1)}$ are the elements of the matrices

$$\bar{\alpha}^{-1} = [(f_i, f_j)]^{-1} \quad \text{and} \quad \bar{\beta}^{-1} = [(\varphi_s, \varphi_k)]^{-1}, \quad i, j, s, k = 1, \ldots, p.$$

For the functionals γ_i and φ_s defined as indicated above, in the case of Hilbert spaces, relation (2.7) takes the form

$$\begin{aligned} \bar{P}_L x &= \sum_{i,j=1}^{p} \bar{\alpha}_{ij}^{(-1)}(f_j, x)\varphi_i, & \bar{P}_L \colon H_1 \to N_1(L^*) \subseteq N(L^*), \\ \bar{P}_{L^*} y &= \sum_{s,k=1}^{p} \bar{\beta}_{sk}^{(-1)}(\varphi_k, y) f_s, & \bar{P}_{L^*} \colon H_2 \to N_1(L) \subseteq N(L). \end{aligned} \tag{3.6}$$

Lemma 3.3. *The orthoprojectors P_L and P_{L^*} and the finite-dimensional operators $\bar{P}_L$ and $\bar{P}_{L^*}$ defined by (3.4) and (3.6) satisfy the following relations:*

$$\begin{aligned}
&(i) \quad P_{L^*}\bar{P}_L = \bar{P}_L P_L = \bar{P}_L,\\
&(ii) \quad P_L\bar{P}_{L^*} = \bar{P}_{L^*}P_{L^*} = \bar{P}_{L^*},\\
&(iii) \quad \bar{P}_L\bar{P}_{L^*}y = \sum_{s,k=1}^{p} \bar{\beta}_{sk}^{(-1)}(\varphi_k, y)\varphi_s,\\
&(iv) \quad \bar{P}_{L^*}\bar{P}_L x = \sum_{i,j=1}^{p} \bar{\alpha}_{ij}^{(-1)}(f_j, x)f_i.
\end{aligned} \tag{3.7}$$

Proof. Let us prove relation (ii). We can write

$$\begin{aligned}
P_L\bar{P}_{L^*}y &= \sum_{i,j=1}^{r} \alpha_{ij}^{(-1)}\Big(f_j, \sum_{s,k=1}^{p} \bar{\beta}_{sk}^{(-1)}(\varphi_k, y)f_s\Big)f_i\\
&= \sum_{s,k=1}^{p} \bar{\beta}_{sk}^{(-1)}(\varphi_k, y)\sum_{i,j=1}^{r} \alpha_{ij}^{(-1)}(f_j, f_s)f_i\\
&= \sum_{s,k=1}^{p} \bar{\beta}_{sk}^{(-1)}(\varphi_k, y)\sum_{i,j=1}^{r} \alpha_{ij}^{(-1)}\tilde{\alpha}_{js}f_i\\
&= \sum_{s,k=1}^{p} \bar{\beta}_{sk}^{(-1)}(\varphi_k, y)f_s,
\end{aligned}$$

which is true because

$$\sum_{j=1}^{r} \alpha_{ij}^{(-1)}\tilde{\alpha}_{js} = \begin{cases} \delta_{is} & \text{for } i, s = 1, \ldots, p,\\ 0 & \text{for } i > p. \end{cases}$$

At the same time, we have

$$\begin{aligned}
\bar{P}_{L^*}P_{L^*}y &= \sum_{s,k=1}^{p} \bar{\beta}_{sk}^{(-1)}\Big(\varphi_k, \sum_{i,j=1}^{d} \beta_{ij}^{(-1)}(\varphi_j, y)\varphi_i\Big)f_s\\
&= \sum_{s,k=1}^{p} \bar{\beta}_{sk}^{(-1)}\sum_{i,j=1}^{d} \beta_{ij}^{(-1)}(\varphi_k, \varphi_i)(\varphi_j, y)f_s\\
&= \sum_{s,k=1}^{p} \bar{\beta}_{sk}^{(-1)}\sum_{i,j=1}^{d} \tilde{\beta}_{ki}\beta_{ij}^{(-1)}(\varphi_j, y)f_s\\
&= \sum_{s,k=1}^{p} \bar{\beta}_{sk}^{(-1)}(\varphi_k, y)f_s
\end{aligned}$$

because

$$\sum_{i=1}^{r} \tilde{\beta}_{ki}\beta_{ij}^{(-1)} = \begin{cases} \delta_{kj} & \text{for } k,j = 1,\dots,p, \\ 0 & \text{for } j > p. \end{cases}$$

The other relations can be checked similarly. Note that if $r = p$, then $P_{L^*}P_L = P_L$ and if $d = p$, then $P_L P_{L^*} = P_{L^*}$.

3.2 An Analog of the Schmidt Lemma for Fredholm Operators

Lemma 2.4 proved in Section 2.2 for Banach spaces remains true (with some modifications) in Hilbert spaces.

Let $L\colon H_1 \to H_2$ be a bounded linear Fredholm operator of arbitrary index ($\operatorname{ind} L = \dim\ker L - \dim\ker L^* = r - d < \infty$).

Lemma 3.4. *The operator $\bar{L} = L + \bar{P}_L$ has the bounded inverse*

$$\bar{L}_{l,r}^{-1} = \begin{cases} (L + \bar{P}_L)_l^{-1} & \textit{left for } r \le d, \\ (L + \bar{P}_L)_r^{-1} & \textit{right for } r \ge d. \end{cases}$$

The proof of this lemma is similar to the proof of Lemma 2.4 and, therefore, we do not present it here. However, it should be emphasized that, unlike the case of Banach spaces, the functionals $\gamma_i(\cdot)$ and $\varphi_s(\cdot)$ are given by relations (3.3) and (3.5), respectively.

To establish the properties of the operator $\bar{L}_{l,r}^{-1}$, we study the action of the operators $\bar{L}$, $\bar{L}^*$, L_l^{-1}, and $(L_r^{-1})^*$ upon the basis vectors of the null spaces $N(L)$ and $N(L^*)$. First, we consider the relationship between the operators $(L_{l,r}^{-1})^*$ and $(L^*)_{r,l}^{-1}$.

Lemma 3.5. *The following relation is true:*

$$(\bar{L}_{l,r}^{-1})^* \sim (\bar{L}^*)_{r,l}^{-1}. \tag{3.8}$$

Proof. By using the obvious relations

$$(\bar{L}\bar{L}_r^{-1})^* = (\bar{L}_r^{-1})^*\bar{L}^* = I_{H_2}, \qquad (\bar{L}^*)_l^{-1}\bar{L}^* = I_{H_2}, \tag{3.9}$$

we conclude that $(\bar{L}_r^{-1})^* \sim (\bar{L}^*)_l^{-1}$. Similarly, it follows from the relations

$$(\bar{L}_l^{-1}\bar{L})^* = \bar{L}^*(\bar{L}_l^{-1})^* = I_{H_1}, \qquad \bar{L}^*(\bar{L}^*)_r^{-1} = I_{H_1} \tag{3.10}$$

that $(\bar{L}_l^{-1})^* \sim (\bar{L}^*)_r^{-1}$. Thus, we get $(L_{l,r}^{-1})^* \sim (\bar{L}^*)_{r,l}^{-1}$.

Note that, unlike the relation $(L^{-1})^* = (L^*)^{-1}$, which holds whenever the inverse operator L^{-1} exists, it is impossible to use the sign of equality in relation (3.8) because the operators $(\bar{L}_{l,r}^{-1})^*$ and $(\bar{L}^*)_{r,l}^{-1}$ are not uniquely determined. Their coincidence can be understood only in the sense of relations (3.9) and (3.10).

Lemma 3.6. *The operators $\bar{L}$, $\bar{L}_l^{-1}$, $\bar{L}^*$, and $(\bar{L}_r^{-1})^*$ satisfy the following relations:*

(i) $\bar{L}f_k = \varphi_k,\ k = 1,\dots,r = p;$

(ii) $\bar{L}_l^{-1}\varphi_k = f_k,\ k = 1,\dots,r = p;$

(iii) $\bar{L}^*\varphi_s = \sum\limits_{j=1}^{d} \xi_{js} f_j,\ s = 1,\dots,d = p;$

(iv) $(\bar{L}_r^{-1})^* f_s = (\bar{L}^*)_l^{-1} f_s = \sum\limits_{j=1}^{d} \xi_{sj}^{(-1)} \varphi_j,\ s = 1,\dots,d = p.$

Here, ξ_{js} are the entries of the symmetric matrix $\Xi = \bar{\alpha}^{-1}\beta$, and $\xi_{sj}^{(-1)}$ are the entries of the matrix inverse to Ξ.

Proof. Assume that $d \geq r = p$ and $x_0 = \sum\limits_{k=1}^{r} d_k f_k \in N(L)$, where d_k are arbitrary constants and $\{f_k\}_{k=1}^r$ are basis vectors of the null space $N(L)$. Then

$$\bar{L}x_0 = Lx_0 + \bar{P}_L x_0 = \bar{P}_L x_0$$

because

$$Lx_0 = \sum_{k=1}^{r} d_k L f_k = 0.$$

Further,

$$\begin{aligned}
\bar{L}\Big(\sum_{k=1}^{r} d_k f_k\Big) &= \sum_{i,j=1}^{r} \alpha_{ij}^{(-1)} \Big(f_j, \sum_{k=1}^{r} d_k f_k\Big)\varphi_i \\
&= \sum_{i,j=1}^{r} \alpha_{ij}^{(-1)} \sum_{k=1}^{r} d_k (f_j, f_k)\varphi_i \\
&= \sum_{k=1}^{r}\sum_{i,j=1}^{r} d_k \alpha_{ij}^{(-1)} \alpha_{jk}\varphi_i = \sum_{k=1}^{r} d_k\varphi_k,
\end{aligned}$$

whence it follows that

$$\bar{L}f_k = \varphi_k, \quad k = 1,\dots,r = p. \tag{3.11}$$

By virtue of Lemma 3.4, for $r \leq d$, there exists an operator $\bar{L}_l^{-1}$ left inverse to the operator $\bar{L}$. Applying the operator $\bar{L}_l^{-1}$ to both sides of relation (3.11) from the left, we get

$$\bar{L}_l^{-1}\bar{L}f_k = \bar{L}_l^{-1}\varphi_k, \quad k = 1,\dots,r = p,$$

and, hence,

$$\bar{L}_l^{-1}\varphi_k = f_k, \quad k = 1, \ldots, r = p. \tag{3.12}$$

We now consider the action of the operator $\bar{L}^*$ on the basis vectors φ_s of the null space $N(L^*)$. To do this, we find $\bar{P}_L^*$. Let $x \in H_1$ and $y \in H_2$. Then, by virtue of the definition of the adjoint operator acting in a Hilbert space (Definition 1.2), we can write

$$\begin{aligned}(\bar{P}_L x, y) &= \sum_{i,j=1}^{p} \bar{\alpha}_{ij}^{(-1)}(f_j, x)(\varphi_i, y) \\ &= \sum_{i,j=1}^{p} \bar{\alpha}_{ij}^{(-1)}(x, f_j)(\varphi_i, y) \\ &= (x, \sum_{i,j=1}^{p} \bar{\alpha}_{ij}^{(-1)}(\varphi_i, y) f_j) = (x, \bar{P}_L^* y).\end{aligned}$$

Since $\bar{\alpha}_{ij}^{(-1)} = \bar{\alpha}_{ji}^{(-1)}$, we have

$$\bar{P}_L^* y = \sum_{i,j=1}^{p} \bar{\alpha}_{ji}^{(-1)}(\varphi_i, y) f_j.$$

Now let $r \geq d = p$ and $y_0 = \sum_{s=1}^{d} c_s \varphi_s \in N(L^*)$. Then

$$\bar{L}^* y_0 = L^* y_0 + \bar{P}_L^* y_0 = \bar{P}_L^* y_0$$

because

$$\bar{L}^* y_0 = \sum_{s=1}^{d} c_s L^* \varphi_s = 0.$$

Further,

$$\begin{aligned}\bar{L}^*\Big(\sum_{s=1}^{d} c_s \varphi_s\Big) &= \sum_{i,j=1}^{d} \bar{\alpha}_{ji}^{(-1)}\Big(\varphi_i, \sum_{s=1}^{d} c_s \varphi_s\Big) f_j \\ &= \sum_{i,j=1}^{d} \bar{\alpha}_{ji}^{(-1)} \sum_{s=1}^{d} c_s (\varphi_i, \varphi_s) f_j \\ &= \sum_{i,j=1}^{d} \sum_{s=1}^{d} c_s \bar{\alpha}_{ji}^{(-1)} \beta_{is} f_i = \sum_{s,j=1}^{d} c_s \xi_{js} f_j,\end{aligned}$$

where $\xi_{js} = \sum_{i=1}^{d} \bar{\alpha}_{ji}^{(-1)} \beta_{is}$ are the entries of the symmetric nonsingular $d \times d$ matrix $\Xi = \bar{\alpha}^{-1}\beta$.

Thus, we get

$$\sum_{s=1}^{d} c_s \bar{L}^* \varphi_s = \sum_{s=1}^{d} c_s \sum_{j=1}^{d} \xi_{js} f_j,$$

whence

$$\bar{L}^* \varphi_s = \sum_{j=1}^{d} \xi_{js} f_j, \quad s = 1, \ldots, d = p. \tag{3.13}$$

By virtue of Lemma 3.4, the right inverse operator $\bar{L}_r^{-1}$ exists for $d \leq r$. Moreover, by virtue of Lemma 3.5, we have $(L_r^{-1})^* \sim (\bar{L}^*)_l^{-1}$. Hence, applying the operator $(L_r^{-1})^*$ to both sides of relation (3.13) from the left, we obtain

$$(\bar{L}_r^{-1})^* \bar{L}^* \varphi_s = (\bar{L}^*)_l^{-1} \bar{L}^* \varphi_s = (\bar{L}_r^{-1})^* \sum_{j=1}^{d} \xi_{js} f_j = (\bar{L}^*)_l^{-1} \sum_{j=1}^{d} \xi_{js} f_j,$$

and, therefore,

$$(\bar{L}_r^{-1})^* f_s = (\bar{L}^*)_l^{-1} f_s = \sum_{j=1}^{p} \xi_{sj}^{-1} \varphi_j, \quad s = 1, \ldots, d = p, \tag{3.14}$$

where ξ_{sj}^{-1} are the entries of the symmetric matrix $\Xi^{-1} = \beta^{-1} \bar{\alpha}$ inverse to the matrix Ξ.

Lemma 3.7. *The operator* $\bar{L}_{l,r}^{-1}$ *satisfies the following relations:*

$$\begin{aligned} &(i) \quad L\bar{L}_r^{-1} = I_{H_2} - P_{L^*}, \\ &(ii) \quad \bar{L}_l^{-1} L = I_{H_1} - P_L. \end{aligned} \tag{3.15}$$

Proof. According to the definition of right inverse operator, we have

$$\bar{L}\bar{L}_r^{-1} = I_{H_2},$$

On the other hand, $\bar{L} = L + \bar{P}_L$ and, hence, $L\bar{L}_r^{-1} = I_{H_2} - \bar{P}_L \bar{L}_r^{-1}$. To show that $\bar{P}_L \bar{L}_r^{-1} = P_{L^*}$, we write

$$\begin{aligned} \bar{P}_L(\bar{L}_r^{-1} y) &= \sum_{i,j=1}^{d} \bar{\alpha}_{ij}^{(-1)} (f_j, \bar{L}_r^{-1} y) \varphi_i \\ &= \sum_{i,j=1}^{d} \bar{\alpha}_{ij}^{(-1)} ((\bar{L}_r^{-1})^* f_j, y) \varphi_i \\ &= \sum_{i,j=1}^{d} \bar{\alpha}_{ij}^{(-1)} \Big(\sum_{k=1}^{d} \xi_{kj}^{(-1)} \varphi_k, y \Big) \varphi_i \\ &= \sum_{i,k=1}^{d} \beta_{ik}^{(-1)} (\varphi_k, y) \varphi_i = P_{L^*} y. \end{aligned}$$

This is true because

$$\sum_{j=1}^{d} \bar{\alpha}_{ij}^{(-1)} \xi_{jk}^{(-1)} = \beta_{ik}^{(-1)}.$$

Thus, we arrive at the equality

$$L\bar{L}_r^{-1} = I_{H_2} - P_{L^*}.$$

According to the definition of left inverse operator, we have

$$\bar{L}_l^{-1}\bar{L} = I_{H_1}.$$

This yields

$$\bar{L}_l^{-1}(L + \bar{P}_L) = \bar{L}_l^{-1}L + \bar{L}_l^{-1}\bar{P}_L = I_{H_1} - \bar{L}_l^{-1}\bar{P}_L.$$

To show that $\bar{L}_l^{-1}\bar{P}_L = P_L$, we write

$$\begin{aligned}
\bar{L}_l^{-1}\bar{P}_L &= \bar{L}_l^{-1}\Big(\sum_{i,j=1}^{r} \bar{\alpha}_{ij}^{(-1)}(f_j, x)\varphi_i\Big) \\
&= \sum_{i,j=1}^{r} \alpha_{ij}^{(-1)}(f_j, x)\bar{L}_l^{-1}\varphi_i \\
&= \sum_{i,j=1}^{r} \alpha_{ij}^{(-1)}(f_j, x) f_i = P_L x.
\end{aligned}$$

Thus, we arrive at the required equality

$$\bar{L}_l^{-1}L = I_{H_1} - P_L.$$

3.3 Left and Right Pseudoinverse Operators for Bounded Linear Fredholm Operators

The definition of the operator L^+ pseudoinverse to an operator L in the Moore–Penrose sense [109, 117] can be found in Chapter 1. According to this definition, the operator L^+ has the following properties:

$$\begin{aligned}
&\text{(i)} \quad LL^+L = L, \\
&\text{(ii)} \quad L^+LL^+ = L^+, \\
&\text{(iii)} \quad (LL^+)^* = LL^+ = I_{H_2} - P_{L^*}, \\
&\text{(iv)} \quad (L^+L)^* = L^+L = I_{H_1} - P_L.
\end{aligned} \tag{3.16}$$

This operator is unique [112, 123]. Since the operators P_L and P_{L^*} have unit norm, properties (iii) and (iv) imply that the solution $x = L^+y$ of the equation $Lx = y$ minimizes the norm of the residual

$$\|Lx - y\|_{H_2} = \|P_{L^*}y\|_{H_2},$$

and the solution $f = (L^*)^+g = (L^+)^*g$ of the conjugate equation $L^*f = g$ minimizes the norm of the residual

$$\|L^*f - g\|_{H_1} = \|P_L g\|_{H_1}.$$

Furthermore, since $(L^*)^+ = (L^+)^*$, these minima are attained for the same operator L^+. However, we have already noted that each minimum can be separately attained with the use of different operators. Indeed, if the generalized inverse operator L^- possesses properties (i)–(iii), then it gives a solution $x = L^-y$ minimizing the residual solely of the original equation. At the same time, if L^- has properties (i), (ii), and (iv), then the residual is minimized only for the conjugate equation. In this connection, we introduce the following definitions:

Definition 3.2. The generalized inverse operator $L^-: H_2 \to H_1$ that satisfies the conditions

$$LL^-L = L, \quad L^-LL^- = L^-, \quad (LL^-)^* = LL^- = I_{H_2} - P_{L^*} \tag{3.17}$$

is called the right pseudoinverse operator for an operator L and is denoted by L_r^+.

Definition 3.3. The generalized inverse operator $L^-: H_2 \to H_1$ satisfying the conditions

$$LL^-L = L, \quad L^-LL^- = L^-, \quad (L^-L)^* = L^-L = I_{H_1} - P_L, \tag{3.18}$$

is called the left pseudoinverse operator for an operator L and is denoted by L_l^+.

Clearly, an operator that is both left and right pseudoinverse is pseudoinverse in the Moore–Penrose sense.

The operator $\bar{L}_{l,r}^{-1}$ specified by Lemma 3.4 satisfies the relation $L\bar{L}_{l,r}^{-1}L = L$ but does not possess the second property of generalized inverse operators, namely, $\bar{L}_{l,r}^{-1}L\bar{L}_{l,r}^{-1} \neq \bar{L}_{l,r}^{-1}$. However, if the operator L_0^- is such that $LL_0^-L = L$ but $L_0^-LL_0^- \neq L_0^-$, then the operator $L_0^-LL_0^-$ satisfying both these conditions can be chosen as the generalized inverse operator L^- [63]. If the role of L^- is played by the operator $\bar{L}_{l,r}^{-1}L\bar{L}_{l,r}^{-1}$, then we get so-called unilateral pseudoinverse operators.

Theorem 3.4. *The operator*

$$L_{l,r}^{+} = \begin{cases} (I_{H_1} - P_L)\bar{L}_l^{-1} & \text{for } r \leq d, \\ \bar{L}_r^{-1}(I_{H_2} - P_{L^*}) & \text{for } r \geq d \end{cases} \tag{3.19}$$

is unilateral pseudoinverse to a bounded linear Fredholm operator L.

Proof. Let $r \geq d$. By virtue of Lemma 3.4, the operator $\bar{L}$ possesses the bounded left inverse operator $\bar{L}_r^{-1}$. It is necessary to verify relations (3.17). It follows from relation (b) in Lemma 2.5 that $L\bar{L}_r^{-1} = I_{H_2} - P_{L^*}$. Further, by using Lemma 3.2, we obtain

$$LL_r^{+}L = L\bar{L}_r^{-1}(I_{H_2} - P_{L^*})L = (I_{H_2} - P_{L^*})(I_{H_2} - P_{L^*})L = L$$

and

$$L_r^{+}LL_r^{+} = \bar{L}_r^{-1}(I_{H_2} - P_{L^*})L\bar{L}_r^{-1}(I_{H_2} - P_{L^*}) = \bar{L}_r^{-1}(I_{H_2} - P_{L^*}) = L_r^{+}.$$

Finally, we get

$$LL_r^{+} = L\bar{L}_r^{-1}(I_{H_2} - P_{L^*}) = I_{H_2} - P_{L^*} = (LL_r^{+})^*$$

because $I_{H_2}^* = I_{H_2}$ and $P_{L^*}^* = P_{L^*}$. Similarly, for $r \geq d$, the operator $L_r^{+} = \bar{L}_r^{-1}(I_{H_2} - P_{L^*})$ is right pseudoinverse to L.

3.4 Pseudoinverse Operators for Bounded Linear Fredholm Operators

For a bounded linear Fredholm operator L in a Banach space, the structure of the generalized inverse operator L^- specified by (2.14) does not determine the pseudoinverse operator in the case of Hilbert spaces.

At the same time, the structure of the unilateral pseudoinverse operator $L_{l,r}^{+}$ enables us to deduce a formula for the pseudoinverse operator.

Theorem 3.5. *The operator*

$$L^{+} = \begin{cases} L_l^{+}(I_{H_2} - P_{L^*}) & \text{for } r \leq d, \\ (I_{H_1} - P_L)L_r^{+} & \text{for } r \geq d \end{cases} \tag{3.20}$$

is the unique pseudoinverse operator for a bounded linear Fredholm operator L.

Proof. Since

$$L(I_{H_1} - P_L) = L \quad \text{and} \quad (I_{H_2} - P_{L^*})L = L,$$

relations (i) and (ii) in (3.16) are true. It is necessary to check relations (iii) and (iv). Since L_r^+ is the right pseudoinverse operator, we have

$$LL^+ = L(I_{H_1} - P_L)L_r^+ = LL_r^+ = I_{H_2} - P_{L^*} = (LL^+)^*.$$

Further,

$$L^+L = (I_{H_1} - P_L)L_r^+L = (I_{H_1} - P_L)\bar{L}_r^{-1}(I_{H_2} - P_{L^*})L$$
$$= (I_{H_1} - P_L)\bar{L}_r^{-1}L = (I_{H_1} - P_L)(I_{H_1} - \mathcal{P}_L),$$

where $\mathcal{P}_L$ is a projector to $N(L)$. Indeed, since

$$\bar{L}_r^{-1}L\bar{L}_r^{-1}L = \bar{L}_r^{-1}(I - P_{L^*})L = \bar{L}_r^{-1}L,$$

the operator $\bar{L}_r^{-1}L$ is a projector onto $R(L^*)$. Therefore, $I_{H_1} - \bar{L}_r^{-1}L = \mathcal{P}_L$ is also a projector. By virtue of Lemma 3.2, we have $P_L\mathcal{P}_L = \mathcal{P}_L$ and, hence,

$$(I_{H_1} - P_L)(I_{H_1} - \mathcal{P}_L) = I_{H_1} - P_L - \mathcal{P}_L + P_L\mathcal{P}_L = I_{H_1} - P_L.$$

Therefore,

$$L^+L = I_{H_1} - P_L = (L^+L)^*$$

because $I_{H_1}^* = I_{H_1}$ and $P_L^* = P_L$.

It is now necessary to check properties (iii) and (iv) for the second part of relation (3.20). We have

$$L^+L = L_l^+(I_{H_2} - P_{L^*})L = L_l^+L = I_{H_1} - P_L = (L^+L)^*$$

because L_l^+ is the left pseudoinverse operator. Further,

$$LL^+ = LL_l^+(I_{H_2} - P_{L^*}) = L(I_{H_1} - P_L)\bar{L}_l^{-1}(I_{H_2} - P_{L^*})$$
$$= L\bar{L}_l^{-1}(I_{H_2} - P_{L^*}) = (I_{H_2} - \mathcal{P}_{L^*})(I_{H_2} - P_{L^*}),$$

where $\mathcal{P}_{L^*}$ is the projector of the operator L^* to the null space $N(L^*)$. By virtue of Lemma 3.2, we have $\mathcal{P}_{L^*}P_{L^*} = \mathcal{P}_{L^*}$ and, therefore,

$$(I_{H_2} - \mathcal{P}_{L^*})(I_{H_2} - P_{L^*}) = I_{H_2} - \mathcal{P}_{L^*} - P_{L^*} + \mathcal{P}_{L^*}P_{L^*} = I_{H_2} - P_{L^*},$$

whence

$$LL^+ = I_{H_2} - P_{L^*} = (LL^+)^*.$$

This completes the proof.

We can also deduce another formula for the operator pseudoinverse to a Fredholm operator in a Hilbert space. Since any Hilbert space coincides with its dual space to within an isomorphism [129], the following compositions of operators

$L\colon H_1 \to H_2$ and $L^*\colon H_2 \to H_1$ are meaningful:

$$L^*L\colon H_1 \to H_1 \qquad \text{and} \qquad LL^*\colon H_2 \to H_2.$$

Thus, we have introduced the operator $S = L^*L\colon H_1 \to H_1$. This operator is self-adjoint because

$$S^* = (L^*L)^* = L^*L^{**} = L^*L = S.$$

Finally, by using relations (3.4), we construct the orthoprojector

$$P_S\colon H_1 \to N(S).$$

Lemma 3.8. *The operators P_L, P_S, and P_{S^*} satisfy the relation*

$$P_L = P_S = P_{S^*}.$$

Proof. As earlier, let $\{f_i\}_{i=1}^r$ be a basis in the null space $N(L)$. Since $Lf_i = 0$ for $i = 1, \dots, r$ and the operators L^* and L are linear, we get

$$Sf_i = L^*Lf_i = 0,$$

i.e., $N(L) \subseteq N(S)$. Let us show that the basis f_i, $i = 1, \dots, r$, of the kernel $N(L)$ cannot be supplemented by adding a linearly independent element $x_0 \in H_1$ such that $x_0 \in N(S)$, i.e., $Sx_0 = 0$. Assume the contrary, i.e., let there exist an element $x_0 \in H_1$ such that $Lx_0 = \psi \neq 0$ and $Sx_0 = L^*\psi = 0$. In this case, we arrive at a contradiction:

$$0 = (Lx_0, 0) = (x_0, L^*\psi) = (Lx_0, \psi) = (\psi, \psi) \neq 0.$$

Hence, the bases of the null spaces $N(L)$ and $N(S)$ coincide and, therefore, $P_L = P_S$. Since the operator S is self-adjoint, we conclude that $N(S) = N(S^*)$ and $P_S = P_{S^*}$.

Theorem 3.6. *The operator*

$$L^+ = (L^*L + P_L)^{-1}L^* = L^*(LL^* + P_{L^*})^{-1} \tag{3.21}$$

is the bounded pseudoinverse operator for a bounded linear Fredholm operator L.

Proof. We prove the relation $L^+ = (L^*L + P_L)^{-1}L^*$. The relation $L^+ = L^*(LL^* + P_{L^*})^{-1}$ is proved similarly.

The operator S is a Fredholm operator of index zero. Hence, the Schmidt lemma [148] is true for this operator, and the operator $S + P_S$ possesses the bounded inverse. The operator $S^+ = (S + P_S)^{-1} - P_S$ is the generalized inverse of S and satisfies the relation $S^+S = I_{H_1} - P_S$. Moreover, since P_S is an orthoprojector, the operator S^+ is not only the generalized inverse but also the unique pseudoinverse operator satisfying (1.8). Indeed, we have

(i) $SS^+S = S(I - P_S) = S - SP_S = S$,

(ii) $S^+SS^+ = (I - P_S)S^+ = S^+ - P_SS^+ = S^+$ (3.22)

because $S^+P_S = 0$ and $P_S(S + P_S)^{-1} = P_S$. Moreover,

(iii) $(S^+S)^* = (I_{H_1} - P_S)^* = I_{H_1} - P_S = S^+S$,

(iv) $(SS^+)^* = (I_{H_1} - P_{S^*})^* = I_{H_1} - P_{S^*} = SS^+$ (3.23)

because $P_S = P_{S^*}$, $I^*_{H_1} = I_{H_1}$, and $P_S = P^*_S$.

By using relations (3.22) and (3.23), we can show that the operator $L^+ = S^+L^*$ satisfies relations (3.16) and, hence, is the pseudoinverse operator for the operator L.

Since $P_L = P_S$, $LP_L = 0$, $S^+S = I_{H_1} - P_S$, and $P_SS^+ = 0$, we can write

(i) $LL^+L = LS^+L^*L = L(I_{H_1} - P_S) = L(I_{H_1} - P_L) = L$,

(ii) $L^+LL^+ = S^+L^*LL^+ = (I_{H_1} - P_S)L^+ = L^+ - P_SS^+L^* = L^+$.

Further, since [129]

$$L^{**} = L, \quad ((S + P_S)^{-1})^* = (S + P_S)^{-1}$$

and, hence,

$$(S^+)^* = ((S + P_S)^{-1} - P_S)^* = ((S + P_S)^{-1})^* - P^*_S = S^+,$$

we conclude that

(iii) $(L^+L)^* = (S^+L^*L)^* = (S^+S)^* = S^+S = S^+L^*L = L^+L$,

(iv) $(LL^+)^* = (LS^+L^*)^* = L^{**}(LS^+)^* = L(S^+)^*L^*LS^+L^* = LL^+$.

Finally, since $P_LL^* = (LP_L)^* = 0$ and $P_S = P_L = P^*_L$, we can write

$$L^+ = S^+L^* = [(S + P_S)^{-1} - P_S]L^* = (L^*L + P_L)^{-1}L^*.$$

Remark 3.1. In the case of a singular square matrix L (a simple example of a Fredholm operator of index zero), relation (3.21) for the pseudoinverse matrix L^+ was apparently deduced for the first time in [147]. In [19], this relation was generalized to the case of rectangular matrices.

Remark 3.2. Relations (3.21) enable one to compute the operators pseudoinverse to n- and d-normal operators in Hilbert spaces [88].

Note that if an operator L is such that $\ker L = 0$ or $\ker L^* = 0$, then relation (3.21) coincides with the formulas

$$L^+ = (L^*L)^{-1}L^* \quad (P_L \equiv 0),$$

$$L^+ = L^*(LL^*)^{-1} \quad (P_{L^*} \equiv 0)$$

presented in [112].

Remark 3.3. Relations (3.21) can be used to determine the pseudoinverse operator L^+ for a closed densely defined operator L.

3.5 Inverse Operators for Fredholm Operators of Index Zero

Relations (3.20) and (3.21) specifying the operators pseudoinverse to Fredholm operators remain valid for Fredholm operators of index zero.

To construct the operator pseudoinverse to a Fredholm operator of index zero in a Hilbert space, one can use an approach based on the Schmidt lemma.

Thus, let L be a Fredholm operator of index zero ($\operatorname{ind} L = 0,\ r = d = p$) and let $\{f_i\}_{i=1}^r$ and $\{\varphi_s\}_{s=1}^r$ be bases in the spaces $\ker L$ and $\ker L^*$, respectively.

Using relations (3.4) and (3.6), we construct the orthoprojectors

$$P_L \colon H_1 \to N(L) \qquad \text{and} \qquad P_{L^*} \colon H_2 \to N(L^*)$$

and finite-dimensional operators

$$\bar{P}_L \colon H_1 \to N(L^*) \qquad \text{and} \qquad \bar{P}_{L^*} \colon H_2 \to N(L).$$

Lemma 3.9. *The operators P_L, P_{L^*}, $\bar{P}_L$, and $\bar{P}_{L^*}$ satisfy the relations*

$$\begin{aligned} &(i) \quad P_{L^*}\bar{P}_L = \bar{P}_L P_L = \bar{P}_L; \\ &(ii) \quad P_L\bar{P}_{L^*} = \bar{P}_{L^*}P_{L^*} = \bar{P}_{L^*}; \\ &(iii) \quad \bar{P}_L\bar{P}_{L^*} = P_{L^*}; \\ &(iv) \quad \bar{P}_{L^*}\bar{P}_L = P_L. \end{aligned} \tag{3.24}$$

Proof. Lemma 3.9 follows from Lemma 3.3. Relations (iii) and (iv) in (3.24) follow from relations (iii) and (iv) in (3.7) because, for $r = d = p$, the expressions

$$\sum_{s,k=1}^{p} \beta_{sk}^{(-1)}(\varphi_k, y)\varphi_s \qquad \text{and} \qquad \sum_{i,j=1}^{p} \alpha_{ij}^{(-1)}(f_j, x)f_i$$

are, in fact, the orthoprojectors P_{L^*} and P_L, respectively.

Lemma 3.10. *The operator $L + \bar{P}_L$ possesses the bounded inverse operator.*

Proof. It is necessary and sufficient to show that

$$\ker(L + \bar{P}_L) = \{0\} \qquad \text{and} \qquad \ker(L + \bar{P}_L)^* = \{0\}.$$

Assume that there exists $x_0 \neq 0$, $x_0 \in H_1$, such that

$$(L + \bar{P}_L)x_0 = 0.$$

This means that

$$Lx_0 = -\sum_{i,j=1}^{r} \alpha_{ij}^{(-1)}(f_j, x_0)\varphi_i.$$

Multiplying both sides of the last equality scalarly by the functional φ_s, $s = 1, \dots, r$, we obtain

$$\begin{aligned} 0 &= (L^*\varphi_s, x_0) = (\varphi_s, Lx_0) \\ &= -\sum_{i,j=1}^{r} \alpha_{ij}^{(-1)}(f_j, x_0)(\varphi_s, \varphi_i) = \sum_{i,j=1}^{r} \alpha_{ij}^{(-1)}\beta_{si}(f_j, x_0). \end{aligned}$$

Since the $r \times r$ matrices α^{-1} and β are nonsingular, the last equality can be true only in the case where $(f_s, x_0) = 0$, $s = 1, \dots, r$. At the same time, in view of the fact that the system of basis vectors f_s is linearly independent, the equality $(f_s, x_0) = 0$ implies that $x_0 \equiv 0$. This contradiction proves that $\ker(L + \bar{P}_L) = \{0\}$.

We now show that $\ker(L + \bar{P}_L)^* = \{0\}$. The operator $\bar{P}_{L^*}$ adjoint to the operator $\bar{P}_L$ is determined in the proof of Lemma 3.6 and has the form

$$\bar{P}_L^* y = \sum_{i,j=1}^{r} \alpha_{ij}^{(-1)}(\varphi_i, y) f_j.$$

Assume that there exists $y_0 \neq 0$, $y_0 \in H_2 = H_2^*$, such that

$$(L + \bar{P}_L)^* y_0 = 0$$

or

$$L^* y_0 = -\sum_{i,j=1}^{r} \alpha_{ij}^{(-1)}(\varphi_i, y_0) f_j.$$

Multiplying both sides of the last equality scalarly by an element f_k, $k = 1, \dots, r$, we get

$$\begin{aligned} 0 &= (L^* y_0, f_k) = (y_0, Lf_k) \\ &= -\sum_{i,j=1}^{r} \alpha_{ij}^{(-1)}(y_0, \varphi_i)(f_j, f_k) = -\sum_{i,j=1}^{r} \alpha_{ij}^{(-1)}\alpha_{jk}(y_0, \varphi_i). \end{aligned}$$

Since the basis vectors φ_i, $i = 1, \dots, r$, are linearly independent, the equality $(y_0, \varphi_i) = 0$ holds only for $y_0 \equiv 0$. Thus, $\ker(L + \bar{P}_L)^* = \{0\}$. It follows from the proof that the operator $(L + \bar{P}_L)$ establishes a one-to-one correspondence between the spaces H_1 and H_2. By virtue of the Banach inverse-operator theorem [146], the operator $(L + \bar{P}_L)^{-1}$ exists and is bounded.

Theorem 3.7. *The operator*

$$L^+ = (L + \bar{P}_L)^{-1} - \bar{P}_{L^*} \tag{3.25}$$

is the bounded pseudoinverse operator for a bounded Fredholm operator L of index zero.

Proof. Theorem 3.7 follows from Theorem 3.4. Indeed, since $r = d = p$, the operator $(L + \bar{P}_L)$ possesses both left $(L + \bar{P}_L)_l^{-1}$ and right $(L + \bar{P}_L)_r^{-1}$ inverse operators. This means that the inverse operator $(L + \bar{P}_L)^{-1}$ exists. Therefore, by analogy with the proof of Theorem 3.4, we arrive at the required assertion of Theorem 3.7.

3.6 Criterion of Solvability and the Representation of Solutions of Fredholm Linear Operator Equations

The relations deduced for the operators pseudoinverse to bounded linear Fredholm operators in Hilbert spaces enable us to obtain, as in the case of Banach spaces, explicit expressions for the solutions of linear operator equations

$$Lx = y. \tag{3.26}$$

Unlike the case of Banach spaces, the expression $x = L^+y$ gives a unique special solution of equation (3.26), provided that this solution exists.

The relation

$$P_{L^*}y = 0, \tag{3.27}$$

where P_{L^*} is the orthoprojector to the kernel $\ker L^*$ of the operator L^*, guarantees the existence of a solution of equation (3.26).

The general solution of the homogeneous equation $Lx = 0$ corresponding to (3.26) can be represented in the form

$$x = [f_1, \dots, f_r]c_r, \tag{3.28}$$

where $\{f_i\}_{i=1}^r$ are basis vectors of the kernel $\ker L$.

Expression (3.28) for the general solution of the homogeneous equation can be rewritten in the following equivalent form:

$$x = P_L x$$

Indeed, substituting the last equality in the equation $Lx = 0$, we arrive at the identity

$$LP_L x = L \sum_{i,j=1}^{r} \alpha_{ij}^{(-1)}(f_j, x)f_i = \sum_{i,j=1}^{r} \alpha_{ij}^{(-1)}(f_j, x)Lf_i \equiv 0.$$

Theorem 3.8. *The Fredholm operator equation (3.26) is solvable if and only if $y \in H_2$ satisfies condition (3.27). In this case, equation (3.26) possesses an r-parameter ($r = \dim\ker L$) family of solutions, which can be represented in the form of a direct orthogonal sum as follows:*

$$x = P_L x \oplus L^+ y = [f_1, \dots, f_r] c_r \oplus L^+ y. \tag{3.29}$$

The first term of this sum is the general solution of the corresponding homogeneous equation and the second term is the unique solution of the operator equation (3.26) orthogonal to any solution of the homogeneous equation.

The assertion of the theorem follows from the equality

$$\begin{aligned} Lx = L(P_L x + L^+ y) &= LP_L x + LL^+ y \\ &= LP_L x + (I_{H_2} - P_{L^*}) y = I_{H_2} y = y, \end{aligned}$$

provided that condition (3.27) is satisfied.

For $P_{L^*} y \neq 0$, the problem is ill posed [145]. In this case, y does not belong to the image $R(L)$ of the operator L, and the operator equation (3.26) is unsolvable.

However, this equation has a so-called pseudosolution [145], which minimizes the residual $\|Lx - y\|_{H_2}$.

In the set of pseudosolutions specified by (3.29), one can find a unique pseudosolution $x_+ \in H_1^+ \subset H_1$ orthogonal to the kernel $\ker L$ of the operator L:

$$x_+ = L^+ y. \tag{3.30}$$

This pseudosolution can be regarded as the best possible approximate solution of equation (3.26) minimizing the norm of the residual

$$\|Lx_+ - y\|_{H_2} = \min_{x \in H_1} \|Lx - y\|_{H_2}.$$

The vector x_+ has the smallest norm

$$\|x_+\|_{H_1} = \min_{x \in H_1^+} \|x\|$$

in the set of all vectors for which the norm of the residual attains its minimum. Furthermore, the norm of the residual is equal to the norm of the expression on the left-hand side of the criterion (3.27) of solvability of equation (3.26):

$$\|Lx_+ - y\|_{H_2} = \|LL^+ y - y\|_{H_2} = \|(I - P_{L^*}) y - y\|_{H_2} = \|P_{L^*} y\|_{H_2}.$$

Thus, in the case where relation (3.27) is true, the norm of the residual is equal to zero and pseudosolution (3.30) turns into a solution of equation (3.26).

Example 1. We now establish a condition for the solvability of the following operator equation [92]:

$$(Lx)(t) = \dot{x}(t) + Ax(t) - \frac{1}{2\pi}\int_0^{2\pi} Bx(s)ds = f(t), \tag{3.31}$$

where $x(t) = \text{col}\,[x_1(t), x_2(t)]$, $f(t) = \text{col}\,[f_1(t), f_2(t)]$,

$$A = \begin{bmatrix} 0 & -1 \\ 1 & 0 \end{bmatrix}, \quad \text{and} \quad B = \begin{bmatrix} 0 & 0 \\ 1 & 0 \end{bmatrix}.$$

The operator $L\colon D_2[0, 2\pi] \to L_2[0, 2\pi]$ is a Fredholm operator. The homogeneous equation $(Lx)(t) = 0$ has three linearly independent solutions

$$x_1(t) = \begin{bmatrix} 1 \\ 0 \end{bmatrix}, \quad x_2(t) = \begin{bmatrix} \cos t \\ -\sin t \end{bmatrix}, \quad x_3(t) = \begin{bmatrix} \sin t \\ \cos t \end{bmatrix}.$$

The conjugate equation

$$(L^*y)(t) = \dot{y}(t) - A^T y(t) + \frac{1}{2\pi}\int_0^{2\pi} B^T y(s)ds$$

has a single solution $y_1(t) = \text{col}\,[-\sin t,\ 1 - \cos t]$.

The orthoprojector P_{L^*} is constructed by using relation (3.4). Since the conjugate homogeneous equation is uniquely solvable, the Gram matrix β turns into the number

$$\beta = \int_0^{2\pi} y_1^T(t) y_1(t)dt = \int_0^{2\pi} [\sin^2 t + (1 - \cos t)^2]dt = 4\pi$$

and, therefore,

$$\begin{aligned}(P_{L^*}y)(t) &= \frac{1}{4\pi}\begin{bmatrix} -\sin t \\ 1 - \cos t \end{bmatrix}\int_0^{2\pi} \begin{pmatrix} -\sin s & 1 - \cos s \end{pmatrix} \begin{bmatrix} y_1(s) \\ y_2(s) \end{bmatrix} ds \\ &= \frac{1}{4\pi}\begin{bmatrix} -\sin t \\ 1 - \cos t \end{bmatrix}\int_0^{2\pi} ((-\sin s)y_1(s) + (1 - \cos s)y_2(s))ds.\end{aligned}$$

Hence, the condition of solvability of the integrodifferential equation (3.31) takes the form

$$\frac{1}{4\pi}\begin{bmatrix} -\sin t \\ 1 - \cos t \end{bmatrix}\int_0^{2\pi} ((-\sin s)f_1(s) + (1 - \cos s)f_2(s))ds = 0.$$

Note that this condition is equivalent to the condition

$$\int_0^{2\pi} ((-\sin s) f_1(s) + (1 - \cos s) f_2(s)) ds = 0$$

obtained by a different method in [92] for the case $f_1(s) = 0$.

Example 2. We now find a solvability condition and the general solution of the equation [8]

$$(Lx)(t) = \dot{x}(t) - x(1) + x(0) = f(t), \quad t \in [0, 2]. \tag{3.32}$$

This functional differential equation can be rewritten in the following equivalent form:

$$(Lx)(t) \equiv \dot{x}(t) - \int_0^1 \dot{x}(s) ds = f(t).$$

The equation conjugate to (3.32) has the form

$$(L^* y)(t) = y(t) - \chi_{[0,1]}(t) \int_0^2 y(s) ds.$$

The operator $L\colon D_2[0,2] \to L_2[0,2]$ is a Fredholm operator ($\operatorname{\mathbf{ind}} L = 1$). The homogeneous equations $Lx = 0$ and $L^* y = 0$ have the linearly independent solutions

$$x_1(t) = t, \quad x_2(t) = 1 \quad \text{and} \quad y(t) = \chi_{[0,1]}(t),$$

respectively, where

$$\chi_{[0,1]}(t) = \begin{cases} 1 & \text{for } \ t \in [0, 1], \\ 0 & \text{for } \ t \in]1, 2] \end{cases}$$

is the characteristic function of the segment $[0, 1]$.

The orthoprojectors P_L and P_{L^*} given by relations (3.4) have the form

$$(P_L x)(t) = \frac{3}{2} t \int_0^2 (s - 1) x(s) ds + \int_0^2 \left(2 - \frac{3}{2} s\right) x(s) ds$$

and

$$(P_{L^*} y)(t) = \chi_{[0,1]}(t) \int_0^2 \chi_{[0,1]}(s) y(s) ds.$$

Thus, in view of the definition of the characteristic function $\chi_{[0,1]}(t)$, the condition of solvability for equation (3.32)

$$(P_{L^*} f)(t) = 0$$

takes the following equivalent form:

$$\int_0^1 f(s)ds = 0.$$

According to Theorem 3.9, for the general solution of equation (3.32) we can write

$$x(t) = c_1 t + c_2 + (L^+ f)(t),$$

where the unique pseudoinverse operator L^+ given by relation (3.21) admits the following representation:

$$(L^+ f)(t) = \int_0^t f(s)ds - t\chi_{[0,1]}(t)\int_0^2 \chi_{[0,1]}(s)f(s)ds + \frac{t}{2}\int_0^2 (\chi_{[0,1]}(s)-1)f(s)ds.$$

3.7 Integral Fredholm Equations with Degenerate Kernels in the Critical Cases

In the present section, we consider integral Fredholm equations with degenerate kernels

$$(Lx)(t) = x(t) - \sum_{i=1}^{n} \psi_i(t)\int_a^b \varphi_i(s)x(s)\,ds = f(t) \tag{3.33}$$

and the corresponding conjugate equations

$$(L^* y)(s) = y(s) - \sum_{i=1}^{n} \varphi_i(s)\int_a^b \psi_i(t)y(t)dt = g(s), \tag{3.34}$$

where

$$L\colon L_2[a,b] \to L_2[a,b] \qquad \text{and} \qquad L^*\colon L_2[a,b] \to L_2[a,b].$$

To find the bases of the kernels of the operators L and L^*, it is necessary to solve the homogeneous equations

$$(Lx)(t) = 0 \qquad \text{and} \qquad (L^* y)(s) = 0. \tag{3.35}$$

The solutions of these equations are sought in the form

$$x(t) = \sum_{i=1}^{n} \psi_i(t)c_i \qquad \text{and} \qquad y(s) = \sum_{i=1}^{n} \varphi_i(s)d_i. \tag{3.36}$$

Substituting solutions (3.36) in equations (3.35), we arrive at the following algebraic systems:

$$(I - A)c = 0 \qquad \text{and} \qquad (I - A^T)d = 0, \tag{3.37}$$

where

$$A = \Big(\int_a^b \varphi_i(t)\psi_j(t)dt \Big)$$

is an $n \times n$ matrix with constant entries,

$$c = \operatorname{col}[c_1, \ldots, c_n], \qquad \text{and} \qquad d = \operatorname{col}[d_1, \ldots, d_n].$$

The case where $\det(I - A) = \det(I - A^T) \neq 0$ (noncritical) was studied in [193]. In this case, systems (3.37) and, hence, the homogeneous equations (3.35) have only trivial solutions. By the Fredholm alternative, the inhomogeneous equations (3.33) and (3.34) are solvable for any $f(t)$ and $g(s)$, and their solutions can be represented in the explicit form.

We now consider the critical case where

$$\det(I - A) = 0 \qquad \text{and} \qquad \operatorname{rank}(I - A) = \operatorname{rank}(I - A^T) = n_1.$$

In this case, the solution of system (3.37) admits the representation

$$c = Fc_r, \quad d = F_1 d_r, \quad c_r, d_r \in R^r, \quad r = n - n_1,$$

where F and F_1 are $n \times r$ matrices formed by basis vectors of the null spaces $N(I - A)$ and $N(I - A^T)$:

$$F = [f^{(1)}, \ldots, f^{(r)}] \quad \text{and} \quad F_1 = [f_1^{(1)}, \ldots, f_1^{(r)}].$$

As a result, solutions (3.36) take the form

$$x(t) = \sum_{i=1}^{n} \bar{\psi}_i(t)c_i, \qquad y(s) = \sum_{i=1}^{n} \bar{\varphi}_i(s)d_i,$$

where

$$\bar{\psi}_i(t) = [\psi_1(t), \ldots, \psi_n(t)]f^{(i)} \qquad \text{and} \qquad \bar{\varphi}_i(s) = [\varphi_1(s), \ldots, \varphi_n(s)]f_1^{(i)}.$$

Thus, the complete systems of functions $\{\bar{\psi}_i(t)\}_{i=1}^{r}$ and $\{\bar{\varphi}_i(s)\}_{i=1}^{r}$ form the bases of $\ker L$ and $\ker L^*$, respectively.

Since L is a Fredholm operator of index zero, its pseudoinverse operator L^+ is given by relation (3.25).

We first find the orthoprojector P_{L^*} and the operators $\bar{P}_L$ and $\bar{P}_{L^*}$. We have

$$(P_{L^*}y)(t) = \sum_{i,j=1}^{r} \beta_{ij}^{(-1)} \bar{\varphi}_j(t) \int_a^b \bar{\varphi}_i(s)y(s)ds = \sum_{i=1}^{r} \tilde{\varphi}_i(t) \int_a^b \bar{\varphi}_i(s)y(s)ds,$$

$$(\bar{P}_L x)(t) = \sum_{i,j=1}^{r} \alpha_{ij}^{(-1)} \bar{\varphi}_j(t) \int_a^b \bar{\psi}_i(s)x(s)ds = \sum_{i=1}^{r} \tilde{\varphi}_i(t) \int_a^b \bar{\psi}_i(s)x(s)ds,$$

$$(\bar{P}_{L^*} y)(t) = \sum_{i,j=1}^{r} \beta_{ij}^{(-1)} \bar{\psi}_j(t) \int_a^b \bar{\varphi}_i(s)y(s)ds = \sum_{i=1}^{r} \tilde{\psi}_i(t) \int_a^b \bar{\varphi}_i(s)y(s)ds,$$

where

$$\tilde{\varphi}_i(t) = \sum_{j=1}^{r} \beta_{i,j}^{(-1)} \bar{\varphi}_j(t), \; \hat{\varphi}_i(t) = \sum_{j=1}^{r} \alpha_{i,j}^{(-1)} \bar{\varphi}_j(t),$$

$$\tilde{\psi}_i(t) = \sum_{j=1}^{r} \beta_{i,j}^{(-1)} \bar{\psi}_j(t).$$

By virtue of Lemma 3.10, the operator

$$[(L + \bar{P}_L)x](t) = x(t) - \sum_{i=1}^{n} \psi_i(t) \int_a^b \varphi_i(s)x(s)ds + \sum_{i=1}^{r} \tilde{\varphi}_i(t) \int_a^b \bar{\psi}_i(s)x(s)ds \tag{3.38}$$

has the bounded inverse.

To construct the operator $(L + \bar{P}_L)^{-1}$ inverse to the operator $(L + \bar{P}_L)$, we rewrite relation (3.38) in the form

$$[(L + \bar{P}_L)x](t) = x(t) - \sum_{k=1}^{n+r} \psi_k^{(1)}(t) \int_a^b \varphi_k^{(1)}(s)x(s)ds,$$

where

$$\psi_k^{(1)}(t) = [\psi_1(t), \ldots, \psi_n(t), \tilde{\varphi}_1(t), \ldots, \tilde{\varphi}_r(t)],$$

$$\varphi_k^{(1)}(t) = [\varphi_1(t), \ldots, \varphi_n(t), \bar{\psi}_1(t), \ldots, \bar{\psi}_r(t)]$$

for $k = 1, \ldots, n + r$.

The operator inverse to the operator $L + \bar{P}_L$ has the form [38]

$$[(L + \bar{P}_L)^{(-1)} f](t) = f(t) + \sum_{k,s=1}^{n+r} \frac{M_{ks}}{D} \psi_k^{(1)}(f) \int_a^b \varphi_s^{(1)}(\tau) f(\tau) d\tau,$$

where

$$D = \det(I - S) \neq 0, \quad S = \left(\int_a^b \varphi_s^{(1)}(t) \psi_k^{(1)}(t) dt \right)$$

for $s, k = 1, \ldots, n + r$, and M_{ks} are cofactors of the matrix $I - S$.

Hence, the pseudoinverse operator $L^+ \colon L_2[a,b] \to L_2[a,b]$ takes the form

$$(L^+f)(t) = \Big[[(L+\bar{P}_L)^{-1} - \bar{P}_{L^*}]f\Big](t)$$

$$= f(t) + \sum_{k,s=1}^{n+r} \frac{M_{ks}}{D}\psi_k^{(1)} \int_a^b \varphi_s^{(1)}(\tau) f(\tau) d\tau - \sum_{i=1}^{r} \tilde{\psi}_i(t) \int_a^b \bar{\varphi}_i(\tau) f(\tau) d\tau.$$

Therefore, in order that equation (3.33) be solvable, it is necessary and sufficient that $f(t) \in L_2[a,b]$ satisfy the condition

$$(P_{L^*}f)(t) = \sum_{i=1}^{r} \tilde{\varphi}_i(t) \int_a^b \bar{\varphi}_i(\tau) f(\tau) d\tau = 0.$$

In this case, the solution of equation (3.33) has the form

$$x(t) = \Big[\bar{\psi}_1(t), \ldots, \bar{\psi}_r(t)\Big] c_r + (L^+f)(t),$$

where the operator L^+ is defined earlier.

3.8 Pseudoinverse Matrices

In the present section, to illustrate the procedures of construction of pseudoinverse operators proposed in the previous sections, we give some auxiliary facts from the linear algebra and matrix theory frequently used in what follows. Thus, we consider the problem of construction of pseudoinverse matrices for rectangular matrices, establish conditions for the solvability of linear algebraic systems in the general case where the number of unknowns is, generally speaking, not equal to the number of equations, and study the representations of solutions of these systems.

In what follows, we restrict ourselves to the case of real domains and, thus, the conjugate matrix always coincides with the transpose.

Let Q be an $m \times n$ matrix with constant entries and let $\operatorname{rank} Q = n_1 \le \min(n,m)$. Identifying matrices with operators in a fixed space R^n, we obtain

$$R^n = R(Q^*) \oplus N(Q) \qquad \text{and} \qquad R^m = R(Q) \oplus N(Q^*).$$

The $n \times n$ and $m \times m$ matrices that play the role of orthoprojectors projecting the spaces R^n and R^m onto the null spaces $N(Q)$ and $N(Q^*)$ of the matrices Q and Q^* are denoted by P_Q and P_{Q^*}, respectively:

$$P_Q \colon R^n \to N(Q), \quad N(Q) = P_Q R^n$$

$$P_{Q^*} \colon R^m \to N(Q^*), \quad N(Q^*) = P_{Q^*} R^m.$$

Since the dimension of the null space $N(Q)$ is equal to the defect of the matrix Q and $\operatorname{rank} Q = n_1$, we have [151]

$$\dim N(Q) = n - \operatorname{rank} Q = n - n_1 = r.$$

In view of the fact that $\operatorname{rank} Q = \operatorname{rank} Q^*$, for the dimension of the null space $N(Q^*)$ we get

$$\dim N(Q^*) = m - \operatorname{rank} Q = m - n_1 = d.$$

Therefore, $\operatorname{rank} P_Q = r$ and $\operatorname{rank} P_{Q^*} = d$. This means that the matrix P_Q consists of r linearly independent columns and the matrix P_{Q^*} consists of d linearly independent rows.

To construct the matrices (orthoprojectors) P_Q and P_{Q^*}, we can use the following procedure [147]:

Denote the bases of the null spaces $N(Q)$ and $N(Q^*)$ by $\{f_i\}_{i=1}^r$ and $\{\varphi_s\}_{s=1}^d$, respectively. These vectors are used for the construction of nonsingular $(n \times n)$- and $(m \times m)$-dimensional Gram matrices:

$$\alpha = \{\alpha_{ij}\} = \{(f_i, f_j)\} \qquad \text{and} \qquad \beta = \{\beta_{sk}\} = \{(\varphi_s, \varphi_k)\},$$

where $(\cdot,\cdot)$ are the inner products in the corresponding Euclidean spaces. The orthoprojectors P_Q and P_{Q^*} are given by the formulas

$$\begin{aligned} P_Q &= \sum_{i,j=1} \alpha_{ij}^{(-1)} f_i f_j^T, \\ P_{Q^*} &= \sum_{s,k=1} \beta_{sk}^{(-1)} \varphi_s \varphi_k^T, \end{aligned} \tag{3.39}$$

where $\alpha_{ij}^{(-1)}$ and $\beta_{sk}^{(-1)}$ are the entries of the matrices inverse to the symmetric Gram matrices α and β, respectively.

The Moore–Penrose pseudoinverse $n \times m$ matrix Q^+ is given (according to (3.21)) by the formula

$$Q^+ = (Q^T Q + P_Q)^{-1} Q^T = Q^T (Q Q^T + P_{Q^*})^{-1}$$

or, according to [160, 184], by the formula

$$Q^+ := \lim_{\varepsilon \to 0} (Q^T Q + \varepsilon I_n)^{-1} Q^T = \lim_{\varepsilon \to 0} Q^T (Q Q^T + \varepsilon I_m)^{-1}.$$

Thus, we get

$$P_Q = I_n - Q^+ Q \qquad \text{and} \qquad P_{Q^*} = I_m - Q Q^+. \tag{3.40}$$

We now establish a solvability criterion and construct solutions for a linear algebraic equation

$$Qc = b, \tag{3.41}$$

where Q is a known $m \times n$ matrix of rank n_1, $\operatorname{rank} Q = n_1 \le \min(m, n)$, b is a known column vector from the space R^m, and c is an unknown column vector (from the space R^n).

Since any rectangular matrix is a simple Fredholm operator, by using Theorem 3.9, we conclude that the algebraic system (3.41) is solvable if and only if its right-hand side b belongs to the orthogonal complement $N^{\perp}(Q) = R(Q)$ of the subspace $N(Q^*)$, i.e., whenever

$$P_{Q^*} b = 0. \tag{3.42}$$

In this case, the general solution of system (3.41) takes the form

$$c = Q^+ b + \bar{c},$$

where $\bar{c}$ is a vector from the null space $N(Q)$, namely,

$$\bar{c} = P_Q c = P_Q \bar{c} \in N(Q).$$

Since $\operatorname{rank} P_{Q^*} = d$, condition (3.42) contains d linearly independent relations and the $m \times m$ matrix P_{Q^*} in the analyzed condition can be replaced by the $d \times m$ matrix $P_{Q_d^*}$ formed by a complete system of d linearly independent rows of the matrix P_{Q^*}. The column vector $\bar{c} = P_Q c \in N(Q)$ can be rewritten as $\bar{c} = P_{Q_r} c_r$, where P_{Q_r} is the $n \times r$ matrix formed by a complete system of r linearly independent columns of the matrix P_Q. Thus, we have proved the following theorem:

Theorem 3.9. *If* $\operatorname{rank} Q = n_1 \le \min(m, n)$, *then the algebraic system (3.41) is solvable if and only if the column vector* $b \in R^m$ *satisfies the condition*

$$P_{Q_d^*} b = 0, \quad d = m - n_1. \tag{3.43}$$

In this case, the system has an r*-parameter* $(r = n - n_1)$ *family of solutions of the form*

$$c = P_{Q_r} c_r + Q^+ b \quad \forall c_r \in R^r. \tag{3.44}$$

Corollary 3.2. *If* $\operatorname{rank} Q = n_1 = n$, *then system (3.41) is solvable if and only if the column vector* $b \in R^m$ *satisfies the condition*

$$P_{Q_d^*} b = 0, \quad d = m - n.$$

In this case, the system has a unique solution of the form

$$c = Q^+ b.$$

Indeed, since $r = 0$, we have $P_{Q_r} = 0$ and $P_{Q_r} c_r = 0$.

Corollary 3.3. *If* $\operatorname{rank} Q = n_1 = m$, *then system (3.41) is solvable for any* $b \in R^m$ *and its solution has the form*

$$c = P_{Q_r} c_r + Q^+ b, \quad r = n - m.$$

Indeed, since $\operatorname{rank} Q = m$, we have $d = m - m = 0$ and $P_{Q_d^*} \equiv 0$. Hence, condition (3.42) is always satisfied.

If condition (3.43) is not satisfied, then we get a simple ill-posed problem [60, 145]. In this case, the algebraic system (3.41) is inconsistent and unsolvable. However, there exists a so-called pseudosolution that minimizes the norm of the residual $\|Qc - b\|$ in the space R^n.

The set R_+ of pseudosolutions is specified by relation (3.44). In this set, one can find a single normal pseudosolution $c_+ \in R_+ \subset R^n$ orthogonal to the null space $N(Q)$:

$$c_+ = Q^+ b. \tag{3.45}$$

This solution is an approximate solution of system (3.41) (which can be obtained by the least-squares method) that minimizes the norm of the residual, i.e.,

$$\|Qc_+ - b\| = \min_{c \in R^n} \|Qc - b\|,$$

and has the smallest length in the set of all vectors for which this minimum is attained:

$$\|c_+\| = \min_{c \in R_+} \|c\|, \quad \|c\| = \left(\sum_{i=1}^{n} |c_i|^2 \right)^{1/2}.$$

Note that the norm of the residual is equal to the norm of the expression on the left-hand side of the solvability criterion (3.42) for system (3.41):

$$\|Qc_+ - b\| = \|QQ^+ b - b\| = \|P_{Q^*} b\|.$$

Thus, in the case where condition (3.42) is satisfied, the residual is equal to zero and pseudosolution (3.45) turns into the solution of system (3.41).

Example 1. To illustrate the proposed algorithm for the investigation of linear algebraic systems, we consider the problem of existence and construction of solutions for the matrix equation

$$Qc = b, \quad Q = \begin{bmatrix} 1 & 0 \\ 0 & 1 \\ 1 & 1 \end{bmatrix}, \quad b = \begin{bmatrix} 0 \\ 0 \\ 1 \end{bmatrix}, \quad c = \begin{bmatrix} c_1 \\ c_2 \end{bmatrix} c \in R^2. \tag{3.46}$$

First, we construct the matrices

$$Q^+ = \frac{1}{3} \begin{bmatrix} 2 & -1 & 1 \\ -1 & 2 & 1 \end{bmatrix},$$

$$P_Q = 0, \quad P_{Q^*} = \frac{1}{3}\begin{bmatrix} 1 & 1 & -1 \\ 1 & 1 & -1 \\ -1 & -1 & 1 \end{bmatrix}, \quad P_{Q_d^*} = \frac{1}{3}[1\ 1\ -1].$$

Further, we check the solvability criterion (3.43) for system (3.46):

$$P_{Q_d^*} b = \frac{1}{3}(1,\ 1,\ -1)(0,\ 0,\ 1)^T = -\frac{1}{3} \neq 0.$$

Thus, equation (3.46) is unsolvable but we can find its pseudosolution

$$c_+ = Q^+ b = \frac{1}{3}\begin{bmatrix} 2 & -1 & 1 \\ -1 & 2 & 1 \end{bmatrix}\begin{bmatrix} 0 \\ 0 \\ 1 \end{bmatrix} = \frac{1}{3}\begin{bmatrix} 1 \\ 1 \end{bmatrix},$$

which minimizes the norm of the residual.

Chapter 4

BOUNDARY-VALUE PROBLEMS FOR OPERATOR EQUATIONS

The constructions of generalized inverse operators L^- in Banach spaces and pseudoinverse operators L^+ in Hilbert spaces suggested in the previous chapters enable us to propose two approaches to the analysis of linear Fredholm boundary-value problems

$$(Lz)(t) = \varphi(t), \qquad lz = \lambda \in R^m. \tag{4.1}$$

The first approach uses the available data about the linear operator L of the original operator equation and is based on the construction of the generalized Green operator for a semihomogeneous ($\lambda = 0$) linear boundary-value problem corresponding to (4.1). It is shown that this approach is applicable not only to the analysis of boundary-value problems for classical operator equations, such as the systems of ordinary differential equations and differential equations with delay (Chapter 5) or impulsive differential systems (Chapter 6) characterized by the common property that the original operator equation $Lz = \varphi$ is everywhere solvable [88] but also to the construction of generalized Green operators for boundary-value problems in the case where L is a normally resolvable operator for which L^- or L^+ can be constructed according to the scheme presented above. We study basic properties of generalized Green operators of this sort and generalized inverse operators Λ^- for the operators

$$\Lambda = \text{col}\,[L, l]$$

of the original linear boundary-value problems. These systems with normally solvable operator L, which is not everywhere solvable, are studied in what follows, namely, we consider the boundary-value problems for integrodifferential equations (Theorem 4.6) and for systems of ordinary differential equations with degenerate matrix at the derivative (Section 5.7).

The second approach is based on the interpretation of the boundary-value problem (4.1) as an operator equation

$$(\Lambda z)(t) = f(t) \tag{4.2}$$

suggested in [153]. Here, $\Lambda = \text{col}\,[L, l]$ is a normally resolvable bounded linear operator for which the formulas of generalized inversion presented in Chapters 2 and 3 are true and

$$f(t) = \text{col}\,[\varphi(t), \lambda].$$

4.1 Linear Boundary-Value Problems for Fredholm Operator Equations

Let B_1 and B_2 be Banach spaces of vector functions $z\colon [a,b] \to R^n$ and $\varphi\colon [a,b] \to R^n$, where $-\infty < a \le t \le b < +\infty$, let $L\colon B_1 \to B_2$ be a bounded linear operator, and let $l\colon B_1 \to R^m$ be a bounded linear vector functional.

Definition 4.1. A system of linear operator equations

$$(Lz)(t) = \varphi(t), \tag{4.3}$$

$$lz = \lambda \tag{4.4}$$

is called an inhomogeneous linear boundary-value problem for the equation $Lz = \varphi$, and the equations $l_i z = \lambda_i$, $i = 1,\ldots,m$, are called the boundary conditions of this problem.

By using the approach proposed in [153], we represent the boundary-value problem (4.3), (4.4) in the form of an operator equation:

$$\Lambda z = \begin{bmatrix} L \\ l \end{bmatrix} z = \begin{bmatrix} \varphi \\ \lambda \end{bmatrix} = f,$$

where $\Lambda = \operatorname{col}[L, l]$ is a linear operator acting from the Banach space B_1 into the direct product $B_2 \times R^m$ of the Banach space B_2 and the Euclidean space R^m, i.e., $\Lambda\colon B_1 \to B_2 \times R^m$.

As shown in [8], this operator is bounded if the norm in the space $B_2 \times R^m$ is introduced as follows:

$$\|(\varphi,\lambda)\|_{B_2 \times R^m} = \|\varphi\|_{B_2} + \|\lambda\|_{R^m}, \qquad \varphi \in B_2, \quad \lambda \in R^m.$$

Parallel with the inhomogeneous boundary-value problem (4.3), (4.4), we consider the following homogeneous boundary-value problem:

$$(Lz)(t) = 0, \tag{4.5}$$

$$lz = 0. \tag{4.6}$$

By analogy with the classification of periodic boundary-value problems known in the theory of oscillations [5, 14, 101], we introduce the following definition:

Definition 4.2. A boundary-value problem for which the problem conjugate to the corresponding homogeneous linear boundary-value problem (4.5), (4.6) does not possess (possesses) nontrivial solutions is called noncritical (critical).

In other words, noncritical (critical) boundary-value problems satisfy the condition $\mathcal{P}_{\Lambda^*} = 0$ $(\mathcal{P}_{\Lambda^*} \neq 0)$, where $\mathcal{P}_{\Lambda^*}$ is the projector defined in what follows.

Consider the problem of conditions necessary and sufficient for the solvability of the inhomogeneous linear boundary-value problem (4.3), (4.4) and the structure of the set of its solutions $z(t) \in B_1$ under the assumption that $L: B_1 \to B_2$ is a bounded linear Fredholm operator ($\operatorname{ind} L = \dim\ker L - \dim\ker L^* = s - k < \infty$).

Theorem 2.3 shows that the Fredholm operator equation (4.3) is solvable if and only if its right-hand $\varphi(t) \in B_2$ satisfies the condition

$$(\mathcal{P}_{L^*}\varphi)(t) = 0, \tag{4.7}$$

where $\mathcal{P}_{L^*}: B_2 \to Y_0$ is the projector specified by relation (2.4). In this case, the general solution of equation (4.3) takes the form

$$z(t) = X(t)c + (L^-\varphi)(t), \quad c \in R^s, \tag{4.8}$$

where $X(t) = [f_1(t), f_2(t), \ldots, f_s(t)]$ is an $n \times s$ matrix formed by a complete system of linearly independent vectors from $\ker L$, and $L^-: B_2 \to B_1$ is the bounded generalized inverse operator (2.14) for the operator L.

Indeed, let $\mathcal{P}_L: B_1 \to N(L)$ be the projector (2.4) onto the null space of the operator L. Then every solution of the homogeneous equation $Lz = 0$ admits a representation

$$z(t) = (\mathcal{P}_L z)(t).$$

Since

$$(\mathcal{P}_L z)(t) = \sum_{i=1}^{s} \gamma_i(z) f_i(t) = [f_1(t), f_2(t), \ldots, f_s(t)] \begin{bmatrix} \gamma_1(z) \\ \gamma_2(z) \\ \vdots \\ \gamma_s(z) \end{bmatrix} = X(t)c,$$

where

$$c = \operatorname{col}[\gamma_1(z), \gamma_2(z), \ldots, \gamma_s(z)], \qquad \gamma_i(f_j) = \delta_{ij}, \quad i, j = 1, .., s,$$

the general solution of equation (4.3) can be represented in the form (4.8).

The solution (4.8) of the inhomogeneous operator equation (4.3) is a solution of the boundary-value problem (4.3), (4.4) if and only if the vector $c \in R^s$ satisfies the algebraic system of equations

$$lX(\cdot)c + l(L^-\varphi)(\cdot) = \lambda$$

obtained as a result of the substitution of solution (4.8) in the boundary condition (4.4).

Denote by $Q = lX(\cdot)$ an $m \times s$ matrix with constant entries. Also let $\mathcal{P}_Q: R^s \to N(Q)$ be the $s \times s$ matrix projector (2.22) and let $\mathcal{P}_{Q^*}: R^m \to N(Q^*)$ be the $m \times m$ matrix representing (2.22). The $s \times m$ matrix inverse to Q in the generalized sense is denoted by Q^-.

By using the algebraic equation

$$Qc = \lambda - l(L^{-}\varphi), \tag{4.9}$$

we find a constant $c \in R^s$ for which the solution (4.8) of equation (4.3) [this solution exists if condition (4.7) is satisfied] is a solution of the boundary-value problem (4.3), (4.4). It follows from Theorem 2.3 that equation (4.9) is solvable if and only if the following condition is satisfied:

$$\mathcal{P}_{Q_d^*}(\lambda - l(L^{-})(\cdot)) = 0 \quad (d = m - \operatorname{rank} Q).$$

In this case, the indicated equation possesses an r-parameter ($r = s - \operatorname{rank} Q$) family of solutions of the form (3.43):

$$c = \mathcal{P}_{Q_r} c_r + Q^{-}\{\lambda - l(L^{-}\varphi)(\cdot)\}, \tag{4.10}$$

where $\mathcal{P}_{Q_d^*}$ ($\mathcal{P}_{Q_r}$) is the $d \times m$ ($s \times r$) matrix formed by a complete system of d (r) linearly independent rows (columns) of the matrix $\mathcal{P}_{Q^*}$ ($\mathcal{P}_Q$).

Substituting relation (4.10) in expression (4.8), we obtain the solution of the boundary-value problem (4.3), (4.4) in the following general form:

$$z(t) = X_r(t)c_r + (G\varphi)(t) + X(t)Q^{-}\lambda,$$

where $X_r(t) = X(t)\mathcal{P}_{Q_r}$ is the $n \times r$ matrix whose columns form a complete system of r linearly independent solutions of the homogeneous boundary-value problem (4.5), (4.6). In other words, the columns of the matrix $X_r(t)$ form a basis in the kernel Λ of the operator $\Lambda = \operatorname{col}[L, l]$.

Definition 4.3. A bounded linear operator $G\colon B_2 \to B_1$ given by the formula

$$(G\varphi)(t) = (L^{-}\varphi)(t) - X(t)Q^{-}l(L^{-}\varphi)(\cdot) \tag{4.11}$$

is called the generalized Green operator of the inhomogeneous boundary-value problem (4.3), (4.4).

Thus, the following statement is true for the inhomogeneous boundary-value problem (4.3), (4.4):

Theorem 4.1. *The homogeneous boundary-value problem (4.5), (4.6) has exactly* $r = s - \operatorname{rank} Q$, $\operatorname{rank} Q \le \min(m, s)$, *linearly independent solutions. The inhomogeneous boundary-value problem (4.3), (4.4) with Fredholm operator* $L\colon B_1 \to B_2$ *is solvable if and only if* $\varphi \in B_2$ *and* $\lambda \in R^m$ *satisfy the conditions*

$$\begin{gathered} (\mathcal{P}_{L^*}\varphi)(t) = 0, \\ \mathcal{P}_{Q_d^*}\{\lambda - l(L^{-}\varphi)(\cdot)\} = 0 \quad (d = m - \operatorname{rank} Q). \end{gathered} \tag{4.12}$$

In this case, the analyzed problem possesses the r*-parameter family of solutions*

$$z_0(t, c_r) = X_r(t)c_r + (G\varphi)(t) + X(t)Q^{-}\lambda. \tag{4.13}$$

For a broad variety of boundary-value problems, the linear operator equation $Lz = \varphi$ is everywhere solvable [88]. Thus, this is true, e.g., for linear systems of ordinary differential equations, systems of linear differential equations with delay, some classes of systems of integral equations, etc. In [8], for these classes of functional differential equations, it was shown that if the operator equation (4.3) is solvable for any $\varphi \in B_2$, then the dimension s of the kernel of the operator L is equal to the dimension n of the system ($s = n$), and the boundary-value problem (4.3), (4.4) is a Fredholm problem with index $n - m$: $\operatorname{ind}[L, l] = n - m$. Moreover, the analyzed boundary-value problem is a Fredholm problem of index zero ($\operatorname{ind}\Lambda = 0$) if and only if $n = m$.

We now consider the boundary-value problem (4.3), (4.4) under the assumption that the operator equation (4.3) is everywhere solvable [88]. This means that $R(L) = B_2$. As follows from relation (1.3), the null space of the adjoint operator L^* consists of the zero element, i.e, $N(L^*) = \{0\}$ and $\mathcal{P}_{L^*} \equiv 0$. According to Corollary 2, the operator equation $Lz = \varphi$ is solvable for any $\varphi \in B_2$ and its solution takes the form

$$z(t) = X(t)c + (L_r^{-1}\varphi)(t), \quad c \in R^n, \tag{4.14}$$

where $X(t)$ is an $n \times n$ fundamental matrix of the homogeneous equation and L_r^{-1} is the bounded linear right inverse operator (according to Theorem 2.2, this operator exists). If the operator L acts from the space of absolutely continuous functions D_p^n into the space of summable functions L_p^n, then the operator $L_r^{-1}\colon L_p^n \to D_p^n$ admits an integral representation [79]

$$(L_r^{-1}\varphi)(t) = \int_a^b K(t,s)\varphi(s)ds,$$

whose kernel is called the Cauchy matrix.

Since the equation $Lz = \varphi$ is everywhere solvable, Theorem 4.1 can be simplified and formulated as follows:

Theorem 4.2. *The homogeneous boundary-value problem (4.5), (4.6) has exactly $r = n - \operatorname{rank} Q$, $\operatorname{rank} Q \le \min(m, n)$, linearly independent solutions. If the operator equation $Lz = \varphi$ is everywhere solvable, then the corresponding inhomogeneous boundary-value problem (4.3), (4.4) is solvable if and only if $\varphi \in B_2$ and $\lambda \in R^m$ satisfy the condition*

$$\mathcal{P}_{Q_d^*}\{\lambda - l(L_r^{-1}\varphi)(\cdot)\} = 0, \quad d = m - \operatorname{rank} Q.$$

In this case, the analyzed problem possesses the r-parameter family of solutions

$$z_0(t, c_r) = X_r(t)c_r + (G\varphi)(t) + X(t)Q^-\lambda.$$

Remark 4.1. If $m=n$ and $\operatorname{rank} Q=n$, then $\det Q=\det(lX(\cdot)) \neq 0$ ($\mathcal{P}_Q* \equiv 0$ and $\mathcal{P}_Q \equiv 0$) and $Q^- = Q^{-1}$. In this case, the boundary-value problem (4.3), (4.4) is solvable both everywhere and uniquely. Hence, the generalized Green operator turns into the well-known Green operator [8]

$$(G\varphi)(t) = (L^{-1}\varphi)(t) - X(t)(lX)^{-1}l(L^{-1}\varphi)(\cdot).$$

4.2 Generalized Green Operator

The generalized Green operator $G\colon B_2 \to B_1$ constructed in the previous section [see (4.11)] plays an important role in the analysis of linear and weakly nonlinear boundary-value problems for operator equations. Thus, by using this operator, one can construct the general solution of the semihomogeneous ($\lambda \equiv 0$) boundary-value problem

$$Lz = \varphi,$$

$$lz = 0,$$

provided that the solvability conditions (4.12) are satisfied. The following property of the generalized Green operator is frequently used in our presentation:

Lemma 4.1. *The generalized Green operator (4.11) satisfies the relation*

$$\Lambda G = \operatorname{col}[I_n - \mathcal{P}_{L^*}, \quad \mathcal{P}_{Q^*}(lL^-)]. \tag{4.15}$$

Proof. Indeed, we have

$$\Lambda G = \begin{bmatrix} L \\ l \end{bmatrix} [L^- - XQ^-lL^-] = \begin{bmatrix} L(L^- - XQ^-lL^-) \\ l(L^- - XQ^-lL^-) \end{bmatrix}$$

$$= \begin{bmatrix} LL^- - LXQ^-lL^- \\ lL^- - lXQ^-lL^- \end{bmatrix} = \begin{bmatrix} LL^- \\ (I_m - QQ^-)lL^- \end{bmatrix} = \begin{bmatrix} (I - \mathcal{P}_{L^*}) \\ \mathcal{P}_{Q^*}lL^- \end{bmatrix}$$

because $LX = 0$ and $I - QQ^- = \mathcal{P}_{Q^*}$.

Remark 4.2. In the case where the equation $Lz = \varphi$ is everywhere solvable, relation (4.15) takes the form

$$\Lambda G = \operatorname{col}[I_n, \quad \mathcal{P}_{Q^*}lL_r^{-1}]$$

because $R(L) = B_2$ and $\mathcal{P}_{L^*} \equiv 0$.

As shown in Section 4.1, under condition (4.12) the boundary-value problem

$$\Lambda z = \begin{bmatrix} L \\ l \end{bmatrix} z = \begin{bmatrix} \varphi \\ \lambda \end{bmatrix}$$

is solvable and its solution has the form (4.13).

The bounded generalized inverse operator $[L,l]^-\colon B_2\times R^m\to B_1$ admits the representation

$$[L,l]^-\begin{bmatrix}\varphi\\ \lambda\end{bmatrix}=G\varphi+XQ^-\lambda,$$

where $G\varphi=[L,l]^-\begin{bmatrix}\varphi\\ 0\end{bmatrix}$ is the solution of the problem

$$Lz=\varphi,$$
$$lz=0,$$

and $XQ^+\lambda=[L,l]^-\begin{bmatrix}0\\ \lambda\end{bmatrix}$ is the solution of the problem

$$Lz=0,$$
$$lz=\lambda.$$

Thus, it is now possible to prove the following statement:

Theorem 4.3. *The operator*

$$\Lambda^-=[G,XQ^-] \tag{4.16}$$

is the bounded generalized inverse operator for the bounded linear operator

$$\Lambda=\operatorname{col}(L,l).$$

Proof. To prove the theorem, it is necessary and sufficient to show that the operator Λ^- satisfies relation (1.7) specifying generalized inverse operators. As indicated in [63], the second relation in (1.7) follows from the first relation. Therefore, it is necessary to prove that the operator Λ^- satisfies the relation $\Lambda^-\Lambda\Lambda^-=\Lambda^-$. Since $LX\equiv 0$ and $lX=Q$, we obtain

$$\Lambda\Lambda^-=\begin{bmatrix}L\\ l\end{bmatrix}[G,\,XQ^-]=\begin{bmatrix}LG & LXQ^-\\ lG & lXQ^-\end{bmatrix}=\begin{bmatrix}I_n-\mathcal{P}_{L^*} & 0\\ \mathcal{P}_{Q^*}lL^- & QQ^-\end{bmatrix}.$$

Note that relations (2.8) and (2.14) imply that $L^-\mathcal{P}_{L^*}=0$, $Q^-\mathcal{P}_{Q^*}=0$, and $G\mathcal{P}_{L^*}=0$. Hence, we can write

$$\begin{aligned}\Lambda^-\Lambda\Lambda^-&=[G,XQ^-]\begin{bmatrix}I_n-\mathcal{P}_{L^*} & 0\\ \mathcal{P}_{Q^*}lL^- & QQ^-\end{bmatrix}\\ &=[G(I_n-\mathcal{P}_{L^*})+XQ^-\mathcal{P}_{Q^*}lL^-,\ XQ^-QQ^-]\\ &=[G-G\mathcal{P}_{L^*}+XQ^-\mathcal{P}_{Q^*}lL^-,\ XQ^-]=[G,\ XQ^-].\end{aligned}$$

The fact that the operator Λ^- is bounded follows from the boundedness of the operator L^- and the fact that the operators G and XQ^- are finite-dimensional (and, hence, bounded) [11]. Consequently, the operator $\Lambda^-\colon B_2 \times R^m \to B_1$ is the bounded generalized inverse operator for the operator Λ of the original boundary-value problem.

By using expression (2.15) for the generalized inverse operator Λ^- via the projector $\mathcal{P}_\Lambda\colon B_2 \times R^m \to \ker\Lambda$ and relation (4.15) for the generalized Green operator, we determine the projector $\mathcal{P}_{\Lambda^*}$ characterized by the block structure. The equality $\Lambda\Lambda^- = I - \mathcal{P}_{\Lambda^*}$ yields

$$\mathcal{P}_{\Lambda^*} = I_{n+m} - \Lambda\Lambda^- = \begin{bmatrix} I_n & 0 \\ 0 & I_m \end{bmatrix} - \begin{bmatrix} L \\ l \end{bmatrix} [G, XQ^-]$$

$$= \begin{bmatrix} I_n & 0 \\ 0 & I_m \end{bmatrix} - \begin{bmatrix} I_n - \mathcal{P}_{L^*} & 0 \\ \mathcal{P}_{Q^*} l L^- & QQ^- \end{bmatrix}$$

$$= \begin{bmatrix} \mathcal{P}_{L^*} & 0 \\ -\mathcal{P}_{Q^*} l L^- & (I_m - QQ^-) \end{bmatrix} = \begin{bmatrix} \mathcal{P}_{L^*} & 0 \\ -\mathcal{P}_{Q^*} l L^- & \mathcal{P}_{Q^*} \end{bmatrix}.$$

The operator $\mathcal{P}_{\Lambda^*}$ is indeed a projector because it possesses the property $\mathcal{P}^2_{\Lambda^*} = \mathcal{P}_{\Lambda^*}$. Thus, we can write

$$\mathcal{P}^2_{\Lambda^*} = \begin{bmatrix} \mathcal{P}_{L^*} & 0 \\ -\mathcal{P}_{Q^*} l L^- & \mathcal{P}_{Q^*} \end{bmatrix} \begin{bmatrix} \mathcal{P}_{L^*} & 0 \\ -\mathcal{P}_{Q^*} l L^- & \mathcal{P}_{Q^*} \end{bmatrix}$$

$$= \begin{bmatrix} \mathcal{P}^2_{L^*} & 0 \\ -\mathcal{P}_{Q^*} l L^- \mathcal{P}_{L^*} - \mathcal{P}^2_{Q^*} l L^- & \mathcal{P}^2_{Q^*} \end{bmatrix}$$

$$= \begin{bmatrix} \mathcal{P}_{L^*} & 0 \\ -\mathcal{P}_{Q^*} l L^- & \mathcal{P}_{Q^*} \end{bmatrix} = \mathcal{P}_{\Lambda^*}.$$

It is clear that the condition $\mathcal{P}_{\Lambda^*} \begin{bmatrix} \varphi \\ \lambda \end{bmatrix} = 0$ is equivalent to the conditions of solvability (4.12) for the boundary-value problem (4.3), (4.4).

Remark 4.3. If the operator equation (4.3) in the analyzed boundary-value problem is everywhere solvable, then the projector $\mathcal{P}_{\Lambda^*}$ has the form

$$\begin{bmatrix} 0 & 0 \\ -\mathcal{P}_{Q^*} l L_r^{-1} & \mathcal{P}_{Q^*} \end{bmatrix}$$

because, in this case, we have $\mathcal{P}_{L^*} \equiv 0$ and $L^- = L_r^{-1}$.

Remark 4.4. For the integral representation [79] of the right inverse operator $L_r^{-1}\colon L_p^n \to D_p^n$, the generalized Green operator takes the form

$$(G\varphi)(t) = \int_a^b K(t,s)\varphi(s)ds - X(t)Q^{-}l\int_a^b K(\cdot,s)\varphi(s)ds.$$

Moreover, if the vector functional $l\colon D_p^n \to R^m$ satisfies the condition [8, 85]

$$l\int_a^b K(\cdot,s)\varphi(s)ds = \int_a^b lK(\cdot,s)\varphi(s)\,ds,$$

then the generalized Green operator admits the representation

$$(G\varphi)(t) = \int_a^b G(t,s)\varphi(s)ds,$$

whose kernel

$$G(t,s) = K(t,s) - X(t)Q^{-}lK(\cdot,s).$$

is called a *generalized Green matrix.*

We now consider the boundary-value problem (4.3), (4.4) in the case where B_1 and B_2 are Hilbert spaces, i.e., $L\colon H_1 \longrightarrow H_2$ and $l\colon H_1 \longrightarrow R^m$.

Let $P_\Lambda\colon H_1 \longrightarrow N(\Lambda)$ and $P_{\Lambda^*}\colon H_2 \longrightarrow N(\Lambda^*)$ be the orthoprojectors from the spaces H_1 and H_2 onto the null spaces of the operators $\Lambda = \mathrm{col}\,[L,l]$ and $\Lambda^* = [L^*, l^*]$. In this case, Theorems 3.9 and 4.1 yield the following formula for the generalized Green operator:

$$G = L^+ - XQ^+lL^+, \tag{4.17}$$

where L^+ is the unique pseudoinverse operator for the operator L minimizing the norm of the residual for the Fredholm equation $Lz = \varphi$. However, the generalized inverse operator Λ^- specified by relation (4.16), where G is given by (4.17), is not the unique pseudoinverse operator for the operator Λ and does not minimize the norm of the residual of the equation $\Lambda z = f$.

As follows from Theorems 3.4 and 3.6, the operator

$$\Lambda^+ = (I - P_\Lambda)\Lambda^-(I - P_{\Lambda^*}) = (I - P_\Lambda)[G, XQ^+](I - P_{\Lambda^*})$$

possesses the required property.

4.3 Examples

We now illustrate the efficiency of the proposed algorithms for the investigation of boundary-value problems for functional differential operators by analyzing special cases of finite difference and integrodifferential operators as examples.

1. Finite Difference Operators. As a rule (see [61, 69, 102, 134]), the boundary-value problems for systems of difference equations with discrete variables are studied under the assumption that their linear part contains a Fredholm operator of index zero and, moreover, that this operator is invertible (e.g., periodic problems in the noncritical case).

We now consider a general inhomogeneous linear Fredholm boundary-value problem for a difference equation, namely,

$$x(k+1) = A(k)x(k) + f(k+1), \tag{4.18}$$

$$lx(\cdot) = \alpha. \tag{4.19}$$

In this problem, the number of boundary conditions m does not coincide with the order of the difference system n and, hence, this problem includes, as special cases, both underdetermined and overdetermined noncritical and critical difference boundary-value problems.

Our aim is to construct a criterion for the existence of bounded solutions $x\colon [n_0, N] \to R^n$ of the difference system (4.18) with boundary conditions (4.19) specified using a bounded linear m-dimensional vector functional

$$l = \operatorname{col}(l_1, \ldots, l_m)\colon \{x\colon [n_0, N] \to R^n\} \to R^m$$

defined in the space of vector functions bounded on $[n_0, N]$, namely,

$$l_i\colon \{x\colon [n_0, N] \to R^n\} \to R.$$

In the problem formulated above, $A(k)$ is a bounded $n \times n$ matrix, $f(k)$ is an n-dimensional column vector, $k \in [n_0, N] \subset Z^+$ is a set of natural numbers, and $\alpha \in R^m$. For the sake of convenience, we represent the solution of equation (4.18) in the form

$$x(k) = X(k, n_0)c + \sum_{j=n_0+1}^{N} K(k, j) f(j) \qquad \text{for any} \quad c \in R^n, \tag{4.20}$$

where

$$X(k, n_0) = \prod_{j=n_0+1}^{k} A(k + n_0 - j), \quad X(n_0, n_0) = E \quad [69],$$

and

$$K(k, j) = \begin{cases} X(k, j) & \text{for } n_0 \le j \le k \le N, \\ 0 & \text{for } j > k \end{cases}$$

is the Cauchy matrix of the Cauchy problem for system (4.18). If $\det A(k) \neq 0$, then relation (4.20) gives the general solution of system (4.18) on the segment $[n_0, N]$.

The solution $x(k)$ of the Cauchy problem for the difference system (4.18) given by (4.20) is a solution of the boundary-value problem (4.18), (4.19) if and only if

$$lX(\cdot, n_0)c + \sum_{j=n_0+1}^{N} lK(\cdot, j)f(j) = \alpha. \tag{4.21}$$

As earlier, let $Q = lX(\cdot, n_0)$ be an $m \times n$ matrix such that $\operatorname{rank} Q = n_1 \le \min(m, n)$, let P_Q and P_{Q^*} be $n \times n$ and $m \times m$ matrices (orthoprojectors) that project R^n and R^m onto the null spaces $N(Q)$ and $N(Q^*)$, respectively, let $P_{Q_d^*}$ be the $d \times m$ matrix formed by a complete system of d linearly independent rows of the matrix P_{Q^*}, let P_{Q_r} be the $n \times r$ matrix formed by a complete system of r linearly independent columns of the matrix P_Q, and let Q^+ be the $n \times m$ matrix pseudoinverse to Q. According to Theorem 3.10, in order that system (4.21) be solvable with respect to $c \in R^n$, it is necessary and sufficient that

$$P_{Q_d^*}\{\alpha - \sum_{j=n_0+1}^{N} lK(\cdot, j)f(j)\} = 0 \quad (d = m - n_1). \tag{4.22}$$

Under the indicated condition, system (4.21) has a solution

$$c = -Q^+\{\alpha - \sum_{j=n_0+1}^{N} lK(\cdot, j)f(j)\} + P_{Q_r}c_r \quad (r = n - n_1) \ \forall c_r \in R^r. \tag{4.23}$$

Substituting relation (4.23) in (4.20), we get

$$\begin{aligned} x(k) &= x^0(k, c_r) \\ &= X(k, n_0)P_{Q_r}c_r \\ &\quad + X(k, n_0)Q^+\left\{\alpha - \sum_{j=n_0+1}^{N} lK(\cdot, j)f(j)\right\} + \sum_{j=n_0+1}^{N} K(k, j)f(j) \\ &= X_r(k, n_0)c_r + X(k, n_0)Q^+\alpha + \sum_{j=n_0+1}^{N} G(k, j)f(j), \end{aligned} \tag{4.24}$$

where

$$G(k, j) = K(k, j) - X(k, n_0)Q^+ lK(\cdot, j)$$

is the generalized Green matrix of the boundary-value problem (4.18), (4.19) for the system of difference equations $X_r(k, n_0) = X(k, n_0)P_{Q_r}$ (see Remark 4.4).

For the difference boundary-value problem (4.18), (4.19), Theorem 4.2 can be reformulated as follows:

Theorem 4.4. *Let* rank $[Q = lX(\cdot, n_0)] = n_1 \leq \min(n, m)$. *Then the homogeneous* ($f(j) = 0$, $\alpha = 0$) *boundary-value problem (4.18), (4.19) has exactly* r *linearly independent solutions. The inhomogeneous boundary-value problem (4.18), (4.19) is solvable if and only if the inhomogeneities* $f(j) \in R^n$ *and* $\alpha \in R^m$ *satisfy condition (4.22). In this case, the problem possesses the* r*-parameter family of solutions specified by (4.24).*

2. ***Integrodifferential Operators.*** We now consider the boundary-value problem for the system of equations [138]

$$\dot{x} - \Phi(t) \int_a^b [A(s)x(s) + B(s)\dot{x}(s)]ds = f(t) \tag{4.25}$$

with boundary conditions

$$lx = \alpha, \tag{4.26}$$

where $\Phi(t)$, $A(t)$, $B(t)$, and $f(t)$ are $n \times m$, $m \times n$, and $n \times 1$ matrices, respectively, with entries from the space $L_2[a, b]$, $l = \mathrm{col}\,(l_1, \ldots, l_p)$ is a bounded linear vector functional defined in the space $D_2[a, b]$ of n-dimensional functions absolutely continuous on $[a, b]$, $\alpha = \mathrm{col}\,(\alpha_1, \ldots, \alpha_p) \in R^p$, and $t \in [a, b]$. The columns of the matrix $\Phi(t)$ are assumed to be linearly independent on $[a, b]$. The solution of problem (4.25), (4.26) is sought in the class of vector functions $x(t)$ such that $x(t) \in D_2[a, b]$ and $\dot{x}(t) \in L_2[a, b]$.

We first consider the problem of solvability of system (4.25) and study the structure of the set of its solutions. It is easy to see that

$$\dot{x}(t) = f(t) + \Phi(t)c_0,$$

$$c_0 = \int_a^b [A(s)x(s) + B(s)\dot{x}(s)]ds \in R^m, \tag{4.27}$$

$$x = \tilde{f}(t) + \Psi(t)c_0 + \tilde{c} = \tilde{f}(t) + \Psi_0(t)c,$$

where

$$\tilde{f}(t) = \int_a^t f(s)ds, \quad \Psi(t) = \int_a^t \Phi(s)ds, \quad \tilde{c} = \mathrm{col}\,(c_{m+1}, \ldots, c_{m+n}),$$

$\Psi_0(t) = [\Psi(t), I_n]$ is an $n \times (m+n)$ matrix, I_n is the identity matrix of order n, and $c = \mathrm{col}\,(c_0, \tilde{c}) \in R^{m+n}$. Substituting relation (4.27) in equations (4.25), we arrive at the following system of algebraic equations for c:

$$Dc = \tilde{b}, \tag{4.28}$$

where

$$D = \left[I_m - \int_a^b [A(s)\Psi(s) + B(s)\Phi(s)]\,ds, \; -\int_a^b A(s)ds\right]$$

is an $m \times (m+n)$ matrix and

$$\tilde{b} = \int_a^b \left[A(s)\tilde{f}(s) + B(s)f(s)\right] ds.$$

System (4.28) is solvable if and only if the vector $\tilde{b}$ satisfies the condition

$$P_{D^T_{d_1}} \tilde{b} = 0, \quad d_1 = m - n_1, \quad n_1 = \operatorname{rank} D, \tag{4.29}$$

where $P_{D^T_{d_1}}$ is a $d_1 \times m$ matrix formed by a complete system of d_1 linearly independent rows of the matrix projector P_{D^T}. If condition (4.29) is satisfied, then system (4.28) possesses the following r_1-parameter family of solutions:

$$c = P_{D_{r_1}} c_{r_1} + D^+ \tilde{b}, \quad c_{r_1} \in R^{r_1} \tag{4.30}$$

where $P_{D_{r_1}}$ is an $(m+n) \times r_1$ matrix formed by a complete system of r_1 linearly independent columns of the matrix orthoprojector P_D, and the matrix D^+ is pseudoinverse (in the Moore–Penrose sense) to the matrix D.

Hence, we can now formulate the following criterion of solvability for system (4.25):

Theorem 4.5. *Let* $\operatorname{rank} D = n_1$. *The system of integrodifferential equations (4.25) is solvable if and only if the function* $f(t)$ *satisfies condition (4.29). If this condition is satisfied, then system (4.25) possesses an* r_1*-parameter* ($r_1 = m + n - n_1$) *family of solutions*

$$x = \Psi_0(t) P_{D_{r_1}} c_{r_1} + F(t), \tag{4.31}$$

where

$$F(t) = \tilde{f}(t) + \Psi_0(t) D^+ \tilde{b}. \tag{4.32}$$

Theorem 4.5 implies that the homogeneous system of equations ($f = 0$) always has $r_1 = m + n - n_1 > 0$ linearly independent solutions. If $\operatorname{rank} D = n_1 = m$, then $d_1 = 0$, $P_{D^T_{d_1}} = 0$, and condition (4.29) is satisfied for all $f(t)$. This is a characteristic feature of Fredholm problems.

We now consider the problem of solvability of the boundary-value problem (4.25), (4.26) and study the structure of the set of its solutions. Assume that the solvability condition (4.29) is satisfied. Substituting solution (4.31) in the boundary conditions (4.26), we obtain

$$lx = l(F(\cdot)) + l(\Psi_0(\cdot) P_{D_{r_1}} c_{r_1}) = \alpha. \tag{4.33}$$

This yields the following algebraic system for the vector c_{r_1}:

$$Q c_{r_1} = \alpha - l(F(\cdot)), \tag{4.34}$$

where $Q = l(\Psi_0(\cdot))P_{D_{r_1}}$ is a $p \times r_1$ matrix. By using Theorem 4.5, together with the criterion of solvability of system (4.34), we arrive at a condition necessary and sufficient for the boundary-value problem (4.25), (4.26) to be solvable. This condition improves the corresponding results established in Theorem 4.1.

Theorem 4.6. *Let* $\operatorname{rank} Q = n_2 \leq \min(p, r_1)$. *The homogeneous problem (4.25), (4.26)* ($f = 0$, $\alpha = 0$) *possesses exactly* r_2 ($r_2 = r_1 - n_2$) *linearly independent solutions of the form*

$$x = \Psi_0(t)P_{D_{r_1}}P_{Q_{r_2}}c_{r_2}, \quad c_{r_2} \in R^{r_2}. \tag{4.35}$$

The inhomogeneous problem is solvable if and only if

$$P_{D^T_{d_1}}\tilde{b} = 0, \quad P_{Q^T_{d_2}}(\alpha - l(F(\cdot))) = 0, \tag{4.36}$$
$$d_1 = m - \operatorname{rank} D, \quad d_2 = p - \operatorname{rank} Q.$$

If condition (4.36) is satisfied, then the boundary-value problem (4.25), (4.26) possesses an r_2*-parameter family of solutions*

$$x = \Psi_0(t)P_{D_{r_1}}P_{Q_{r_2}}c_{r_2} + \Psi_0 P_{D_{r_1}}Q^+(\alpha - l(F(\cdot))) + F(t),$$

where Q^+ *is pseudoinverse to* Q, $F(t)$ *is given by (4.32), and*

$$r_1 = m + n - \operatorname{rank} D, \quad r_2 = m + n - \operatorname{rank} D - \operatorname{rank} Q.$$

As a consequence of Theorem 4.6, we can mention the fact that the boundary-value problem (4.25), (4.26) is solvable for all $f(t)$ provided that $\operatorname{rank} D = m$ and $\operatorname{rank} Q = p$, $p \leq n$. Moreover, for $p = n$, the solution of this problem is unique.

Remark 4.5. The system of integrodifferential equations

$$\dot{x} + R(t)x - \Phi(t)\int_a^b [A(s)x(s) + B(s)\dot{x}(s)]ds = f(t), \tag{4.37}$$

where $R(t)$ is an $n \times n$ matrix continuous on $[a, b]$, can be reduced to a system of equations similar to (4.25) by the change of variables $x = X(t)y$, where $X(t)$ is the fundamental matrix of the system $\dot{x} = -R(t)x$.

The general system

$$\dot{x} - \sum_{i=1}^{l}\Phi_i(t)\int_a^b [A_i(s)x(s) + B_i(s)\dot{x}(s)]ds = f(t) \tag{4.38}$$

can be also reduced to a system of the form (4.25). Here, $\Phi_i(t)$, $i = 1, \ldots, l$, are $n \times q$ matrices, and $A_i(t)$ and $B_i(t)$, $i = 1, \ldots, l$, are $q \times n$ matrices.

Indeed, the second term in system (4.38) can be rewritten in the form

$$\dot{x} - f(t) = \sum_{i=1}^{l} \sum_{j=1}^{q} \phi_j^i(t) \int_a^b [a_{ij}(s)x(s) + b_{ij}(s)\dot{x}(s)]ds, \tag{4.39}$$

where $\phi_j^i(t)$ is the jth column of the matrix $\mathbf{\Phi}_i(t)$, and a_{ij} and b_{ij} are the jth rows of the matrices $A_i(t)$ and $B_i(t)$, respectively.

Let

$$\mathbf{\Phi}(t) = [\phi_1(t), \ldots, \phi_m(t)]$$

be an $n \times m$ matrix whose columns form a complete system of linearly independent columns of the $n \times ql$ matrix $\bar{\mathbf{\Phi}}(t) = [\phi_1^1(t), \ldots, \phi_q^l(t)]$. If we express the vector functions $\phi_j^i(t)$ in equations (4.39) via the quantities $\phi_1(t), \ldots, \phi_m(t)$, then we get a system of integrodifferential equations of the form (4.25).

Chapter 5

BOUNDARY-VALUE PROBLEMS FOR SYSTEMS OF ORDINARY DIFFERENTIAL EQUATIONS

In the case where the operators specifying the analyzed systems of equations can be extended to the entire space C of continuous functions (systems of ordinary differential equations can be regarded as a typical example of operators of this sort [51]), it is natural to seek solutions of boundary-value problems for these systems of equations in the space $C^1[a, b]$ of continuously differentiable functions. This enables one to study the specific features of nonlinear boundary-value problems connected with the Fredholm nature of their linear parts without any additional assumptions concerning the spaces where the required solutions are sought. The indicated boundary-value problems are investigated in the present chapter.

5.1 Linear Boundary-Value Problems. Criterion of Solvability

Consider the problem of necessary and sufficient conditions for the solvability of an inhomogeneous linear boundary-value problem

$$\dot{z} = A(t)z + \varphi(t), \qquad t \in [a, b], \tag{5.1}$$

$$lz = \alpha \tag{5.2}$$

and the structure of the set of its solutions in the space of n-dimensional continuously differentiable vector functions

$$z = z(t) = \text{col}\,(z_1(t), \ldots, z_n(t)) \colon z(t) \in C^1[a, b].$$

Here, $A(t)$ is an $n \times n$ matrix whose entries are real functions continuous on $[a, b]$, i.e., $A(t) \in C[a, b]$, $\varphi(t) = \text{col}\,(\varphi_1(t), \varphi_2(t), \ldots, \varphi_n(t))$ is an n-dimensional column vector whose components are real functions continuous on $[a, b]$, i.e., $\varphi(t) \in C[a, b]$, α is a constant column vector from the m-dimensional real Euclidean vector space R^m, l is a linear vector functional defined in the space of n-dimensional vector functions continuous on $[a, b]$, $l = \text{col}\,(l_1, \ldots, l_m)$, $l \colon C[a, b] \to R^m$, and $-\infty < a \le t \le b < \infty$. For different specific forms of the functional, we get different boundary-value problems, some aspects of the theory of which have been studied in the previous chapters.

Example 1. The Fredholm case ($n \neq m$) of the multipoint boundary-value problem with $lz = \sum_{i=1}^{k} M_i z(t_i)$, where M_i are $m \times n$ matrices and $-\infty < a = t_1 \leq \ldots \leq t_i \ldots \leq t_k = b < \infty$, was studied for the first time by R. Conti in [48], and the case of Fredholm operator of index zero ($n = m$) was analyzed in [13, 56, 126, 127, 140]. For $k = 2$, $m = n$, $M_1 = -M_2$, $\det M_1 \neq 0$, and $\alpha = 0$, we arrive at frequently encountered periodic boundary-value problems (see [42, 54, 101, 154]). We can choose lz in the form

$$lz = \int_a^b d\Phi(t) z(t),$$

where $\Phi(t)$ is an $m \times n$ matrix whose entries are functions of bounded variation on $[a, b]$, and the integral used to represent linear functionals [52, 84] in the space $C[a, b]$ is understood in the Riemann–Stieltjes sense.

Parallel with the inhomogeneous boundary-value problem (5.1), (5.2), we consider the corresponding homogeneous problem:

$$\dot{z} = A(t)z, \tag{5.3}$$

$$lz = 0. \tag{5.4}$$

As earlier, the boundary-value problems for which $\operatorname{rank} Q = m$ or $\operatorname{rank} Q < m$, where $Q = lX(\cdot)$ is an $m \times n$ matrix obtained as a result of the substitution of the normal fundamental matrix $X(t) = X(t, a)$, $X(a) = I_n$, of system (5.3) in the boundary condition, are called noncritical and critical, respectively.

It is known that the general solution of the system of linear differential equations (5.1) has the form

$$z(t) = X(t)c + \bar{z}(t), \tag{5.5}$$

where $\bar{z}(t)$ is an arbitrary special solution of the inhomogeneous system (5.1). As a result of the direct substitution of the proposed solution in (5.2), we conclude that the required special solution $\bar{z}(t)$ can be chosen in the form

$$\bar{z}(t) = \int_a^b K(t, \tau) \varphi(\tau)\, d\tau, \tag{5.6}$$

where $K(t, \tau)$ is an $n \times n$ Cauchy matrix of the form

$$K(t, \tau) = \frac{1}{2} X(t) X^{-1}(\tau) \operatorname{sign}(t - \tau), \tag{5.7}$$

or

$$K(t, \tau) = \begin{cases} X(t) X^{-1}(\tau), & a \leq \tau \leq t \leq b, \\ 0, & a \leq t < \tau \leq b, \end{cases}$$

or

$$K(t,\tau) = \begin{cases} -X(t)X^{-1}(\tau), & a \le t \le \tau \le b, \\ 0, & a \le \tau < t \le b. \end{cases}$$

For the solution $z(t)$ of system (5.1) [given by relation (5.5)] to be a solution to the boundary-value problem (5.1), (5.2), it is necessary to satisfy the boundary condition (5.2). Substituting relation (5.5) in the boundary condition (5.2), we arrive at the following algebraic system for $c \in R^n$:

$$Qc + l\bar{z} = \alpha, \tag{5.8}$$

where Q is an $m \times n$ matrix and

$$l\bar{z} = l \int_a^b K(\cdot,\tau)\varphi(\tau)\, d\tau.$$

By using Theorem 3.10, we establish the condition of solvability of system (5.8) and, hence, of the boundary-value problem (5.1), (5.2). Moreover, we find a constant $c \in R^n$ for which the solution (5.5) of the differential system (5.1) is a solution of the analyzed boundary-value problem.

Critical Problems. Let $\operatorname{rank} Q = n_1$. Then, by virtue of Theorem 3.10, system (5.8) is solvable if and only if

$$P_{Q_d^*}[\alpha - l\bar{z}] = 0 \qquad (d = m - n_1). \tag{5.9}$$

In this case, the general solution of system (5.8) has the form

$$c = P_{Q_r} c_r + Q^+(\alpha - l\bar{z}).$$

Substituting the constant obtained as a result in (5.5), we arrive at the following expression for the general solution of the boundary-value problem (5.1), (5.2):

$$z_0(t, c_r) = X_r(t)c_r + \int_a^b K(t,\tau)\varphi(\tau)d\tau$$

$$- X(t)Q^+ l \int_a^b K(\cdot,\tau)\varphi(\tau)d\tau + X(t)Q^+\alpha, \tag{5.10}$$

where $X_r(t)$ is an $n \times r$ matrix whose columns form a complete system of r linearly independent solutions of the homogeneous boundary-value problem (5.3), (5.4), i.e., $X_r(t) = X(t)P_{Q_r}$, and P_{Q_r} is the $n \times r$ matrix defined in Section 3.8.

Thus, it is easy to see that the following theorem is true:

Theorem 5.1. *If* $\operatorname{rank} Q = n_1$, *then the homogeneous boundary-value problem (5.3), (5.4) has exactly* $r = n - n_1$ *linearly independent solutions. The inhomoge-*

neous boundary-value problem (5.1), (5.2) is solvable if and only if $\varphi(t) \in C[a,b]$ *and* $\alpha \in R^m$ *satisfy condition (5.9):*

$$P_{Q_d^*}\left[\alpha - l\int_a^b K(\cdot,\tau)\varphi(\tau)d\tau\right] = 0 \quad (d = m - n_1).$$

In this case, the problem possesses an r*-parameter family of solutions*

$$z_0(t, c_r) = X_r(t)c_r + (G\varphi)(t) + X(t)Q^+\alpha \ \in C^1[a,b], \tag{5.11}$$

where $(G\varphi)(t)$ *is a generalized Green operator acting upon an arbitrary vector function* $\varphi(t)$ *from* $C[a,b]$ *as follows:*

$$(G\varphi)(t) \stackrel{\text{def}}{=} \int_a^b K(t,\tau)\varphi(\tau)d\tau - X(t)Q^+l\int_a^b K(\cdot,\tau)\varphi(\tau)d\tau. \tag{5.12}$$

If l is a vector functional such that [8, 85]

$$l\int_a^b K(\cdot,\tau)\varphi(\tau)d\tau = \int_a^b lK(\cdot,\tau)\varphi(\tau)d\tau, \tag{5.13}$$

then

$$(G\varphi)(t) = \int_a^b G(t,\tau)\varphi(\tau)d\tau,$$

where $G(t,\tau)$ is the $n \times n$ generalized Green matrix of the boundary-value problem (5.1), (5.2):

$$G(t,\tau) = K(t,\tau) - X(t)Q^+lK(\cdot,\tau). \tag{5.14}$$

If $\operatorname{rank} Q = n_1 = n$, then the homogeneous boundary-value problem (5.3), (5.4) possesses only the trivial solution. The inhomogeneous boundary-value problem (5.1), (5.2) is solvable if and only if $\varphi(t) \in C[a,b]$ and $\alpha \in R^m$ satisfy the condition

$$P_{Q_d^*}\left[\alpha - l\int_a^b K(\cdot,\tau)\varphi(\tau)d\tau\right] = 0 \quad (d = m - n). \tag{5.15}$$

In this case, the analyzed problem possesses a unique solution

$$z_0(t) = (G\varphi)(t) + X(t)Q^+\alpha \ \in C^1[a,b]. \tag{5.16}$$

Noncritical Problems. Theorem 5.1 yields the following assertion for the noncritical boundary-value problems:

Theorem 5.2. *If* $\operatorname{rank} Q = n_1 = m$, *then the homogeneous boundary-value problem (5.3), (5.4) possesses exactly* $r = n - m$ *linearly independent solutions.*

The inhomogeneous boundary-value problem (5.1), (5.2) possesses an r-parameter family of solutions (5.11) for arbitrary inhomogeneities $\varphi(t) \in C[a,b]$ and $\alpha \in R^m$.

Periodic Boundary-Value Problems. If, instead of the boundary-value problem (5.1), (5.2), we consider a periodic boundary-value problem

$$\frac{dz}{dt} = A\,(t)\,z + \varphi\,(t)\,, \qquad t \in [0,T]$$

$$l\,z\,(\cdot) := z(0) - z(T) = 0 \in R^n,$$

then, for problem (5.1), (5.1), Theorem 5.1 yields the following assertion:

Corollary 5.1. *If $rank\{Q := X(0) - X(T) = I - X(T)\} = n_1 \le n$, then the homogeneous problem*

$$\frac{dz}{dt} = A\,(t)\,z, \;\; z(0) = z(T); \qquad t \in [0,T]$$

has exactly r (where $r = n - n_1$) linearly independent solutions in the space $C^1[0,T]$. The inhomogeneous periodic problem (5.1), (5.1) is solvable in the space $C^1[0,T]$ if and only if $\varphi(t) \in C[0,T]$ satisfies r linearly independent conditions

$$P_{Q_r^*} \int\limits_0^T K(T,\tau)\varphi\,(\tau)\,d\tau = 0$$

and has an r-parameter family of linearly independent solutions represented in the form

$$z_0(t,c_r) = X_r(t)c_r + (G\varphi)(t) \in C^1[0,T],$$

where

$$(G\varphi)(t) = \int\limits_0^T G(t,\tau)\varphi\,(\tau)\,d\tau, \;\; G(t,\tau) = K(t,\tau) + X(t)Q^+K(T,\tau).$$

Note that, in view of the fact that $m = n$ and, hence, $d = r$, the homogeneous $(\varphi\,(t) = 0)$ periodic problem (5.1) and the conjugate periodic problem

$$\frac{dz}{dt} = -A^*\,(t)\,z, \;\; z(0) = z(T); \qquad t \in [0,T]$$

have the same number of linearly independent solutions. Thus, in the theory of Fredholm problems of index zero in which the number of boundary conditions m is equal to the number of unknowns n (as a simple example of these problems, we can mention periodic boundary-value problems), the critical (or resonance) boundary-value problems are defined (see I. G. Malkin [101], V. A. Yakubovich and V. M. Starzhinsky [154], and Yu. A. Ryabov [65]) as problems for which the corresponding homogeneous boundary-value problem has nontrivial solutions.

In the case of general overdetermined or underdetermined problems when the number of boundary conditions m is not equal to the number of unknowns n, the homogeneous boundary-value problem and the problem conjugate to this problem have unequal numbers of linearly independent solutions. In the corresponding spaces, these are Fredholm problems with nonzero index. In this case, the *critical or resonance* problems are defined (Definition 4.2) as problems for which the conjugate homogeneous problem has nontrivial solutions, i.e., whenever $P_{Q^*} \neq 0$ and, hence, $P_{Q_d^*} \neq 0$. Our aim is to study these critical (resonance) problems. We also note that, in the case where the coefficients of the system are periodic with period T, we deal with periodic solutions of problem (5.1), (5.1).

Generalized Inverse Operator. By analogy with Chapter 4, we can interpret the facts presented above as follows: The boundary-value problem (5.1), (5.2) is regarded as an operator system

$$\Lambda z = f, \tag{5.17}$$

where $\Lambda\cdot = \operatorname{col}[d\cdot/dt - A\cdot, l\cdot]$ is a linear operator that maps the real space $X^n = C^1[a,b]$ of n-dimensional vector functions continuously differentiable on $[a,b]$ into the real space $Y^{n+m} = C[a,b] \times R^m$ of $(n+m)$-dimensional vector functions f:

$$\Lambda\colon X^n \to Y^{n+m}, \quad f = \operatorname{col}(\varphi, \alpha).$$

Theorem 2.3 implies that the linear operator system (5.17) is solvable if and only if $f \in R(\Lambda)$, i.e., in the case where the condition $\mathcal{P}_{\Lambda^*} f = 0$ is satisfied, where $\mathcal{P}_{\Lambda^*}$ is a projector decomposing the space Y^{n+m} into a direct sum of subspaces

$$Y^{n+m} = R(\Lambda) \oplus Y_0, \quad \text{namely} \quad \mathcal{P}_{\Lambda^*} Y^{n+m} = Y_0. \tag{5.18}$$

By Λ^- we denote the operator

$$\Lambda^- * = [G*, XQ^+ *]. \tag{5.19}$$

By direct calculation, we can show that the operator given by relation (5.19) satisfies relations (1.7) specifying the generalized inverse operator for the operator Λ. With the help of the operators Λ and Λ^-, the projector $\mathcal{P}_{\Lambda^*}$ creating partition (5.18) can be represented in the form

$$\mathcal{P}_{\Lambda^*} = \mathcal{P}_{\Lambda^*}^2 = I_{n+m} - \Lambda\Lambda^-$$
$$= \begin{bmatrix} I_n & 0 \\ 0 & I_m \end{bmatrix} - \begin{bmatrix} I_n & 0 \\ P_{Q^*} l \int\limits_a^b K(\cdot,\tau) * d\tau & QQ^+ * \end{bmatrix}$$

$$= \begin{bmatrix} 0 & 0 \\ -P_{Q^*} l \int_a^b K(\cdot, \tau) * d\tau & P_{Q^*} * \end{bmatrix}. \tag{5.20}$$

By using projector (5.20), we can rewrite the criterion of solvability of the operator system (5.17) in the form $\mathcal{P}_{\Lambda^*} f = 0$. This is equivalent to criterion (5.9) because $\operatorname{rank} P_{Q^*} = d = m - n$.

Quasisolutions of Boundary-Value Problems. Assume that the inhomogeneities $\varphi(t) \in C[a, b]$ and $\alpha \in R^m$ of the boundary-value problem (5.1), (5.2) do not satisfy condition (5.9). Theorem 5.1 implies that the boundary-value problem (5.1), (5.2) is unsolvable in the class of vector functions $z(t) \in C^1[a, b]$. In this case, the operator system (5.17) equivalent to the original boundary-value problem belongs to the class of ill-posed problems [145]. These boundary-value problems are solved by using special regularization methods, e.g., by adding linear perturbing terms to the differential system with an aim to make the analyzed problem well posed (see Section 6.3). In the case where $f \bar{\in} R(\Lambda)$ (i.e., $\mathcal{P}_{\Lambda^*} f \neq 0$), the operator system (5.17) is unsolvable but possesses so-called quasisolutions. The set X^n of quasisolutions $z^-(t, c_r)$ of the operator system (5.17) is defined as

$$z_0^-(t, c_r) = X_r(t) c_r + (\Lambda^- f)(t). \tag{5.21}$$

Moreover, the norm of the residual is equal to the norm of the expression on the left-hand side of the criterion of solvability (5.9) of the boundary-value problem (5.1), (5.2):

$$||\Lambda z_0^- - f|| = \left\| P_{Q^*} \left(\alpha - l \int_a^b K(\cdot, \tau) \varphi(\tau) d\tau \right) \right\|. \tag{5.22}$$

The norm $||z||$ of an element $z \in X^n$ and the norm $||\Lambda||$ of an operator Λ compatible with the indicated norm of elements are introduced in the ordinary way [84].

As a result, we can propose the following procedure for the investigation of boundary-value problems of the form (5.1), (5.2):

1. First, we find the normal fundamental matrix $X(t) = X(t, a)$ of the differential system (5.3). This is the strongest requirement in the analyzed procedure guaranteeing, in principle, the possibility of construction of the efficient criterion of solvability of the original inhomogeneous boundary-value problem and its solution.

2. Second, we construct an $m \times n$ matrix $Q = lX(\cdot)$, determine its rank and projectors, and find the generalized Green operator $(G*)(t)$.

3. Third, we check the validity of the criterion of solvability (5.9) for the analyzed boundary-value problem.

4. For $f = \operatorname{col}(\varphi, \alpha) \in Y^{n+m}$ satisfying criterion (5.9), we construct the r-parameter $(0 \le r = n - n_1 \le n)$ family of solutions (5.11) of the boundary-value problem (5.1), (5.2).

5. For $f = \operatorname{col}(\varphi, \alpha) \in Y^{n+m}$ that do not satisfy (5.9), relation (5.11) specifies a family $z_0^-(t, c_r)$ of quasisolutions of the boundary-value problem (5.1), (5.2). In this case, the analyzed boundary-value problem is ill posed.

Note that the proposed procedure of investigation of boundary-value problems can also be realized in the case where the fundamental matrix $X(t)$ is given numerically by using the methods described in [9, 44, 62].

Generalized Green Matrix. The generalized Green matrix $G(t, \tau)$ for the boundary-value problem (5.1), (5.2) given by relation (5.14) plays an important role in the analysis of linear and nonlinear boundary-value problems for systems of ordinary differential equations.

The generalized Green matrix $G(t, \tau)$ of the boundary-value problem (5.1), (5.2) possesses the following basic properties:

(i) for $t \neq \tau$, each column of the matrix is a continuously differentiable solution of the homogeneous differential system (5.3)

$$\frac{\partial G(t,\tau)}{\partial t} = A(t)G(t,\tau), \qquad t \neq \tau; \tag{5.23}$$

(ii) for $t = \tau$, the matrix possesses a discontinuity of the first kind as a function of t:

$$G(\tau + 0, \tau) - G(\tau - 0, \tau) = I_n; \tag{5.24}$$

(iii) the matrix satisfies the boundary condition

$$lG(\cdot, \tau) = P_{Q^*} lK(\cdot, \tau).$$

These properties can easily be verified. Moreover, they can be used as a definition of the generalized Green matrix for the boundary-value problem (5.1), (5.2). Note that, according to properties (i) and (ii), the investigated matrix satisfies the system of ordinary differential equations (5.23), (5.24) with impulsive action [135, 139]. For the noncritical Fredholm multipoint boundary-value problems of index zero $(m = n)$, the Green matrices were constructed in [65] and [140] as follows:

$$G(t, \tau) = K(t, \tau) - X(t)Q^{-1}lK(\cdot, \tau).$$

In this case, property (iii) takes the form $lG(\cdot, \tau) = 0$ because $P_{Q^*} = 0$. For the critical Fredholm boundary-value problems of index zero, the corresponding matrices were also constructed in [154] (for periodic problems) and in [41, 54, 127] (for two-point problems).

Example 2. Consider a two-dimensional differential system

$$\dot{z} = \varphi(t), \quad t \in [a,b], \quad \varphi(t) \in C[t], \quad n = 2 \tag{5.25}$$

equipped with a two-point boundary condition

$$lz = M_1 z(a) + M_2 z(b) = \alpha, \tag{5.26}$$

where M_i $(i = 1,2)$ are rectangular 3×2 matrices of the form

$$M_1 = \begin{bmatrix} m_1 & 0 \\ 0 & m_2 \\ 0 & m_3 \end{bmatrix} \quad \text{and} \quad M_2 = \begin{bmatrix} m_1 & 0 \\ 0 & m_2 \\ 0 & -m_3 \end{bmatrix},$$

$0 \neq m_j \in R^1$, $j = 1,2,3$, and clearly, $\alpha = \operatorname{col}(\alpha_1, \alpha_2, \alpha_3) \in R^3$ $(m = 3)$. After elementary transformations, we conclude that

$$X(t) = I_2 = \begin{bmatrix} 1 & 0 \\ 0 & 1 \end{bmatrix}$$

is the normal fundamental matrix of the system $\dot{z} = 0$,

$$Q = lX = 2 \begin{bmatrix} m_1 & 0 \\ 0 & m_2 \\ 0 & 0 \end{bmatrix}$$

is a 3×2 matrix,

$$P_Q = \begin{bmatrix} 0 & 0 \\ 0 & 0 \end{bmatrix}, \quad Q^* = 2 \begin{bmatrix} m_1 & 0 & 0 \\ 0 & m_2 & 0 \end{bmatrix}$$

is a 2×3 matrix,

$$P_{Q^*} = \begin{bmatrix} 0 & 0 & 0 \\ 0 & 0 & 0 \\ 0 & 0 & 1 \end{bmatrix}, \quad P_{Q_d^*} = [0\ 0\ 1], \quad K(t,\tau) = \frac{\operatorname{sign}(t-\tau)}{2} I_2,$$

and

$$lK(\cdot,\tau) = \frac{1}{2}(-M_1 + M_2) = -m_3 \begin{bmatrix} 0 & 0 \\ 0 & 0 \\ 0 & 1 \end{bmatrix}.$$

Since $\operatorname{rank} Q = n = 2$ $(r = 0,\ d = m - n = 1)$, it follows from Definition 4.2 that this case is critical. By virtue of Theorem 5.1, the criterion of solvability (5.9) of the boundary-value problem (5.25), (5.26) takes the form

$$[0\ 0\ 1]\left\{\alpha + m_3 \begin{bmatrix} 0 & 0 \\ 0 & 0 \\ 0 & 1 \end{bmatrix} \int_a^b \varphi(\tau)d\tau \right\} = 0.$$

Thus, problem (5.25), (5.26) is solvable for all $\alpha_1, \alpha_2 \in R^1$ and $\varphi_1(t) \in C[a,b]$ if and only if $\alpha_3 \in R^1$ and $\varphi_2(t) \in C[a,b]$ satisfy the relation

$$\alpha_3 + m_3 \int_a^b \varphi_2(\tau)\, d\tau = 0. \tag{5.27}$$

In this case, the solution of the problem is unique and takes the form

$$z_0(t) = X(t)Q^+\alpha + \int_a^b G(t,\tau)\varphi(\tau)\, d\tau,$$

where

$$Q^+ = \frac{1}{2}\begin{bmatrix} \frac{1}{m_1} & 0 & 0 \\ 0 & \frac{1}{m_2} & 0 \end{bmatrix},$$

$$G(t,\tau) = K(t,\tau) - X(t)Q^+ lK(\cdot,\tau) = K(t,\tau).$$

Therefore, the unique solution of the boundary-value problem (5.25), (5.26) has the form

$$z_0(t) = \frac{1}{2}\begin{bmatrix} \frac{1}{m_1} & 0 & 0 \\ 0 & \frac{1}{m_2} & 0 \end{bmatrix}\alpha + \frac{1}{2}\left\{\int_a^t \varphi(\tau)d\tau - \int_t^b \varphi(\tau)d\tau\right\}, \tag{5.28}$$

provided that condition (5.27) is satisfied.

If $\alpha_3 \in R^1$ and $\varphi_2(t) \in C[a,b]$ do not satisfy (5.27), then relation (5.28) specifies a quasisolution of the analyzed boundary-value problem and, in addition,

$$||\Lambda z_0 - f|| = \left|\alpha_3 + m_3 \int_a^b \varphi_2(\tau)d\tau\right|.$$

Example 3. We now consider a boundary-value problem

$$\dot{z} = \varphi(t), \quad lz = M_1 z(a) + M_2 z(b) = \alpha, \quad t \in [a,b], \tag{5.29}$$

where φ is a two-dimensional column vector ($n = 2$), $\varphi = \varphi(t) \in C[a,b]$, M_1 and M_2 are 1×2 matrices, $M_1 = [m_1\ 0]$, $M_2 = [0\ m_2]$ ($m_i \in R^1$, $m_i \neq 0$, $i = 1, 2$), and $\alpha \in R^1$ is a scalar ($m = 1$).

In this case,

$$X(t) = I_2, \quad Q = lX = [m_1 \; m_2], \quad Q^* = Q^T = \begin{bmatrix} m_1 \\ m_2 \end{bmatrix}, \quad \text{and} \quad P_{Q^*} = 0.$$

Since $\mathrm{rank}\, Q = n_1 = m = 1 < n = 2$, it follows from Definition 4.2 that the analyzed case is noncritical:

$$X_r(t) = \begin{bmatrix} -m_2 \\ m_1 \end{bmatrix},$$

$$P_Q = \frac{1}{m_1^2 + m_2^2} \begin{bmatrix} -m_2 \\ m_1 \end{bmatrix} [-m_2 \; m_1] = \frac{1}{m_1^2 + m_2^2} \begin{bmatrix} m_2^2 & -m_1 m_2 \\ -m_1 m_2 & m_1^2 \end{bmatrix},$$

$$Q^+ = (Q^* Q + P_Q)^{-1} Q^* = \frac{1}{m_1^2 + m_2^2} \begin{bmatrix} m_1 \\ m_2 \end{bmatrix}.$$

According to relation (5.14), the generalized Green matrix of problem (5.29) has the form

$$G(t, \tau) = \frac{\mathrm{sign}\,(t - \tau)}{2} I_2 - \frac{1}{2(m_1^2 + m_2^2)} \begin{bmatrix} -m_1^2 & m_1 m_2 \\ -m_1 m_2 & m_2^2 \end{bmatrix}.$$

By virtue of Theorem 5.2, the boundary-value problem (5.29) possesses a one-parameter family $(r = n - n_1 = n - m = 1)$ of solutions of the form

$$z_0(t, c_1) = \begin{bmatrix} -m_2 \\ m_1 \end{bmatrix} c_1 + \int_a^b G(t, \tau) \varphi(\tau) d\tau + \frac{1}{m_1^2 + m_2^2} \begin{bmatrix} m_1 \\ m_2 \end{bmatrix} \alpha \quad \forall c_1 \in R^1$$

for any $\varphi(t) = \mathrm{col}\,(\varphi_1(t), \varphi_2(t)) \in C[a, b]$ and $\alpha \in R^1$ $(P_{Q^*} = 0)$.

Example 4. Consider a boundary-value problem of the form (5.1), (5.2), namely,

$$\dot{x} = A(t)x + \varphi(t), \quad lx = \alpha, \quad t \in [a, b], \tag{5.30}$$

and assume that $A(t)$ is an $n \times n$ matrix satisfying the Lappo-Danilevskii condition

$$A(t) \int_a^t A(s) ds = \int_a^t A(s) ds A(t). \tag{5.31}$$

In this case, the normal fundamental matrix of the homogeneous differential system (5.30) admits the following explicit representation:

$$X(t) = \exp \int_a^t A(s)\, ds.$$

This enables us to perform efficient analysis of the boundary-value problem (5.30) for any specific $A(t) \in C[a,b]$ satisfying condition (5.31), any $\varphi(t) \in C[a,b]$ and $\alpha \in R^m$, and an arbitrary form of the linear functional l, i.e., to establish conditions of solvability and construct a family of solutions of the problem in the analytic form. Thus, if $lX = X(a) - X(b) = \alpha \in R^m$ $(m = n)$, then

$$Q = l\left[\exp \int_a^{(\cdot)} A(s)\, ds\right] = I_n - \exp \int_a^b A(s)\, ds.$$

For $\det Q \neq 0$ and arbitrary inhomogeneities $\varphi(t) \in C[a,b]$ and $\alpha \in R^n$, it follows from Theorem 5.2 that the boundary-value problem (5.30) possesses a unique solution

$$x_0(t) = \int_a^b G(t,\tau)\varphi(\tau)d\tau + X(t)Q^{-1}\alpha,$$

where

$$G(t,\tau) = \frac{1}{2}e^{\int_\tau^t A(s)ds} \operatorname{sign}(t-\tau)$$
$$+\frac{1}{2}e^{\int_a^t A(s)ds} Q^{-1}\left[e^{\int_\tau^b A(s)ds} + e^{\int_\tau^a A(s)ds}\right].$$

Similarly, Theorem 5.1 enables us to construct the solution of the boundary-value problem (5.30) in the critical case ($\det Q = 0$).

5.2 Weakly Nonlinear Boundary-Value Problems

Statement of the Problem. We consider boundary-value problems for nonlinear systems of ordinary differential equations containing a small nonnegative parameter ε of the form

$$\dot{z} = A(t)z + \varphi(t) + \varepsilon Z(z,t,\varepsilon), \quad lz = \alpha + \varepsilon J(z(\cdot,\varepsilon),\varepsilon), \tag{5.32}$$

where $A(t)$ is an $n \times n$ matrix from $C[a,b]$, $\varphi(t)$ is an n-dimensional vector function from $C[a,b]$, $Z(z,t,\varepsilon)$ is an n-dimensional vector function nonlinear as a function of z, continuously differentiable with respect to z in a certain neighborhood of the generating solution, and continuous as a function of t and ε, i.e., $Z(\cdot,t,\varepsilon) \in C^1[z]$, $||z - z_0|| \leq q$, $Z(z,\cdot,\varepsilon) \in C[t]$, $t \in [a,b]$, and

$Z(z,t,\cdot) \in C[\varepsilon]$, $\varepsilon \in [0,\varepsilon_0]$, and q and ε_0 are sufficiently small constants. Finally, $J(z(\cdot,\varepsilon),\varepsilon)$ is a nonlinear bounded m-dimensional vector functional continuously differentiable with respect to z (in a sense of Frechêt) and continuous as a function of ε in a certain neighborhood of the generating solution. Our aim is to establish conditions for the existence and develop an algorithm for the construction of the solution $z = z(t,\varepsilon)$ of the boundary-value problem (5.32) such that $z(\cdot,\varepsilon) \in C^1[a,b]$ and $z(t,\cdot) \in C[\varepsilon]$, which turns into the solution of the boundary-value problem

$$\dot{z} = A(t)z + \varphi(t), \qquad lz = \alpha, \tag{5.33}$$

for $\varepsilon = 0$. Problem (5.33) is obtained from problem (5.32) by setting $\varepsilon = 0$ and is called generating for (5.32).

The general method used for the analysis of the posed problem is based on the transition from the original boundary-value problem (5.32) to a certain operator system performed by using the generalized Green operator of problem (5.33) (constructed in Chapter 4) with subsequent application of the Lyapunov–Poincaré method of small parameter and its iterative analog (the method of simple iteration). In our subsequent presentation, it becomes clear that the dimension of the indicated operator system is different for critical and noncritical boundary-value problems.

Let $\operatorname{rank} Q = n$. In other words, we assume that the homogeneous boundary-value problem (5.3), (5.4) possesses only the trivial solution. Thus, according to Theorem 5.1, the generating boundary-value problem (5.33) possesses a unique solution $z_0(t)$ given by relation (5.16) if and only if $\varphi(t) \in C[a,b]$ and $\alpha \in R^m$ satisfy condition (5.15). The solution $z_0(t)$ is called the generating solution for problem (5.32). Our aim is to construct a solution $z(t,\varepsilon)$ of the boundary-value problem (5.32) continuous as a function of ε which turns into the generating solution $z_0(t)$ given by relation (5.16) for $\varepsilon = 0$. By the change of variables

$$z(t,\varepsilon) = z_0(t) + x(t,\varepsilon) \tag{5.34}$$

in (5.32), we arrive at the following boundary-value problem for the deviation $x = x(t,\varepsilon)$ from the generating solution:

$$\dot{x} = A(t)x + \varepsilon Z(z_0 + x, t, \varepsilon), \quad lx = \varepsilon J(z_0(\cdot) + x(\cdot,\varepsilon),\varepsilon). \tag{5.35}$$

We seek a condition for the existence and an algorithm for the construction of a solution $x = x(t,\varepsilon)$ such that $x(\cdot,\varepsilon) \in C^1[a,b]$ and $x(t,\cdot) \in C[\varepsilon]$, $\varepsilon \in [0,\varepsilon_0]$, vanishing for $\varepsilon = 0$. By applying Theorem 5.1 to the boundary-value problem (5.35) and taking into account the nonlinearities $Z(z_0 + x, t, \varepsilon)$ in the system and $J(z_0(\cdot) + x(\cdot,\varepsilon),\varepsilon)$ in the boundary conditions as inhomogeneities, we conclude that problem (5.35) is solvable if and only if the nonlinearities

$Z(z,t,\varepsilon)$ and $J(z(\cdot,\varepsilon),\varepsilon)$ satisfy the condition

$$P_{Q_d^*}\left\{J(z(\cdot),\varepsilon) - l\int_a^b K(\cdot,\tau)Z(z(\tau,\varepsilon),\tau,\varepsilon)\,d\tau\right\} = 0. \tag{5.36}$$

In this case, the solution $x(t,\varepsilon)$ of the boundary-value problem (5.35) admits a unique representation of the form

$$\begin{aligned} x(t,\varepsilon) &= \varepsilon\Lambda^{-}\mathcal{F}(x,\cdot,\varepsilon)(t) \\ &= \varepsilon(GZ(z_0 + x,\tau,\varepsilon))(t) + \varepsilon X(t)Q^{+}J(z_0(\cdot) + x(\cdot,\varepsilon),\varepsilon), \end{aligned} \tag{5.37}$$

where $\mathcal{F} = \operatorname{col}(Z, J)$ and, in addition,

$$\mathcal{F}(0,t,0) = 0 \qquad \text{and} \qquad \frac{\partial\mathcal{F}(0,t,0)}{\partial x} = 0. \tag{5.38}$$

Hence, under the necessary and sufficient condition (5.36), the operator system (5.37) is equivalent to the boundary-value problem (5.32) on the set of functions $x(t,\cdot) \in C[\varepsilon]$, $\varepsilon \in [0,\varepsilon_0]$, vanishing for $\varepsilon = 0$.

System (5.37) belongs to the class of systems [65] solved by the method of simple iterations. For these systems, on the basis of the method of majorizing Lyapunov equations [79], efficient procedures were proposed for the estimation of the range of convergence of lower estimates of the upper bound of the values of the parameter ε for which the iterative process converges to the required solution. The application of the method of simple iterations to system (5.37) gives the following algorithm for finding the solution $x(t,\cdot) \in C[\varepsilon]$, $\varepsilon \in [0,\varepsilon_0]$:

$$x_{k+1}(t,\varepsilon) = \varepsilon\Lambda^{-}\mathcal{F}(x_k,\cdot,\varepsilon)(t), \quad k = 0,1,\ldots, \quad x_0 = 0. \tag{5.39}$$

The procedure used for the estimation of the range of convergence of the iterative process (5.39) and based on the method of Lyapunov majorants is described in [19, 65, 77]. Under the indicated conditions imposed on the operators Λ^{-} and $\mathcal{F}$, one can always find $\varepsilon = \varepsilon_*$ such that the iterative process (5.39) converges for $\varepsilon \in [0,\varepsilon_*]$. Therefore, to establish the conditions of solvability of the boundary-value problem (5.35), it suffices to determine the conditions of its reducibility to the operator system (5.37). Thus, the following theorem is true:

Theorem 5.3. *Assume that the boundary-value problem (5.32) satisfies the conditions presented above and the generating problem (5.33) possesses a unique generating solution $z_0(t)$ given by relation (5.16) provided that* $\operatorname{rank} Q = n$ *and condition (5.15) is satisfied.*

Then the boundary-value problem (5.35) is solvable if and only if the nonlinearities $Z(z,t,\varepsilon) \in C^1[z]$, $Z(z,\cdot,\varepsilon) \in C[t]$, $Z(z,t,\cdot) \in C[\varepsilon]$, and $J(z,\cdot) \in C[\varepsilon]$

satisfy condition (5.36). In this case, for $\varepsilon \in [0, \varepsilon_*]$, the analyzed problem possesses a unique solution $x(t, \varepsilon)$ in the space $C[\varepsilon]$ of functions vanishing for $\varepsilon = 0$. This solution can be found with the help of the iterative process (5.39) convergent for $\varepsilon \in [0, \varepsilon_*]$.

In view of the change of variables (5.34), we can state that the boundary-value problem (5.32) is solvable if and only if condition (5.36) is satisfied. In this case, the problem possesses a unique solution $z(t, \varepsilon)$ in $C[\varepsilon]$ approaching the generating solution $z_0(t)$ given by relation (5.16). The indicated solution is determined by using the recurrence relation

$$z_{k+1}(t, \varepsilon) = z_0(t) + x_{k+1}(t, \varepsilon), \quad k = 0, 1, 2, \ldots, \quad \varepsilon \in [0, \varepsilon_*], \tag{5.40}$$

where $x_{k+1}(t, \varepsilon)$ are given by relations (5.39).

We also note that, according to the definition of the operator Λ^-, the vector functions $x_{k+1}(t, \varepsilon)$, $k = 0, 1, 2, \ldots$, are solutions of the boundary-value problems

$$\dot{x}_{k+1} = A(t)x_{k+1} + \varepsilon Z(z_0 + x_k, t, \varepsilon),$$

$$l x_{k+1} = \varepsilon J(z_0(\cdot) + x_k(\cdot, \varepsilon), \varepsilon), \quad k = 0, 1, 2, \ldots,$$

whose solvability in each step of the iterative process

$$P_{Q_d^*}\left\{ J(z_0(\cdot) + x_k(\cdot, \varepsilon), \varepsilon) - \ell \int_a^b K(\cdot, \tau) Z(z_0(\tau) + x_k(\tau, \varepsilon), \tau, \varepsilon) d\tau \right\} = 0$$

is guaranteed by requirement (5.36) imposed on the nonlinearities $Z(z, t, \varepsilon)$ and $J(z(\cdot, \varepsilon), \varepsilon)$. The vector functions $z_{k+1}(t, \varepsilon)$ are the solutions of the boundary-value problems

$$\dot{z}_{k+1} = A(t)z_{k+1} + \varphi(t) + \varepsilon Z(z_k, t, \varepsilon),$$

$$l z_{k+1} = \alpha + \varepsilon J(z_k(\cdot, \varepsilon), \varepsilon), \qquad k = 0, 1, 2, \ldots.$$

For $m = n$ (noncritical case), similar iterative processes are studied in [16, 17]. In [65, 93, 101], these processes are investigated for the case of periodic boundary-value problems.

Example 1. To illustrate the proposed procedure for the analysis of boundary-value problems of the form (5.32), we consider the boundary-value problem

$$\dot{z} = \varphi(t) + \varepsilon Z(z, t), \qquad l z = M_1 z(a) + M_2 z(b) = \alpha, \tag{5.41}$$

where

$$z = z(t, \varepsilon) = \operatorname{col}(z^{(1)}, z^{(2)}) \qquad (n = 2),$$

$$\alpha = \mathrm{col}\,(\alpha_1, \alpha_2, \alpha_3) \in R^3 \qquad (m = 3),$$
$$Z(z,t) = \mathrm{col}\,(g_1(t)(z^{(1)} + z^{(2)})^2,\ g_2(t)z^{(2)2}),$$
$$\varphi(t),\ g_j(t) \in C[a,b], \quad j = 1, 2,$$
$$M_1 = \begin{bmatrix} m_1 & 0 \\ 0 & m_2 \\ 0 & m_3 \end{bmatrix}, \quad M_2 = \begin{bmatrix} m_1 & 0 \\ 0 & m_2 \\ 0 & -m_3 \end{bmatrix},$$

and $0 \neq m_i \in R^1$ for $i = 1, 2, 3$. We now establish conditions for the existence of a solution $z(t,\varepsilon) \in C[\varepsilon]$ of the boundary-value problem (5.41) that turns into the generating solution $z_0(t)$ for $\varepsilon = 0$.

Since $\mathrm{rank}\, Q = n = 2$, this case is critical. As indicated in Section 5.1, the boundary-value problem (5.25), (5.26) generating for the boundary-value problem (5.41) is obtained from the latter if we set $\varepsilon = 0$ and assume that

$$\alpha_3 + m_3 \int_a^b \varphi_2(s)\, ds = 0.$$

The indicated generating problem possesses a unique solution $z_0(t)$ given by relation (5.28):

$$z_0(t) = \mathrm{col}\,(z_0^{(1)}, z_0^{(2)}) = \frac{1}{2}\begin{bmatrix} \frac{1}{m_1} & 0 & 0 \\ 0 & \frac{1}{m_2} & 0 \end{bmatrix} \alpha + \frac{1}{2}\left\{ \int_a^t \varphi(\tau) d\tau - \int_t^b \varphi(\tau)\, d\tau \right\},$$

$$G(t,\tau) = K(t,\tau) = \frac{1}{2}\mathrm{sign}\,(t-\tau) I_2.$$

By using Theorem 5.2, we conclude that the boundary-value problem (5.41) is solvable if and only if

$$P_{Q_d^*} l \int_a^b K(\cdot,\tau) \begin{bmatrix} g_1(\tau)(z^{(1)}(\tau,\varepsilon) + z^{(2)}(\tau,\varepsilon))^2 \\ g_2(\tau) z^{(2)^2}(\tau,\varepsilon) \end{bmatrix} d\tau$$
$$= -m_3 \int_a^b g_2(\tau) z^{(2)^2}(\tau,\varepsilon)\, d\tau = 0. \tag{5.42}$$

For $g_2(t) = 0$, condition (5.42) is satisfied and the boundary-value problem (5.41) possesses a unique solution $z(t,\cdot) \in C[\varepsilon]$, $z(t,0) = z_0(t)$, $\varepsilon \in [0, \varepsilon_*]$, obtained from the operator system

$$z(t,\varepsilon) = z_0(t) + \varepsilon \int_a^b G(t,\tau) \begin{bmatrix} g_1(\tau)(z^{(1)}(\tau,\varepsilon) + z^{(2)}(\tau,\varepsilon))^2 \\ 0 \end{bmatrix} d\tau$$

by using the iterative process

$$z_{k+1}^{(1)}(t,\varepsilon) = z_0^{(1)}(t) + \frac{\varepsilon}{2}\int_a^b \operatorname{sign}(t-\tau) g_1(\tau)(z_k^{(1)}(\tau,\varepsilon) + z_0^{(2)}(\tau))^2\, d\tau,$$

$$z^{(2)}(t,\varepsilon) = z_0^{(2)}(t), \qquad k = 0,1,2,\ldots,$$

convergent for $\varepsilon \in [0, \varepsilon_*]$. To estimate the range of convergence of the iterative process, i.e., the quantity ε_*, we use the procedure proposed in [19].

Note that condition (5.42) is also satisfied for $m_3 = 0$. In this case, the boundary-value problem (5.41) turns into a Fredholm problem of index zero: $m = n = 2$, $\alpha_3 = 0$, $\alpha = \operatorname{col}(\alpha_1, \alpha_2) \in R^2$, and

$$M_1 = M_2 = \begin{bmatrix} m_1 & 0 \\ 0 & m_2 \end{bmatrix}.$$

In the analyzed noncritical case, for any nonlinearity $Z(z,t)$, problem (5.41) possesses a unique solution $z(t,\cdot) \in C[\varepsilon]$ that turns into the generating solution

$$z_0(t) = \frac{1}{2}\begin{bmatrix} \dfrac{1}{m_1} & 0 \\ 0 & \dfrac{1}{m_2} \end{bmatrix}\alpha + \frac{1}{2}\int_a^b \operatorname{sign}(t-\tau)\varphi(\tau)\, d\tau$$

for $\varepsilon = 0$. This solution can be found with the help of the iterative process

$$z_{k+1}(t,\varepsilon) = z_0(t) + \frac{\varepsilon}{2}\int_a^b \operatorname{sign}(t-\tau)\begin{bmatrix} g_1(\tau)(z_k^{(1)}(\tau,\varepsilon) + z_k^{(2)}(\tau,\varepsilon))^2 \\ g_2(\tau)(z_k^{(2)}(\tau,\varepsilon))^2 \end{bmatrix} d\tau,$$

$$z_0(t,\varepsilon) = z_0(t), \qquad k = 0,1,2,\ldots,$$

convergent for $\varepsilon \in [0, \varepsilon_*]$.

Critical Case. We now consider a boundary-value problem

$$\dot z = A(t)z + \varphi(t) + \varepsilon Z(z,t,\varepsilon), \quad lz = \alpha + \varepsilon J(z(\cdot,\varepsilon),\varepsilon) \tag{5.43}$$

similar to problem (5.32). Unlike the previous section, we study the general critical case where $\operatorname{rank} Q = n_1 < m$. Thus, by virtue of Theorem 5.1, the generating boundary-value problem corresponding to (5.43), i.e.,

$$\dot z = A(t)z + \varphi(t), \qquad lz = \alpha, \tag{5.44}$$

is solvable if and only if $\varphi(t) \in C[a,b]$ and $\alpha \in R^m$ satisfy condition (5.9). In this case, the analyzed problem possesses an r-parameter ($r = n - n_1$) family of solutions given by (5.11). Our aim is to establish conditions for the existence and propose an algorithm for the construction of the solution $z(t,\varepsilon)$ of the boundary-value problem (5.43) from $C[\varepsilon]$, $\varepsilon \in [0,\varepsilon_0]$, which turns into one of the generating solutions of the boundary-value problem (5.44) for $\varepsilon = 0$. Note that periodic $(m = n,\ lz = z(a) - z(b) = \alpha = 0)$ and two-point $\big(m = n,$ $lz = M_1 z(a) + M_2 z(b)$, where M_i are $n \times n$ matrices$\big)$ boundary-value problems were studied in similar statements in [16–18, 65, 93, 99, 101, 142].

Necessary Conditions for the Existence of Solutions. We now establish conditions necessary for the existence of solutions $z(t,\varepsilon)$ of the boundary-value problem (5.43) which turn into the generating solutions $z_0(t,c_r)$ of the boundary-value problem (5.44) given by relation (5.11) for $\varepsilon = 0$.

Theorem 5.4. *Assume that the boundary-value problem (5.43) satisfies the conditions presented above and possesses a solution $z(t,\varepsilon)$ such that $z(\cdot,\varepsilon) \in C^1[a,b]$ and $z(t,\cdot) \in C[\varepsilon]$, $\varepsilon \in [0,\varepsilon_0]$, which turns into the generating solution $z_0(t,c_r)$ given by relation (5.11) with constant $c_r = c_0^*$ for $\varepsilon = 0$. Then the vector $c_0^* \in R^r$ satisfies the equation*

$$P_{Q_d^*}\left\{ J(z_0(\cdot,c_0^*),0) - l\int_a^b K(\cdot,\tau) Z(z_0(\tau,c_0^*),\tau,0)\,d\tau \right\} = 0. \tag{5.45}$$

Proof. Condition (5.9) for the existence of generating solutions $z_0(t,c_r) \in C[a,b]$ given by relation (5.11) is assumed to be satisfied. If we now consider the vector function $Z(z,t,\varepsilon)$ and vector functional $J(z(\cdot,\varepsilon),\varepsilon)$ in relations (5.32) as inhomogeneities and apply Theorem 5.1 to problem (5.32), then we get the following condition of solvability for the boundary-value problem (5.32):

$$P_{Q_d^*}\left\{ J(z(\cdot,\varepsilon),\varepsilon) - l\int_a^b K(\cdot,\tau) Z(z(\tau,\varepsilon),\tau,\varepsilon)\,d\tau \right\} = 0.$$

Passing to the limit as $\varepsilon \to 0$, we arrive at the required condition (5.45).

Note that, for the periodic boundary-value problem (5.43) $(m = n,\ d = r,$ $lz = z(a) - z(b) = \alpha = 0)$, equation (5.45) corresponds to the equation for generating amplitudes well known in the theory of nonlinear oscillations [65, 101]. Indeed, the vector $c_r = c_0^* \in R^r$ in equation (5.45) determines the amplitude of oscillations described by the generating solution (5.11) of the periodic boundary-value problem (5.44).

In general case, we say that

$$F(c_r) = P_{Q_d^*}\left\{ J(z_0(\cdot,c_r),0) - l\int_a^b K(\cdot,\tau) Z(z_0(\tau,c_r),\tau,0)d\tau \right\} = 0 \tag{5.46}$$

is the equation for generating constants of the boundary-value problem (5.43).

If equation (5.46) is solvable, then the vector c_0^* specifies the generating solution $z_0(t, c_0^*)$ associated with a solution $z(t, \cdot) \in C[\varepsilon]$, $z(t, 0) = z_0(t, c_0^*)$, of the original boundary-value problem (5.43). However, in the case where equation (5.46) is unsolvable, the boundary-value problem (5.43) also does not possess the required solution in $C[\varepsilon]$, $\varepsilon \in [0, \varepsilon_0]$. Note that, here and in what follows, all calculations are performed in real numbers and, thus, we seek real roots of the equation (5.46) for generating constants (algebraic or transcendental).

Note that the condition necessary for the existence of a solution of the critical boundary-value problem (5.43) according to which equation (5.46) must have at least one solution $c_r = c_0^* \in R^r$ and the condition necessary for the existence of solutions (5.36) of the boundary-value problem (5.32) have similar forms but, at the same time, they are characterized by the following basic distinctions:

In the critical case, condition (5.46) is satisfied by choosing a proper constant c_r in the r-parameter family of generating solutions (5.11). In the case $\operatorname{rank} Q = n$, there is no choice of constants at all because the generating solution $z_0(t)$ is unique. Therefore, the necessary condition (5.36) is, in fact, a severe restriction imposed on the nonlinearities $Z(z, t, \varepsilon)$ and $J(z(\cdot, \varepsilon), \varepsilon)$ and guaranteeing the existence of solution of the boundary-value problem (5.32).

Sufficient Conditions for the Existence of Solutions. Our aim is to establish a condition guaranteeing the existence of solutions of the boundary-value problem (5.43) in the so-called *critical case of the first order.* In this case (see [65]), the problem of the existence of solutions of the original boundary-value problem (5.43) is solved after the analysis of the boundary-value problem used to find the first approximation to the exact solution.

Assume that a constant vector $c_0^* \in R^r$ satisfies equation (5.46) for generating constants. By the change of variables

$$z(t, \varepsilon) = z_0(t, c_0^*) + x(t, \varepsilon) \tag{5.47}$$

in relations (5.43), we arrive at the following problem: It is necessary to establish conditions for the existence and an algorithm for the construction of a solution $x(t, \varepsilon)$ of the boundary-value problem

$$\begin{aligned} \dot{x} &= A(t)x + \varepsilon Z(z_0(t, c_0^*) + x(t, \varepsilon), t, \varepsilon), \\ lx &= \varepsilon J(z_0(\cdot, c_0^*) + x(\cdot, \varepsilon), \varepsilon) \end{aligned} \tag{5.48}$$

such that $x(\cdot, \varepsilon) \in C^1[a, b]$ and $x(t, \varepsilon) \in C[\varepsilon]$, $\varepsilon \in [0, \varepsilon_0]$; vanishing for $\varepsilon = 0$. By using the fact that the vector function $Z(z, t, \varepsilon)$ and the vector functional $J(z(\cdot, \varepsilon), \varepsilon)$ are continuously differentiable with respect to z in a neighborhood of the point $\varepsilon = 0$, we can separate the term linear as a function of x and the terms of order zero with respect to ε from the vector function $Z(z_0 + x, t, \varepsilon)$ and the vector functional $J(z_0(\cdot, c_0^*) + x(\cdot, \varepsilon), \varepsilon)$. As a result, we get the following expansions:

$$
\begin{aligned}
Z(z_0 + x, t, \varepsilon) &= f_0(t, c_0^*) + A_1(t)x + R(x(t,\varepsilon), t, \varepsilon), \\
J(z_0(\cdot, c_0^*) + x(\cdot, \varepsilon), \varepsilon) &= J_0(z_0(\cdot, c_0^*)) + l_1 x(\cdot, \varepsilon) + R_1(x(\cdot, \varepsilon), \varepsilon),
\end{aligned}
\tag{5.49}
$$

where

$$
f_0(t, c_0^*) = Z(z_0(t, c_0^*), t, 0) \in C[t],
$$

$$
J_0(z_0(\cdot, c_0^*)) = J(z_0(\cdot, c_0^*), 0),
$$

$$
A_1(t) = A_1(t, c_0^*) = \partial Z(z, t, 0)/\partial z \Big|_{z = z_0(t, c_0^*)} \in C[t],
$$

and $l_1 x(\cdot, \varepsilon)$ is the linear part of the vector functional $J(z_0(\cdot, c_0^*) + x(\cdot, \varepsilon), \varepsilon)$. The nonlinear vector function $R(x(t,\varepsilon), t, \varepsilon)$ belongs to the class $C^1[x]$, $C[t]$, $C[\varepsilon]$ in the region $||x|| \leq q$, $t \in [a, b]$, $\varepsilon \in [0, \varepsilon_0]$ and, moreover,

$$
R(0, t, 0) = 0, \qquad \frac{\partial R(0, t, 0)}{\partial x} = 0,
$$

$$
R_1(0, 0) = 0, \qquad \frac{\partial R_1(0, 0)}{\partial x} = 0.
$$

We now want to construct an operator system equivalent to the boundary-value problem (5.48) in the class of functions $C[\varepsilon]$ vanishing for $\varepsilon = 0$. By applying Theorem 5.1 (in which the nonlinearities $Z(z_0 + x, t, \varepsilon)$ and $J(z_0(\cdot, \varepsilon) + x(\cdot, \varepsilon), \varepsilon)$ are formally regarded as inhomogeneities) to the boundary-value problem (5.48), we get

$$
x(t, \varepsilon) = X_r(t)c + x^{(1)}(t, \varepsilon).
$$

In this case, the unknown constant vector $c = c(\varepsilon) \in R^r$ and the unknown vector function $x^{(1)}(t, \varepsilon)$ are determined, respectively, from the condition of solvability of the boundary-value problem (5.48) and the following representation of a special solution of the problem $x^{(1)}(t, \varepsilon)$:

$$
\begin{aligned}
P_{Q_d^*} \Big\{ & J_0(z_0(\cdot, c_0^*)) + l_1(X_r(\cdot)c + x^{(1)}(\cdot, \varepsilon)) + R_1(x(\cdot, \varepsilon), \varepsilon) \\
& - l \int_a^b K(\cdot, \tau) \Big[f_0(\tau, c_0^*) + A_1(\tau)(X_r(\tau)c + x^{(1)}(\tau, \varepsilon)) \\
& \qquad + R(x(\tau, \varepsilon), \tau, \varepsilon) \Big] d\tau \Big\} = 0,
\end{aligned}
$$

$$
\begin{aligned}
x^{(1)}(t, \varepsilon) = {} & \varepsilon \Big(G\big[f_0(\tau, c_0^*) + A_1(\tau)x(\tau, \varepsilon) + R(x(\tau, \varepsilon), \tau, \varepsilon) \big] \Big)(t) \\
& + \varepsilon X(t) Q^+ \Big\{ J_0(z_0(\cdot, c_0^*)) + l_1 x(\cdot, \varepsilon) + R_1(x(\cdot, \varepsilon), \varepsilon) \Big\}.
\end{aligned}
$$

In view of the fact that the vector constant c_0^* satisfies equation (5.45), for the solution $x(t,\cdot) \in C[\varepsilon]$, $x(t,0)=0$, of the boundary-value problem (5.48), we obtain an equivalent operator system:

$$\begin{aligned} x(t,\varepsilon) &= X_r(t)c + x^{(1)}(t,\varepsilon), \\ B_0 c &= -P_{Q_d^*}\Big\{ l_1 x^{(1)}(\cdot,\varepsilon) + R_1(x(\cdot,\varepsilon),\varepsilon) \\ &\quad - l\int_a^b K(\cdot,\tau)[A_1(\tau)x^{(1)}(\tau,\varepsilon) + R(x(\tau,\varepsilon),\tau,\varepsilon)]\,d\tau \Big\}, \\ x^{(1)}(t,\varepsilon) &= \varepsilon\Big(G\big[f_0(\tau,c_0^*) + A_1(\tau)x(\tau,\varepsilon) + R(x,\tau,\varepsilon)\big]\Big)(t) \\ &\quad + \varepsilon X(t)Q^+\Big\{ J_0(z_0(\cdot,c_0^*)) + l_1 x(\cdot,\varepsilon) + R_1(x(\cdot,\varepsilon),\varepsilon)\Big\}, \end{aligned} \tag{5.50}$$

where B_0 is a $d \times r$ matrix,

$$B_0 = P_{Q_d^*}\Big\{ l_1 X_r(\cdot) - l\int_a^b K(\cdot,\tau)A_1(\tau)X_r(\tau)\,d\tau \Big\}.$$

Let P_{B_0} be an $r \times r$ matrix orthoprojector $R^r \to N(B_0)$. The second equation in the operator system (5.50) is solvable if and only if the following condition is satisfied:

$$\begin{aligned} P_{B_0^*}P_{Q_d^*}\Big\{ l_1 x^{(1)}(\cdot,\varepsilon) + R_1(x(\cdot,\varepsilon),\varepsilon) \\ - l\int_a^b K(\cdot,\tau)[A_1(\tau)x^{(1)}(\tau,\varepsilon) + R(x(\tau,\varepsilon),\tau,\varepsilon)]\,d\tau \Big\} = 0, \end{aligned} \tag{5.51}$$

where $P_{B_0^*}$ is a $d \times d$ matrix orthoprojector onto the cokernel of the matrix $B_0\colon R^d \to N(B_0^*)$. If $P_{B_0^*} = 0$, i.e., $\operatorname{rank} B_0 = d$, then the last equality is always true. Hence, $d \le r$, which means that the number m of boundary conditions is not greater than the dimension n of the differential system because $d = m - n_1 \le r = n - n_1$. Hence, for $P_{B_0^*} = 0$, the second equation in the operator system (5.50) is solvable with respect to $c \in R^r$ to within an arbitrary vector constant $P_{B_0}\bar{c}$ $(\forall \bar{c} \in R^r)$ from the null space of the matrix B_0:

$$\begin{aligned} c = -B_0^+ P_{Q_d^*}\Big\{ l_1 x^{(1)}(\cdot,\varepsilon) + R_1(x(\cdot,\varepsilon),\varepsilon) \\ - l\int_a^b K(\cdot,\tau)[A_1(\tau)x^{(1)}(\tau,\varepsilon) + R(x(\tau,\varepsilon),\tau,\varepsilon)]\,d\tau \Big\} + P_{B_0}\bar{c} \quad (\forall \bar{c} \in R^r). \end{aligned}$$

One of solutions $x = x(t,\varepsilon)$ of problem (5.48) such that $x(\cdot,\varepsilon)\colon [a,b] \to R^n$, $x(\cdot,\varepsilon) \in C^1[a,b]$, $x(t,\cdot) \in C[0,\varepsilon_0]$, and $x(t,0) = 0$ can be found from the following operator system:

$$x(t,\varepsilon) = X_r(t)c + x^{(1)}(t,\varepsilon),$$

$$c = -B_0^+ P_{Q_d^*}\Big\{ l_1 x^{(1)}(\cdot,\varepsilon) + R_1(x(\cdot,\varepsilon),\varepsilon) \\ - l\int_a^b K(\cdot,\tau)[A_1(\tau)x^{(1)}(\tau,\varepsilon) + R(x(\tau,\varepsilon),\tau,\varepsilon)]\,d\tau \Big\}, \tag{5.52}$$

$$x^{(1)}(t,\varepsilon) = \varepsilon\Big(G[f_0(\tau,c_0^*) + A_1(\tau)x(\tau,\varepsilon) + R(x,\tau,\varepsilon)]\Big)(t) \\ + \varepsilon X(t)Q^+\Big\{ J_0(z_0(\cdot,c_0^*)) + l_1 x(\cdot,\varepsilon) + R_1(x(\cdot,\varepsilon),\varepsilon)\Big\}.$$

The operator system (5.52) belongs to the class of systems studied in [65]. These systems can be solved by the method of simple iterations. Indeed, system (5.52) can be rewritten in the form

$$y(t,\varepsilon) = L^{(1)}y + Fy, \tag{5.53}$$

where $y = y(t,\varepsilon) = \mathrm{col}\,[x(t,\varepsilon), c(\varepsilon), x^{(1)}(t,\varepsilon)]$ is a $(2n+r)$-dimensional column vector, $L^{(1)}$ and F are, respectively, linear and nonlinear bounded operators, namely,

$$L^{(1)} = \begin{bmatrix} 0 & X_r(t) & I_n \\ 0 & 0 & L_1 \\ 0 & 0 & 0 \end{bmatrix},$$

$$L_1\psi(t) = -B_0^+ P_{Q_d^*}\left[l_1\psi(\cdot) - l\int_a^b K(\cdot,\tau)A_1(\tau)\psi(\tau)\,d\tau \right],$$

$$F = F(y,t,\varepsilon) \\ = \mathrm{col}\,\Big\{ 0, -B_0^+ P_{Q_d^*}\left[R_1(x(\cdot,\varepsilon),\varepsilon) - l\int_a^b K(\cdot,\tau)R(x(\tau,\varepsilon),\tau,\varepsilon)\,d\tau \right], \\ \varepsilon\big(G[f_0(\tau,c_0^*) + A_1(\tau)x(\tau,\varepsilon) + R(x,\tau,\varepsilon)]\big)(t) \\ + \varepsilon X(t)Q^+[J_0(z_0(\cdot,c_0^*)) + l_1 x(\cdot,\varepsilon) + R_1(x(\cdot,\varepsilon),\varepsilon)] \Big\},$$

$$F(0,t,0) = 0, \qquad \frac{\partial F(0,t,0)}{\partial y} = 0,$$

and I_n is the $n \times n$ identity matrix. The upper-triangular block-matrix operator $L^{(1)}$ has zero blocks on the principal diagonal and below. Therefore, the operator system (5.53) admits the following transformation:

$$y(t,\varepsilon) = \tilde{L}Fy(t,\varepsilon), \quad \tilde{L} = (I_s - L^{(1)})^{-1}, \quad s = 2n + r.$$

The system obtained as a result can be solved by the method of simple iterations. Moreover, by the method of Lyapunov majorants, one can show [19] that the iterative process converges in a sufficiently small neighborhood of the generating solution and establish the required estimates for the error of approximation. In addition, the operator $S = \tilde{L}F$ of the analyzed system is a contraction operator in a sufficiently small neighborhood of a point $z_0(t, c_r^*)$, $\varepsilon = 0$ and, hence, one of the versions of the fixed-point principle [87] is applicable to this system for sufficiently small $\varepsilon \in [0, \varepsilon_*]$. Thus, the method of simple iterations enables us to construct the solution of the boundary-value problem (5.48) as a result of the iterative process described below.

Iterative Algorithm. The iterative process used to find the solution $x(t,\cdot) \in C[\varepsilon]$, $x(t,0) = 0$, of the boundary-value problem (5.48) is based on the operator system (5.52). As the first approximation $x_1^{(1)}(t,\varepsilon)$ to the solution $x^{(1)}(t,\varepsilon)$, we take

$$x_1^{(1)}(t,\varepsilon) = \varepsilon\Lambda^- \begin{bmatrix} f_0(\tau, c_0^*) \\ J_0(z_0(\cdot, c_0^*)) \end{bmatrix}(t)$$

$$= \varepsilon(Gf_0(\tau, c_0^*))(t) + \varepsilon X(t)Q^+ J_0(z_0(\cdot, c_0^*)).$$

As follows from the definition of the operator Λ^-, the vector function $x_1^{(1)} = x_1^{(1)}(t,\varepsilon)$ is a special solution of the boundary-value problem

$$\dot{x}_1 = A(t)x_1 + \varepsilon f_0(t, c_0^*), \qquad lx_1 = \varepsilon J_0(z_0(\cdot, c_0^*)).$$

Since $c_0^* \in R^r$ is chosen to satisfy equation (5.46) for the generating amplitudes of this boundary-value problem, the indicated solution exists. Hence, the first approximation $x_1(t,\varepsilon)$ to the exact solution $x(t,\varepsilon)$ of problem (5.48) is assumed to be equal to $x_1^{(1)}(t,\varepsilon)$. For the second approximation $x_2^{(1)}(t,\varepsilon)$ to the solution $x^{(1)}(t,\varepsilon)$, we set

$$x_2^{(1)}(t,\varepsilon) = \varepsilon\Lambda^- \left\{ \begin{matrix} f_0(\tau, c_0^*) + A_1(\tau)[X_r(\tau)c_1 + x_1^{(1)}(\tau,\varepsilon)] + R(x_1(\tau,\varepsilon), \tau, \varepsilon) \\ J_0(z_0(\cdot, c_0^*)) + l_1[X_r(\cdot)c_1 + x_1^{(1)}(\cdot,\varepsilon)] + R_1(x_1(\cdot,\varepsilon), \varepsilon) \end{matrix} \right\}(t).$$

By virtue of the definition of the operator Λ^-, the vector function $x_2^{(1)}(t,\varepsilon)$ is a solution of the boundary-value problem

$$\dot{x}_2 = A(t)x_2 + \varepsilon\Big\{f_0(t, c_0^*) + A_1(t)[X_r(t)c_1 + x_1^{(1)}(t,\varepsilon)] + R(x_1(t,\varepsilon), t, \varepsilon)\Big\},$$

$$lx_2 = \varepsilon\Big\{J_0(z_0(\cdot, c_0^*)) + l_1[X_r(\cdot)c_1 + x_1^{(1)}(\cdot,\varepsilon)] + R_1(x_1(\cdot,\varepsilon), \varepsilon)\Big\}.$$

By using the necessary and sufficient condition of solvability of this problem

$$B_0 c_1 + P_{Q_d^*}\Big\{ l_1 x_1^{(1)}(\cdot,\varepsilon) + R_1(x_1(\cdot,\varepsilon),\varepsilon) - l\int_a^b K(\cdot,\tau)[A_1(\tau)x_1^{(1)}(\tau,\varepsilon) + R(x_1(\tau,\varepsilon),\tau,\varepsilon)]\,d\tau\Big\} = 0,$$

we find the first approximation c_1 to $c(\varepsilon)$.

By virtue of Theorem 3.10, this system is solvable (with respect to $c_1 \in R^r$) if and only if

$$P_{B_0^*} P_{Q_d^*}\Big\{ l x_1^{(1)}(\cdot,\varepsilon) + R_1(x_1(\cdot,\varepsilon),\varepsilon) - l\int_a^b K(\cdot,\tau)[A_1(\tau)x_1^{(1)}(\tau,\varepsilon) + R(x_1(\tau,\varepsilon),\tau,\varepsilon)]\,d\tau\Big\} = 0.$$

Therefore, if $P_{B_0^*} = 0$, then the last condition is always satisfied and the equation presented earlier is solvable with respect to $c \in R^r$ to within an arbitrary vector constant $P_{B_0}c$ ($\forall c \in R^r$) from the null space of the matrix B_0. For one of these solutions, we get the following representation:

$$c_1 = -B_0^+ P_{Q_d^*}\Big\{ l_1 x_1^{(1)}(\cdot,\varepsilon) + R_1(x_1(\cdot,\varepsilon),\varepsilon) - l\int_a^b K(\cdot,\tau)[A_1(\tau)x_1^{(1)}(\tau,\varepsilon) + R(x_1(\tau,\varepsilon),\tau,\varepsilon)]\,d\tau\Big\}.$$

As a result, the second approximation $x_2(t,\varepsilon)$ to the exact solution $x(t,\varepsilon)$ takes the form

$$x_2(t,\varepsilon) = X_r(t)c_1 + x_2^{(1)}(t,\varepsilon).$$

The third approximation $x_3^{(1)}(t,\varepsilon)$ is sought in the form

$$x_3^{(1)}(t,\varepsilon) = \varepsilon\Lambda^-\left\{\begin{array}{c} f_0(\tau,c_0^*) + A_1(\tau)[X_r(\tau)c_2 + x_2^{(1)}(\tau,\varepsilon)] + R(x_2(\tau,\varepsilon),\tau,\varepsilon) \\ J_0(z_0(\cdot,c_0^*)) + l_1[X_r(\cdot)c_2 + x_2^{(1)}(\cdot,\varepsilon)] + R_1(x_2(\cdot,\varepsilon),\varepsilon) \end{array}\right\}(t)$$

as a solution of the boundary-value problem

$$\dot x_3 = A(t)x_3 + \varepsilon\{f_0(t,c_0^*) + A_1(t)[X_r(t)c_2 + x_2^{(1)}(t,\varepsilon)] + R(x_2(t,\varepsilon),t,\varepsilon)\},$$

$$l x_3 = \varepsilon\{J_0(z_0(\cdot,c_0^*)) + l_1[X_r(\cdot)c_2 + x_2^{(1)}(\cdot,\varepsilon)] + R_1(x_2(\cdot,\varepsilon),\varepsilon)\}.$$

The necessary and sufficient condition of solvability of this problem yields the following algebraic system:

$$B_0 c_2 + P_{Q_d^*}\Big\{ l_1 x_2^{(1)}(\cdot,\varepsilon) + R_1(x_2(\cdot,\varepsilon),\varepsilon) - l\int_a^b K(\cdot,\tau)[A_1(\tau)x_2^{(1)}(\tau,\varepsilon) + R(x_2(\tau,\varepsilon),\tau,\varepsilon)]\,d\tau\Big\} = 0, \quad (5.54)$$

whence we get the second approximation c_2 to $c(\varepsilon)$ in the form

$$c_2 = -B_0^+ P_{Q_d^*}\Big\{ l_1 x_2^{(1)}(\cdot,\varepsilon) + R_1(x_2(\cdot,\varepsilon),\varepsilon) - l\int_a^b K(\cdot,\tau)[A_1(\tau)x_2^{(1)}(\tau,\varepsilon) + R(x_2(\tau,\varepsilon),\tau,\varepsilon)]\,d\tau\Big\}.$$

System (5.54) is solvable if and only if

$$P_{B_0^*} P_{Q_d^*}\Big\{ l_1 x_2^{(1)}(\cdot,\varepsilon) + R_1(x_2(\cdot,\varepsilon),\varepsilon) - l\int_a^b K(\cdot,\tau)[A_1(\tau)x_2^{(1)}(\tau,\varepsilon) + R(x_2(\tau,\varepsilon),\tau,\varepsilon)]\,d\tau\Big\} = 0. \quad (5.55)$$

Thus, in the case where $P_{B_0^*} = 0$, the criteria of solvability similar to (5.55) for the corresponding algebraic systems are satisfied in each step of the iterative process. Hence, if we continue the iterative process, then, from the operator system (5.52), we get the following procedure for the construction of a solution $x(t,\cdot) \in C[\varepsilon]$, $\varepsilon \in [0,\varepsilon_0]$, $x(t,0) = 0$, of the boundary-value problem (5.48):

$$c_k = -B_0^+ P_{Q_d^*}\Big\{ l_1 x_k^{(1)}(\cdot,\varepsilon) + R_1(x_k(\cdot,\varepsilon),\varepsilon) - l\int_a^b K(\cdot,\tau)[A_1(\tau)x_k^{(1)}(\tau,\varepsilon) + R(x_k(\tau,\varepsilon),\tau,\varepsilon)]\,d\tau\Big\}, \quad (5.56)$$

$$x_{k+1}^{(1)}(t,\varepsilon) = \varepsilon\Big(G\{f_0(\tau,c_0^*) + A_1(\tau)[X_r(\tau)c_k + x_k^{(1)}(\tau,\varepsilon)] + R(x_k(\tau,\varepsilon),\tau,\varepsilon)\}\Big)(t) + \varepsilon X(t)Q^+\{J_0(z_0(\cdot,c_0^*)) + l_1[X_r(\cdot)c_k + x_k^{(1)}(\cdot,\varepsilon)] + R_1(x_k(\cdot,\varepsilon),\varepsilon)\},$$

$$x_{k+1}(t,\varepsilon) = X_r(t)c_k + x_{k+1}^{(1)}(t,\varepsilon),$$

$$k = 0,1,2,\ldots, \qquad x_0(t,\varepsilon) = x_0^{(1)}(t,\varepsilon) = 0.$$

To prove that the iterative process (5.56) is convergent and establish the required estimates, we can use the procedure of the method of Lyapunov majorants described in [19, 65].

Note that, in order to prove the existence of a solution of the boundary-value problem (5.48), it suffices to establish the conditions required for its reducibility to the operator form (5.52) with the corresponding properties (without constructing the majorizing Lyapunov equations). As already indicated, these majorizing Lyapunov equations and their solutions always exist. Thus, it is necessary to construct and solve these equations only in the case where this is required in order establish specific estimates characterizing the iterative process (5.56). Thus, the following theorem is true:

Theorem 5.5. *Assume that the boundary-value problem (5.43) satisfies the indicated conditions in the critical case* ($\operatorname{rank} Q = n_1 < m$) *and the corresponding generating boundary-value problem (5.44) possesses an r-parameter* ($r = n - n_1$) *family of generating solutions (5.11) provided that condition (5.9) is satisfied* ($d = m - n_1$).

Then, for any value of the vector $c_r = c_0^* \in R^r$ satisfying equation (5.46) for the generating constants, the boundary-value problem (5.48) possesses at least one solution $x(t,\varepsilon)\colon x(\cdot,\varepsilon) \in C[a,b]$, $x(t,\cdot) \in C[0,\varepsilon_*]$, vanishing for $\varepsilon = 0$, provided that

$$P_{B_0^*} = 0 \Leftrightarrow \operatorname{rank} B_0 = d. \tag{5.57}$$

One of these solutions can be obtained as a result of the iterative process (5.56) convergent for $\varepsilon \in [0,\varepsilon_*]$, where ε_* is a proper constant characterizing the range of convergence of the iterative process (5.56). In this case, the boundary-value problem (5.43) possesses at least one solution $z(t,\varepsilon)\colon z(\cdot,\varepsilon) \in C^1[a,b]$, $z(t,\cdot) \in C[0,\varepsilon_*]$, which turns into the generating solution $z_0(t,c_0^*)$ given by relation (5.11) for $\varepsilon = 0$ and can be obtained as a result of the iterative process (5.56) by using the relation $z_k(t,\varepsilon) = z_0(t,c_0^*) + x_k(t,\varepsilon)$, $k = 0,1,2,\ldots$.

For boundary-value problems with Fredholm linear terms of index zero ($m{=}n$), the relation $P_{B_0^*} = 0$ implies that $\det B_0 \neq 0$, which means that the equation for generating constants (5.46) has simple roots [17, 19]. Therefore, Theorem 5.5 yields the following assertion:

Corollary 5.2. *Assume that the boundary-value problem (5.43) satisfies the conditions presented above in the critical case* ($\operatorname{rank} Q = n_1 < m$) *and the corresponding generating boundary-value problem (5.44) possesses an r-parameter* ($r = n - n_1$) *family of generating solutions (5.11) provided that condition (5.9) is satisfied* ($d = m - n_1 = r = n - n_1$).

Then, for any simple ($\det B_0 \neq 0$) *root $c_r = c_0^* \in R^r$ of the equation for generating constants (5.46), the boundary-value problem (5.48) has exactly one solution $x(t,\varepsilon)\colon x(\cdot,\varepsilon) \in C^1[a,b]$, $x(t,\cdot) \in C[0,\varepsilon_*]$, vanishing for $\varepsilon = 0$. This* *solution can be obtained with the help of the iterative process (5.56) convergent for* $\varepsilon \in [0, \varepsilon_*]$, *where* ε_* *is a proper constant characterizing the range of convergence of the iterative process (5.56). In this case, the boundary-value problem (5.43) possesses a unique solution* $z(t,\varepsilon)$: $z(\cdot,\varepsilon) \in C^1[a,b]$, $x(t,\cdot) \in C[0,\varepsilon_*]$, *which turns into the generating solution* $z_0(t,c_0^*)$ *given by relation (5.11) for* $\varepsilon = 0$ *and can be obtained with the help of the iterative process (5.56) and the relation* $z_k(t,\varepsilon) = z_0(t,c_0^*) + x_k(t,\varepsilon)$, $k = 0,1,2,\dots$.

Critical Case of the Second Order. Consider the boundary-value problem (5.48) in the case characterized by the fact that the problem of the existence of solutions of the original boundary-value problem is solved after analyzing the boundary-value problem used to find the second approximation to the solution. Furthermore, the existence of solution to the original boundary-value problem depends on the condition established by using nonlinearities and the first approximation to the required solution. This situation is called the critical case of the second order.

As everywhere in the present chapter, to establish sufficient conditions for the existence of solutions, we assume that $P_Q \neq 0 \Leftrightarrow \operatorname{rank} Q = n_1 < m$ and, unlike the previous cases, impose the condition $P_{B_0^*} \neq 0$ $(\operatorname{rank} B_0 < d)$.

To simplify calculations, we assume that the vector function $R(x,t,\varepsilon)$ and the vector functional $R_1(x,\varepsilon)$ in the expansions of nonlinearities (5.49) are such that

$$R(\cdot,t,\varepsilon) \in C^1[\|x\| \le q], \quad R(x,\cdot,\varepsilon) \in C[a,b], \quad R(x,t,\cdot) \in C[0,\varepsilon_0],$$

$$R_1(\cdot,\varepsilon) \in C^1[\|x\| \le q], \quad R_1(x,\cdot) \in C[0,\varepsilon_0],$$

and the following conditions are satisfied:

$$\begin{aligned} R(0,t,0) = 0, &\quad \frac{\partial R(0,t,\varepsilon)}{\partial x} = 0, \\ R_1(0,0) = 0, &\quad \frac{\partial R_1(0,\varepsilon)}{\partial x} = 0. \end{aligned} \tag{5.58}$$

The second equation in (5.50) is solvable if and only if

$$P_{B_0^*} P_{Q_d^*} \Big\{ l_1 x^{(1)}(\cdot,\varepsilon) + R_1(x^{(1)}(\cdot,\varepsilon),\varepsilon) \\ - l \int_a^b K(\cdot,\tau)[A_1(\tau)x^{(1)}(\tau,\varepsilon) + R(x(\tau,\varepsilon),\tau,\varepsilon)]\, d\tau \Big\} = 0. \tag{5.59}$$

Its solution admits a representation in the form of a direct sum

$$c = c^{(0)} + c^{(1)}, \tag{5.60}$$

where

$$c^{(0)} = -B_0^+ P_{Q_d^*}\Big\{ l_1 x^{(1)}(\cdot,\varepsilon) + R_1(x(\cdot,\varepsilon),\varepsilon) - l\int_a^b K(\cdot,\tau)[A_1(\tau)x^{(1)}(\tau,\varepsilon) + R(x(\tau,\varepsilon),\tau,\varepsilon)]\,d\tau \Big\}$$

and $c^{(1)}$ is an arbitrary r-dimensional constant:

$$c^{(1)} = P_{B_0}c = P_{B_0}c^{(1)} \in N(B_0) \subseteq R^r.$$

In view of representation (5.60), from the third equation in the operator system (5.50), we obtain

$$x^{(1)}(t,\varepsilon) = \varepsilon G_1(t)P_{B_0}c^{(1)} + x^{(2)}(t,\varepsilon), \tag{5.61}$$

where

$$G_1(t) = (G[A_1(\tau)X_r(\tau)])(t) + X(t)Q^+ l_1 X_r(\cdot)$$

is an $n \times r$ matrix and

$$\begin{aligned} x^{(2)}(t,\varepsilon) = \varepsilon\Big(G\big[f_0(\tau,c_0^*) &+ A_1(\tau)\big(X_r(\tau)(I_r - P_{B_0})c^{(0)} + \varepsilon G_1(\tau)P_{B_0}c^{(1)} + x^{(2)}(\tau,\varepsilon)\big) \\ &+ R(x(\tau,\varepsilon),\tau,\varepsilon)\big]\Big)(t) \\ + \varepsilon X(t)Q^+\Big\{ J_0(z_0(\cdot,c_0^*)) &+ l_1\big[X_r(\cdot)(I_r - P_{B_0})c^{(0)} + \varepsilon G_1(\cdot)P_{B_0}c^{(1)} + x^{(2)}(\cdot,\varepsilon)\big] \\ &+ R_1(x(\cdot,\varepsilon),\varepsilon)\Big\}. \end{aligned}$$

In view of representation (5.61) and the condition of solvability (5.59) of the second equation in (5.50), we arrive at the following algebraic system:

$$\varepsilon B_1 c^{(1)} + P_{B_0^*}P_{Q_d^*}\Big\{ l_1 x^{(2)}(\cdot,\varepsilon) + R_1(x(\cdot,\varepsilon),\varepsilon) - l\int_a^b K(\cdot,\tau)[A_1(\tau)x^{(2)}(\tau,\varepsilon) + R(x(\tau,\varepsilon),\tau,\varepsilon)]\,d\tau \Big\} = 0, \tag{5.62}$$

where B_1 is a $d \times r$ matrix, namely,

$$B_1 = P_{B_0^*}P_{Q_d^*}\Big\{ l_1 G_1(\cdot) - l\int_a^b K(\cdot,\tau)A_1(\tau)G_1(\tau)\,d\tau \Big\} P_{B_0}.$$

Let P_{B_1} be an $r \times r$ matrix orthoprojector $R^r \to N(B_1)$ and let $P_{B_1^*}$ be a $d\times d$ matrix orthoprojector $R^d \to N(B_1^*)$. Then, in order that equation (5.62) be solvable, it is necessary and sufficient that

$$P_{B_1^*}P_{B_0^*}P_{Q_d^*}\Big\{l_1x^{(2)}(\cdot,\varepsilon)+R_1(x(\cdot,\varepsilon),\varepsilon)$$
$$-\,l\int_a^b K(\cdot,\tau)[A_1(\tau)x^{(2)}(\tau,\varepsilon)+R(x(\tau,\varepsilon),\tau,\varepsilon)]\,d\tau\Big\}=0. \tag{5.63}$$

If the null spaces $N(B_0^*)$ and $N(B_1^*)$ have empty intersection, i.e., $P_{B_1^*}P_{B_0^*}=0$, then equality (5.63) always holds and system (5.62) is solvable with respect to $\varepsilon c^{(1)}\in N(B_1)$ to within an arbitrary vector constant $P_{B_0}P_{B_1}c$ ($\forall c\in R^r$):

$$\varepsilon c^{(1)}=-B_1^+P_{B_0^*}P_{Q_d^*}\Big\{l_1x^{(2)}(\cdot,\varepsilon)+R_1(x(\cdot,\varepsilon),\varepsilon)$$
$$-\,l\int_a^b K(\cdot,\tau)[A_1(\tau)x^{(2)}(\tau,\varepsilon)+R(x(\tau,\varepsilon),\tau,\varepsilon)]\,d\tau\Big\}+c^{(2)}, \tag{5.64}$$

where $c^{(2)}$ is an arbitrary vector from the set $N(B_0)\cap N(B_1)$:

$$c^{(2)}=P_{B_1}c^{(1)}=P_{B_1}P_{B_0}c=P_{B_0}P_{B_1}c\in N(B_0)\cap N(B_1).$$

If the null spaces $N(B_0)$ and $N(B_1)$ have empty intersection, i.e., $P_{B_0}P_{B_1}=0$, then equation (5.62) possesses the unique solution (5.64) ($c^{(2)}=0$). In any case, under the assumption that $P_{B_1^*}P_{B_0^*}=0$, system (5.62) possesses at least one solution.

Therefore, in the case where

$$P_{B_1^*}P_{B_0^*}=0, \tag{5.65}$$

we pass from system (5.50) to the following operator system:

$$x(t,\varepsilon)=X_r(t)(I_r-P_{B_0})c^{(0)}+\varepsilon G_1(t)P_{B_0}c^{(1)}+x^{(2)}(t,\varepsilon),$$
$$c^{(0)}=-B_0^+P_{Q_d^*}\Big\{l_1[\varepsilon G_1(\cdot)P_{B_0}c^{(1)}+x^{(2)}(\cdot,\varepsilon)]+R_1(x(\cdot,\varepsilon),\varepsilon)$$
$$-\,l\int_a^b K(\cdot,\tau)\Big[A_1(\tau)(\varepsilon G_1(\tau)P_{B_0}c^{(1)}+x^{(2)}(\tau,\varepsilon))$$
$$+\,R(x(\tau,\varepsilon),\tau,\varepsilon)\Big]\,d\tau\Big\},$$
$$\varepsilon c^{(1)}=-B_1^+P_{B_0^*}P_{Q_d^*}\Big\{l_1x^{(2)}(\cdot,\varepsilon)+R_1(x(\cdot,\varepsilon),\varepsilon)$$
$$-\,l\int_a^b K(\cdot,\tau)[A_1(\tau)x^{(2)}(\tau,\varepsilon)+R(x(\tau,\varepsilon),\tau,\varepsilon)]\,d\tau\Big\}, \tag{5.66}$$

$$x^{(2)}(t,\varepsilon) = \varepsilon\Big(G\big[f_0(\tau,c_0^*)$$
$$+ A_1(\tau)(X_r(\tau)(I_r - P_{B_0})c^{(0)} + \varepsilon G_1(\tau)P_{B_0}c^{(1)} + x^{(2)}(\tau,\varepsilon))$$
$$+ R(x(\tau,\varepsilon),\tau,\varepsilon)\big]\Big)(t)$$
$$+ \varepsilon X(t)Q^+\Big\{J_0(z_0(\cdot,c_0^*)) + l_1[X_r(\cdot)(I_r - P_{B_0})c^{(0)}$$
$$+ \varepsilon G_1(\cdot)P_{B_0}c^{(1)} + x^{(2)}(\cdot,\varepsilon)] + R_1(x(\cdot,\varepsilon),\varepsilon)\Big\}.$$

System (5.66) can be rewritten in the following form similar to (5.53):

$$y = L^{(1)}y + Fy, \tag{5.67}$$

where $L^{(1)}$ is a bounded linear block-matrix operator, namely,

$$L^{(1)} = \begin{bmatrix} 0 & X_r(I_r - P_{B_0}) & G_1P_{B_0} & I_n \\ 0 & 0 & S_1 & L_1 \\ 0 & 0 & 0 & L_2 \\ 0 & 0 & 0 & 0 \end{bmatrix},$$

y is a $2(n+r)$-dimensional column vector, namely,

$$y = y(t,\varepsilon) = \text{col}\,[x(t,\varepsilon), (I_r - P_{B_0})c^{(0)}, \varepsilon P_{B_0}c^{(1)}, x^{(2)}(t,\varepsilon)],$$

F is a $2(n+r)$-dimensional bounded nonlinear vector function of the vector variable y from the class $C^{(1)}[y]$, $C[t]$, $C[\varepsilon]$ in the region $||y|| \le q$, $t \in [a,b]$, $\varepsilon \in [0,\varepsilon_0]$, namely,

$$Fy = \text{col}\Big\{0, -B_0^+P_{Q_d^*}\Big[R_1(x(\cdot,\varepsilon),\varepsilon)) - l\int_a^b K(\cdot,\tau)R(x(\tau,\varepsilon),\tau,\varepsilon)\,d\tau\Big],$$
$$- B_1^+P_{B_0^*}P_{Q_d^*}\Big[R_1(x(\cdot,\varepsilon),\varepsilon) - l\int_a^b K(\cdot,\tau)R(x(\tau,\varepsilon),\tau,\varepsilon)\,d\tau\Big],$$
$$\varepsilon\Big(G\big[f_0(\tau,c_0^*)$$
$$+ A_1(\tau)\big(X_r(\tau)(I_r - P_{B_0})c^{(0)} + \varepsilon G_1(\tau)P_{B_0}c^{(1)} + x^{(2)}(\tau,\varepsilon)\big)\big]\Big)(t)$$
$$+ \varepsilon X(t)Q^+\big[J_0(z_0(\cdot,c_0^*))$$
$$+ l_1\big(X_r(\cdot)(I_r - P_{B_0})c^{(0)} + \varepsilon G_1(\cdot)P_{B_0}c^{(1)} + x^{(2)}(\cdot,\varepsilon)\big)$$
$$+ R_1(x(\cdot,\varepsilon),\varepsilon))\big]\Big\},$$

S_1 is an $r \times r$ matrix, namely,

$$S_1 = -B_0^+ P_{Q_d^*} \left[l_1 G_1(\cdot) - l \int_a^b K(\cdot,\tau) A_1(\tau) G_1(\tau) P_{B_0} \, d\tau \right],$$

and L_1 and L_2 are bounded linear matrix integral operators given by the formulas

$$L_1 \psi(\cdot) = -B_0^+ P_{Q_d^*} \left[l_1 \psi(\cdot) - l \int_a^b K(\cdot,\tau) A_1(\tau) \psi(\tau) \, d\tau \right],$$

$$L_2 \psi(\cdot) = -B_1^+ P_{B_0^*} P_{Q_d^*} \left[l_1 \psi(\cdot) - l \int_a^b K(\cdot,\tau) A_1(\tau) \psi(\tau) \, d\tau \right].$$

We see that if $P_{B_0^*} \neq 0$, then the operator system (5.50) does not belong to the class of systems of the form (5.53) whose solution can be found by the method of simple iterations.

As a result of the separation of an additional variable, system (5.50) reduces to system (5.67), i.e., to (5.53), provided that condition (5.65) is satisfied. As a result of the decomposition of an r-dimensional constant into the direct sum (5.60) of two variables defined in different ways, the dimension of the operator system (5.48) increases by $r = \dim N(Q)$. Under condition (5.65), this enables us to reduce the $(2n+r)$-dimensional system (5.48) to the regular $2(n+r)$-dimensional operator system (5.66) whose solution can be found by the method of simple iterations.

Iterative Algorithm. By applying the method of simple iterations to the operator system (5.66), we construct an iterative process aimed at finding the solution $x(t,\cdot) \in C[\varepsilon]$, $x(t,0) = 0$, of the boundary-value problem (5.48). As the first approximation $x_1^{(2)}(t,\varepsilon)$ to the required solution $x^{(2)}(t,\varepsilon)$, we take

$$x_1^{(2)}(t,\varepsilon) = \varepsilon \left(\Lambda^- \begin{bmatrix} f_0(\tau, c_0^*) \\ J_0(z_0(\cdot, c_0^*)) \end{bmatrix} \right)(t)$$

$$= \varepsilon (G f_0(\tau, c_0^*))(t) + \varepsilon X(t) Q^+ J_0(z_0(\cdot, c_0^*)).$$

According to the definition of the operator Λ^-, the vector function $x_1^{(2)}(t,\varepsilon)$ is a solution of the following boundary-value problem:

$$\dot{x}_1 = A(t) x_1 + \varepsilon f_0(t, c_0^*), \quad l x_1 = \varepsilon J_0(z_0(\cdot, c_0^*)).$$

By the choice of $c_0^* \in R^r$ from the equation for generating constants (5.46), the indicated solution exists. We assume that the first approximation $x_1(t,\varepsilon)$ obtained for the boundary-value problem (5.48) is equal to $x_1^{(2)}(t,\varepsilon)$. For the second approximation $x_2^{(2)}(t,\varepsilon)$ to the required solution $x^{(2)}(t,\varepsilon)$, we set

$$x_2^{(2)}(t,\varepsilon) = \varepsilon\Lambda^- \left\{ \begin{array}{c} f_0(\tau, c_0^*) + A_1(\tau)[X_r(\tau)(I_r - P_{B_0})c_1^{(0)} + x_1^{(2)}(\tau,\varepsilon)] \\ + R(x_1^{(2)}(\tau,\varepsilon),\tau,\varepsilon) \\ J_0(z_0(\cdot, c_0^*)) + l_1[X_r(\cdot)(I_r - P_{B_0})c_1^{(0)} + x_1^{(2)}(\cdot,\varepsilon)] \\ + R_1(x_1^{(2)}(\cdot,\varepsilon),\varepsilon) \end{array} \right\}(t).$$

By definition, the vector function $x_2^{(2)}(t,\varepsilon)$ is a special solution of the boundary-value problem

$$\begin{aligned} \dot{x}_2 &= A(t)x_2 + \varepsilon\Big\{ f_0(t, c_0^*) + A_1(t)[X_r(t)(I_r - P_{B_0})c_1^{(0)} + x_1^{(2)}] \\ &\qquad + R(x_1^{(2)}(t,\varepsilon),t,\varepsilon)\Big\}, \\ lx_2 &= \varepsilon\Big\{ J_0(z_0(\cdot, c_0^*)) + l_1[X_r(\cdot)(I_r - P_{B_0})c_1^{(0)} + x_1^{(2)}(\cdot,\varepsilon)] \\ &\qquad + R_1(x_1^{(2)}(\cdot,\varepsilon),\varepsilon)\Big\}. \end{aligned} \tag{5.68}$$

This solution exists due to the choice of the vector constant $c_1^{(0)} \in R^r$ from the condition of solvability of problem (5.68)

$$B_0 c_1^{(0)} + P_{Q_d^*}\Big\{ l_1 x_1^{(2)}(\cdot,\varepsilon) + R_1(x_1^{(2)}(\cdot,\varepsilon),\varepsilon) \\ - l\int_a^b K(\cdot,\tau)[A_1(\tau)x_1^{(2)}(\tau,\varepsilon) + R(x_1^{(2)}(\tau,\varepsilon),\tau,\varepsilon)]\,d\tau\Big\} = 0. \tag{5.69}$$

In turn, system (5.69) is solvable if and only if

$$P_{B_0^*}P_{Q_d^*}\Big\{ l_1 x_1^{(2)}(\cdot,\varepsilon) + R_1(x_1^{(2)}(\cdot,\varepsilon),\varepsilon) \\ - l\int_a^b K(\cdot,\tau)[A_1(\tau)x_1^{(2)}(\tau,\varepsilon) + R(x_1^{(2)}(\tau,\varepsilon),\tau,\varepsilon)]\,d\tau\Big\} = 0. \tag{5.70}$$

Under this condition, we obtain the first approximation to $c^{(0)} \in R^r$ from system (5.69) by the formula

$$c_1^{(0)} = -B_0^+ P_{Q_d^*}\Big\{ l_1 x_1^{(2)}(\cdot,\varepsilon) + R_1(x_1^{(2)}(\cdot,\varepsilon),\varepsilon) \\ - l\int_a^b K(\cdot,\tau)[A_1(\tau)x_1^{(2)}(\tau,\varepsilon) + R(x_1^{(2)}(\tau,\varepsilon),\tau,\varepsilon)]\,d\tau\Big\}.$$

As the second approximation $x_2(t,\varepsilon)$ to the required solution $x(t,\varepsilon)$, we take the quantity

$$x_2(t,\varepsilon) = X_r(t)(I_r - P_{B_0})c_1^{(0)} + x_2^{(2)}(t,\varepsilon).$$

The third approximation is sought as a solution of the boundary-value problem

$$\begin{aligned}\dot{x}_3 = A(t)x_3 + \varepsilon\Big\{ & f_0(t, c_0^*) \\ & + A_1(t)[X_r(t)(I_r - P_{B_0})c_2^{(0)} + \varepsilon G_1(t)P_{B_0}c_1^{(1)} + x_2^{(2)}(t,\varepsilon)] \\ & + R(x_2(t,\varepsilon), t, \varepsilon)\Big\}, \\ lx_3 = \varepsilon\Big\{ & J_0(z_0(\cdot, c_0^*) \\ & + l_1[X_r(\cdot)(I_r - P_{B_0})c_2^{(0)} + \varepsilon G_1(\cdot)P_{B_0}c_1^{(1)} + x_2^{(2)}(\cdot,\varepsilon)] \\ & + R_1(x_2(\cdot,\varepsilon),\varepsilon)\Big\},\end{aligned}$$

where the constants $c_2^{(0)}$ and $c_1^{(1)}$ are determined from the condition of its solvability:

$$\begin{aligned}B_0c_2^{(0)} + P_{Q_d^*}\Big\{ & \varepsilon l_1[G_1(\cdot)P_{B_0}c_1^{(1)} + x_2^{(2)}(\cdot,\varepsilon)] + R_1(x_2(\cdot,\varepsilon),\varepsilon) \\ & - l\int_a^b K(\cdot,\tau)[A_1(\tau)(\varepsilon G_1(t)P_{B_0}c_1^{(1)} + x_2^{(2)}(\tau,\varepsilon)) \\ & + R(x_2(\tau,\varepsilon),\tau,\varepsilon)]\,d\tau\Big\} = 0.\end{aligned}$$

The necessary and sufficient condition for the solvability of the last system with respect to

$$c_2^{(0)} = (I_r - P_{B_0})c_2^{(0)} \in R^r \ominus N(B_0)$$

yields the following algebraic system for $\varepsilon c_1^{(1)} \in N(B_0)$:

$$\begin{aligned}\varepsilon B_1 c_1^{(1)} = -P_{B_0^*}P_{Q_d^*}\Big\{ & l_1x_2^{(2)}(\cdot,\varepsilon) + R_1(x_2(\cdot,\varepsilon),\varepsilon) \\ & - l\int_a^b K(\cdot,\tau)[A_1(\tau)x_2^{(2)}(\tau,\varepsilon) + R(x_2(\tau,\varepsilon),\tau,\varepsilon)]\,d\tau\Big\}.\end{aligned}$$

It is assumed that

$$P_{B_1^*}P_{B_0^*} = 0$$

and, hence, the last two systems enable us to obtain the first approximation $c_1^{(1)}$ to $c^{(1)} = c^{(1)}(\varepsilon)$ and the second approximation $c_2^{(0)}$ to $c^{(0)} = c^{(0)}(\varepsilon)$ given by the formulas

$$\varepsilon c_1^{(1)} = -B_1^+ P_{B_0^*} P_{Q_d^*} \Big\{ l_1 x_2^{(2)}(\cdot,\varepsilon) + R_1(x_2(\cdot,\varepsilon),\varepsilon) - l \int_a^b K(\cdot,\tau)[A_1(\tau) x_2^{(2)}(\tau,\varepsilon) + R(x_2(\tau,\varepsilon),\tau,\varepsilon)]d\tau \Big\},$$

$$c_2^{(0)} = -B_0^+ P_{Q_d^*} \Big\{ l_1(\varepsilon G_1(\cdot) P_{B_0} c_1^{(1)} + x_2^{(2)}(\cdot,\varepsilon)) + R_1(x_2(\cdot,\varepsilon),\varepsilon) - l \int_a^b K(\cdot,\tau)[A_1(\tau)(\varepsilon G_1(\tau) P_{B_0} c_1^{(1)} + x_2^{(2)}(\tau,\varepsilon)) + R(x_2(\tau,\varepsilon),\tau,\varepsilon)]d\tau \Big\}.$$

The third approximation $x_3^{(2)}(t,\varepsilon)$ to $x^{(2)}(t,\varepsilon)$ is sought in the form

$$x_3^{(2)}(t,\varepsilon) = \varepsilon \Lambda^- \left(\begin{bmatrix} f_0(\tau, c_0^*) + A_1(\tau)\Big(X_r(\tau)(I_r - P_{B_0}) c_2^{(0)} + \varepsilon G_1(\tau) P_{B_0} c_1^{(1)} + x_2^{(2)}(\tau,\varepsilon)\Big) + R(x_2(\tau,\varepsilon),\tau,\varepsilon) \\ J_0(z_0(\cdot, c_0^*)) + l_1\Big(X_r(\cdot)(I_r - P_{B_0}) c_2^{(0)} + \varepsilon G_1(\cdot) P_{B_0} c_1^{(1)} + x_2^{(2)}(\cdot,\varepsilon)\Big) + R_1(x_2(\cdot,\varepsilon)\varepsilon) \end{bmatrix} \right)(t),$$

and, for the third approximation $x_3(t,\varepsilon)$ to the required solution $x(t,\varepsilon)$, we get

$$x_3(t,\varepsilon) = X_r(t)(I_r - P_{B_0}) c_2^{(0)} + \varepsilon G_1(t) P_{B_0} c_1^{(1)} + x_3^{(2)}(t,\varepsilon).$$

Acting in this way, we arrive at the following iterative procedure for finding the solution $x(t,\cdot) \in C[0,\varepsilon_0]$, $x(t,0) = 0$, of the boundary-value problem (5.48):

$$\varepsilon c_{k-1}^{(1)} = -B_1^+ P_{B_0^*} P_{Q_d^*} \Big\{ l_1 x_k^{(2)}(\cdot,\varepsilon) + R_1(x_k(\cdot,\varepsilon),\varepsilon) - l \int_a^b K(\cdot,\tau)[A_1(\tau) x_k^{(2)}(\tau,\varepsilon) + R(x_k(\tau,\varepsilon),\tau,\varepsilon)]d\tau \Big\},$$

$$c_k^{(0)} = -B_0^+ P_{Q_d^*}\Big\{ l_1[\varepsilon G_1(\cdot) P_{B_0} c_{k-1}^{(1)} + x_k^{(2)}(\cdot,\varepsilon)] + R_1(x_k(\cdot,\varepsilon),\varepsilon)$$

$$- l \int_a^b K(\cdot,\tau)[A_1(\tau)(\varepsilon G_1(\tau) P_{B_0} c_{k-1}^{(1)} + x_k^{(2)}(\tau,\varepsilon))$$

$$+ R(x_k(\tau,\varepsilon),\tau,\varepsilon)]d\tau \Big\}, \tag{5.71}$$

$$x_{k+1}^{(2)}(t,\varepsilon) = \varepsilon \Lambda^- \left(\begin{bmatrix} f_0(\tau, c_0^*) + A_1(\tau)\Big(X_r(\tau)(I_r - P_{B_0})c_k^{(0)} \\ \qquad + \varepsilon G_1(\tau) P_{B_0} c_{k-1}^{(1)} + x_k^{(2)}(\tau,\varepsilon)\Big) \\ + R(x_k(\tau,\varepsilon),\tau,\varepsilon) \\ J_0(z_0(\cdot, c_0^*)) + l_1\Big(X_r(\cdot)(I_r - P_{B_0})c_k^{(0)} \\ \qquad + \varepsilon G_1(\cdot) P_{B_0} c_{k-1}^{(1)} + x_k^{(2)}(\cdot,\varepsilon)\Big) \\ + R_1(x_k(\cdot,\varepsilon)\varepsilon) \end{bmatrix} \right)(t),$$

$$x_{k+1}(t,\varepsilon) = X_r(t)(I_r - P_{B_0})c_k^{(0)} + \varepsilon G_1(t) P_{B_0} c_{k-1}^{(1)} + x_{k+1}^{(2)}(t,\varepsilon),$$

$$k = 0,1,2,\ldots, \qquad x_0(t,\varepsilon) = x_0^{(2)}(t,\varepsilon) = 0.$$

As above, to show that the iterative process converges, one can use the method of majorizing Lyapunov equations described, e.g., in [19, 65], or [77]. Thus, we have proved the following assertion:

Theorem 5.6. *Assume that the boundary-value problem (5.43) satisfies the conditions imposed above for the critical case* ($\operatorname{rank} Q = n_1 < m$) *and the corresponding generating boundary-value problem (5.44) possesses the r-parameter ($r = n - n_1$) family of generating solutions (5.11) if and only if condition (5.9) is satisfied. If, in addition,*

$$P_{B_1^*} P_{B_0^*} = 0, \tag{5.72}$$

then, for any value of the vector $c_r = c_0^* \in R^r$ satisfying equation (5.46) for generating constants, the boundary-value problem (5.48) possesses at least one solution $x(t,\cdot) \in C[\varepsilon]$, $\varepsilon \in [0,\varepsilon_*]$, vanishing for $\varepsilon = 0$ provided that

$$P_{B_0^*} P_{Q_d^*}\Big\{ l_1 x_1^{(2)}(\cdot,\varepsilon) + R_1(x_1^{(2)}(\cdot,\varepsilon),\varepsilon)$$

$$- l \int_a^b K(\cdot,\tau)[A_1(\tau) x_1^{(2)}(\tau,\varepsilon) + R(x_1^{(2)}(\tau,\varepsilon),\tau,\varepsilon)]d\tau \Big\} = 0. \tag{5.73}$$

One of these solutions can be obtained with the help of the iterative process (5.71) convergent for $\varepsilon \in [0, \varepsilon_*]$. *Furthermore, the boundary-value problem (5.43) possesses at least one solution* $z(t, \varepsilon)$ *such that* $z(t, \cdot) \in C[\varepsilon]$, $\varepsilon \in [0, \varepsilon_*]$, *which turns into the generating solution* $z_0(t, c_0^*)$ *given by relation (5.11) for* $\varepsilon = 0$. *The indicated solution is specified by relation (5.71) and the formula*

$$z(t, \varepsilon) = \lim_{k \to \infty} z_k(t, \varepsilon),$$

where

$$z_k(t, \varepsilon) = z_0(t, c_0^*) + x_k(t, \varepsilon), \quad k = 0, 1, 2, \ldots .$$

In the case where

$$P_{B_0} P_{B_1} = 0,$$

Theorem 5.6 turns into to the following assertion:

Corollary 5.3. *Assume that the boundary-value problem (5.43) satisfies the conditions imposed above for the critical case* (rank $Q = n_1 < n$) *and the corresponding generating boundary-value problem (5.44) possesses the* r*-parameter* ($r = n - n_1$) *family of generating solutions (5.11) if and only if condition (5.9) is satisfied. In this case, if*

$$P_{B_1^*} P_{B_0^*} = 0 \qquad \text{and} \qquad P_{B_0} P_{B_1} = 0, \tag{5.74}$$

then, for any value of the vector $c_r = c_0^* \in R^r$ *satisfying equation (5.46) for generating constants, the boundary-value problem (5.48) possesses a unique solution* $x(t, \cdot) \in C[\varepsilon]$, $\varepsilon \in [0, \varepsilon_*]$, *vanishing for* $\varepsilon = 0$ *provided that relation (5.73) is true. This solution can be obtained with the help of the iterative process (5.71) convergent for* $\varepsilon \in [0, \varepsilon_*]$. *Furthermore, the boundary-value problem (5.43) possesses a unique solution* $z(t, \varepsilon)$ *such that* $z(t, \cdot) \in C[\varepsilon]$, $\varepsilon \in [0, \varepsilon_*]$, *which turns into the generating solution* $z_0(t, c_0^*)$ *given by relation (5.11) for* $\varepsilon = 0$. *This solution is specified by relation (5.71) and the formula*

$$z(t, \varepsilon) = \lim_{k \to \infty} z_k(t, \varepsilon),$$

where

$$z_k(t, \varepsilon) = z_0(t, c_0^*) + x_k(t, \varepsilon), \quad k = 0, 1, 2, \ldots .$$

Concluding Remarks. In the analyzed critical case (5.72), relation (5.73) is a natural condition necessary and sufficient for the existence of the second approximation to the exact solution. If relation (5.73) is not true, then the required solution $x(t, \cdot) \in C[0, \varepsilon_*]$ of the boundary-value problem (5.48) determined by the method of simple iterations does not exist. In the analyzed case, the sufficient condition (5.72) is the only restriction guaranteeing the existence of solution and the possibility of its construction by an iterative analog of the Lyapunov–Poincaré method of small parameter.

Let us now analyze the structure of successive approximations to the solution of the boundary-value problems (5.35) and (5.48) obtained with the help of the iterative processes (5.39), (5.56), and (5.71). Thus, the following estimates are true for the successive approximations $x_k(t,\varepsilon)$ to the required solution $x(t,\cdot) \in C[\varepsilon]$, $\varepsilon \in [0,\varepsilon_*]$, of the boundary-value problem (5.35) [or (5.48)]:

The uniform norm of the difference between the kth and $(k-l)$th approximations is a quantity of order ε^k:

$$|||x_k(t,\varepsilon) - x_{k-1}(t,\varepsilon)||| \sim \varepsilon^k, \quad k = 2,3,\ldots,$$

$$|||x_1(t,\varepsilon)||| \sim \varepsilon.$$

We now replace the sequence $\{x_k(t,\varepsilon)\} \underset{k\to\infty}{\longrightarrow} x(t,\varepsilon)$ by a series

$$x(t,\varepsilon) = x_1(t,\varepsilon) + \sum_{k=2}^{\infty}[x_k(t,\varepsilon) - x_{k-1}(t,\varepsilon)]. \tag{5.75}$$

This series is equivalent, in a sense of the order of smallness of its terms in the corresponding norm, to a series in powers of the small parameter ε. Therefore, algorithms (5.39), (5.56), and (5.71) proposed for the solution of boundary-value problems are iterative versions of the Lyapunov–Poincaré method of small parameter [65, 93, 96, 101, 106] based on the expansion of solutions in powers of the parameter ε. To apply the method of small parameter to the solution of boundary-value problems of the form (5.35) or (5.43), it is necessary to make an additional assumption that the nonlinearities $Z(z,t,\varepsilon)$ and $J(z,\varepsilon)$ appearing in (5.35) or (5.43) are analytic in x and ε in the corresponding region. The conditions guaranteeing the existence of solutions in the form of series are identical to the conditions required for the reducibility of boundary-value problems to equivalent operator systems. To prove that the series converge in the analyzed case of analytic nonlinearities, it is necessary to construct analytic majorants appearing in the Lyapunov majorizing equations for the corresponding equivalent operator systems. It is known [65, 77] that, in the case of nonanalytic majorants, the majorizing equations specify only the region (interval $[0,\varepsilon_*]$) where the required solution of the boundary-value problem exists but not the region where its expansion in a series in powers of the parameter ε converges. At the same time, if the Lyapunov majorants used for the estimation of the range $[0,\varepsilon_*]$ of convergence of the iterative processes similar to (5.39) are analytic, then the indicated region can be regarded as an estimate of the range of convergence of the series in powers of the parameter ε for the required solution of the boundary-value problem.

Note that, from the viewpoint of compactness of the algorithm, the convenience of its realization on computers, and the analysis of convergence of the method and its errors in the case of weaker restrictions imposed on the right-hand side of the system, it is preferable to solve boundary-value problems by using iterative methods rather than the methods based on the expansion of solutions in powers of a small parameter [65].

Example 2. Consider a nonlinear scalar three-point boundary-value problem of the form (5.43):

$$\dot{z} = \varepsilon g(t)z^2 + \varphi(t), \quad lz = \sum_{i=1}^{3} m_i z(t_i) = \alpha, \quad t \in [a, b], \tag{5.76}$$

where $m_1 = m_3 = 0.5$, $m_2 = -1$, $a = t_1 < t_2 < t_3 = b$, $g(t)$, $\varphi(t) \in C[a, b]$, $\alpha \in R^1$, and $n = m = 1$.

We now study the critical case where the generating boundary-value problem

$$\dot{z} = \varphi(t), \quad lz = \alpha \tag{5.77}$$

obtained by setting $\varepsilon = 0$ in problem (5.76) is solvable. By using the notation introduced earlier, we set

$$X(t) = 1, \quad lX = Q = 0, \quad n = r = 1, \quad P_Q = P_{Q^*} = P_{Q_d^*} = 1,$$

$$G(t, \tau) = K(t, \tau) = \frac{1}{2}\text{sign}\,(t - \tau), \quad lK(\cdot, \tau) = -\frac{1}{2}\text{sign}\,(t_2 - \tau).$$

By Theorem 5.1, the criterion of solvability of the generating boundary-value problem has the form

$$\alpha + \frac{1}{2}\left\{\int_a^{t_2} \varphi(\tau)d\tau - \int_{t_2}^{b} \varphi(\tau)d\tau\right\} = 0. \tag{5.78}$$

In this case, problem (5.77) possesses a one-parameter ($r = 1$) family of solutions

$$z_0(t, c_r) = c_r + \int_a^b G(t, \tau)\varphi(\tau)d\tau \qquad \text{for any} \quad c_r \in R^1.$$

We now establish conditions required for the existence of the solution $z(t, \varepsilon)$ of the boundary-value problem (5.76) that turns into one of generating solutions $z_0(t, c_r)$ for $\varepsilon = 0$. To obtain specific conditions, we need more data about the parameters of the problem. Thus, let $t_1 = a = 0$, $t_2 = 1$, $t_3 = b = 2$, and $\varphi(t) = t$. Relation (5.78) implies that the boundary-value problem (5.77) is solvable only for $\alpha = 0.5$. In this case, it possesses a one-parameter family of solutions

$$z_0(t, c_r) = c_r + \frac{t^2}{2} - 1.$$

The equation for generating constants (5.46) corresponding to the boundary-value problem (5.76) has the form

$$F(c_r) = \int_0^2 \operatorname{sign}(1-\tau) g(\tau) z_0^2(\tau, c_r) d\tau = 0. \tag{5.79}$$

Specifically, for $g(t) = t$, equation (5.79) turns into the quadratic equation

$$F(c_r) = c_r^2 + \frac{3}{2} c_r + \frac{1}{12} = 0$$

with two real roots:

$$c_{r1} = c_{01}^* = \frac{-3\sqrt{3} + \sqrt{23}}{4\sqrt{3}} \quad \text{and} \quad c_{r2} = c_{02}^* = \frac{-3\sqrt{3} - \sqrt{23}}{4\sqrt{3}}.$$

Since $B_0 = 2$ $(P_{B_0} = P_{B_0^*} = 0)$, it follows from Theorem 5.5 that the boundary-value problem (5.76) has two solutions $z(t, \cdot) \in C[\varepsilon]$ in a neighborhood of $\varepsilon = 0$ which turn into generating solutions $z_0(t, c_{01}^*)$ and $z_0(t, c_{02}^*)$ [with constants specified by equation (5.79)] for $\varepsilon = 0$.

Example 3. Consider the conditions required for the existence of solution of a two-point boundary-value problem of the form (5.43), where

$$A(t) = 0, \quad lz = Mz(a) + Nz(b) = \alpha = 0,$$

$$M = \operatorname{diag}\{k, M_1\}, \quad N = \operatorname{diag}\{-k, N_1\},$$
$$Z(z, t, \varepsilon) = \operatorname{col}(Z_1(z, t, \varepsilon), \ldots, Z_n(z, t, \varepsilon)), \tag{5.80}$$

k is an arbitrary real constant, and M_1 and N_1 are $(n-1) \times (n-1)$ matrices such that $\operatorname{rank}[M_1 + N_1] = n - 1$. Clearly,

$$X(t) \equiv I_n, \quad m = n,$$

$$Q = M + N = \operatorname{diag}\{0, M_1 + N_1\}, \quad \operatorname{rank} Q = n-1, \quad r = d = 1,$$

$$P_Q = \operatorname{diag}\{1, 0, \ldots, 0\}, \quad X_r(t) = \begin{bmatrix} 1 \\ 0 \\ \vdots \\ 0 \end{bmatrix}, \quad H(\tau) = P_{Q_r^*} l K(\cdot, \tau) = [10 \ldots 0].$$

In this case, the matrix B_0 is a scalar,

$$B_0 = \int_a^b H(\tau) A_1(\tau) X_r(\tau) d\tau = \int_a^b a_{11}(\tau) d\tau,$$

where $a_{11}(\tau)$ is the entry of the matrix $A_1(\tau)$ located at the intersection of the first row with the first column. Thus, if $B_0 \neq 0$, then, for any $c_1 = c_0^* \in R^1$ satisfying the equation for generating constants

$$\int_a^b H(\tau)Z(z_0(\tau, c_0^*), \tau, 0)d\tau = \int_a^b Z_1(z_0(\tau, c_0^*), \tau, 0)d\tau = 0,$$

the two-point boundary-value problem (5.43) with properties (5.80) possesses a unique solution $z(t, \cdot) \in C[\varepsilon]$, $\varepsilon \in [0, \varepsilon_*]$, which turns into the generating solution $z_0(t, c_0^*)$ of the generating problem

$$\dot{z} = \varphi(t), \qquad lz = Mz(a) + Nz(b) = 0$$

for $\varepsilon = 0$.

Example 4. As an illustration of the applicability of the proposed procedure to the solution of boundary-value problems, we now consider a problem from the control theory.

It is known [128] that the minimum of the quadratic functional

$$x^T(b)\Phi x(b) + \int_a^b [x^T(t)M(t)x(t) + u^T(t)N(t)u(t)]dt \tag{5.81}$$

is attained on the solutions of the system

$$\dot{x} = A(t)x + B(t)u, \quad x(a) = x_0 \tag{5.82}$$

for the optimal control

$$u_0(t) = -N^{-1}(t)B^T(t)Z(t)x, \tag{5.83}$$

where Φ and $M(t)$ are given $n \times n$ symmetric nonnegative matrices, $N(t)$ is a given $m \times m$ positive-definite matrix, $t \in [a, b]$, and the unknown $n \times n$ matrix $Z(t)$ is obtained as a solution of the boundary-value problem for the matrix Riccati equation

$$\begin{gathered} \dot{Z} = -ZA(t) - A^T(t)Z + ZB(t)N^{-1}(t)B(t)Z - M(t), \\ lZ = Z(b) = \Phi. \end{gathered} \tag{5.84}$$

If $A(t) = A_0 + \varepsilon A_1(t)$ and $B(t) = B_0 + \varepsilon B_1(t)$, then we can use the theorems proved above to establish conditions required for the existence of solutions of problem (5.84) and construct these solutions. As a result, we obtain conditions required for the existence of the optimal control (5.83) minimizing functional (5.81) on the solutions of system (5.82), together with algorithms used for its construction.

5.3 Autonomous Boundary-Value Problems

Consider the following autonomous system of ordinary differential equations:

$$\frac{dz}{dt} = Az + f + \varepsilon Z(z, \varepsilon), \tag{5.85}$$

Our aim is to establish conditions for the existence of solutions

$$z(t, \varepsilon)\colon\; z(\cdot, \varepsilon) \in C^1[t], \quad t \in [a, b(\varepsilon)], \quad z(t, \cdot) \in C[\varepsilon], \quad \varepsilon \in [0, \varepsilon_0]$$

of this system of equations and construct these solutions satisfying the following boundary condition:

$$lz(\cdot, \varepsilon) = \alpha + \varepsilon J(z(\cdot, \varepsilon), \varepsilon) \tag{5.86}$$

and turning, for $\varepsilon = 0$, into a generating solution

$$z_0(t)\colon\; z_0(\cdot) \in C^1[t], \quad t \in [a, b^*], \quad b^* = b(0),$$

of the generating boundary-value problem

$$\frac{dz_0}{dt} = Az_0 + f, \tag{5.87}$$

$$lz_0(\cdot) = \alpha, \tag{5.88}$$

where A is a constant $n \times n$ matrix and l and J are vector functionals linear and nonlinear in z, respectively. The nonlinear vector function $Z(z, \varepsilon)$ is such that $Z(\cdot, \varepsilon) \in C[z]$ and $Z(z, \cdot) \in C[\varepsilon]$. The functional $J(z(\cdot, \varepsilon), \varepsilon)$ is continuously differentiable with respect to z (in the sense of Frechét) in a vicinity of the generating solution of problem (5.87), (5.88) (see [25, 26]). For constant column vectors f and α, we have $f \in R^n$ and $\alpha \in R^m$. The interval $[a, b^*]$ in which we seek the solution of the generating boundary-value problem is regarded as known. The constant matrix $Q = lX(\cdot)$ is $(m \times n)$-dimensional and $X(t)$ is a normal $(X(a) = I_n)$ fundamental matrix of the homogeneous differential system (5.87). Problem (5.85), (5.86) is a natural generalization of the autonomous periodic problem [101, 149] with

$$\alpha = 0, \quad f = 0, \quad J(z(\cdot, \varepsilon), \varepsilon) = 0,$$

and

$$lz(\cdot, \varepsilon) = z(0, \varepsilon) - z(T_1(\varepsilon), \varepsilon).$$

According to Theorem 5.1, the generating problem (5.87), (5.88) is solvable if and only if

$$P_{Q_d^*}\left\{\alpha - l\int_a^{b^*} K(\cdot, s)ds f\right\} = 0. \tag{5.89}$$

If this condition is satisfied, then the general solution of the generating boundary-value problem admits the following representation:

$$z_0(t, c_r) = X_r(t)c_r + X(t)Q^+\alpha + (Gf)(t), \tag{5.90}$$

where the $n \times r$ matrix $X_r(t)$ is formed by r linearly independent solutions of the homogeneous generating problem, $X_r(t) = X(t)P_{Q_r}$, and $(Gf)(t)$ is the generalized Green operator (5.12).

Let $\operatorname{rank} Q = n_1$. Then $\operatorname{rank} P_{Q^*} = m - n_1 = d$ and $\operatorname{rank} P_Q = n - n_1 = r$. Hence, $P_{Q_d^*}$ is a $d \times m$ matrix whose rows form a complete system of d linearly independent rows of the orthoprojector $P_{Q^*}\colon R^d \to N(Q^*)$ and P_{Q_r} is an $n \times r$ matrix formed by a complete system of r linearly independent columns of the orthoprojector $P_Q\colon R^r \to N(Q)$.

Necessary Condition. Both the autonomous periodic problem [101] and the weakly nonlinear boundary-value problem (5.85), (5.86) are noticeably different from similar nonautonomous boundary-value problems. Unlike the case of nonautonomous problems, the right endpoint of the interval where the solution of problem (5.85), (5.86) is sought is unknown and should be found in the process of construction of the required solution. We seek this endpoint in the form

$$b(\varepsilon) = b^* + \varepsilon(b^* - a)\beta(\varepsilon), \quad \beta(0) = \beta^*. \tag{5.91}$$

By using this relation and the change of variables

$$t = a + (\tau - a)(1 + \varepsilon\beta(\varepsilon)) \tag{5.92}$$

in problem (5.85), (5.86), we arrive at the problem of finding a solution

$$z(\tau, \varepsilon): \ z(\cdot, \varepsilon) \in C^1[\tau], \quad z(\tau, \cdot) \in C[\varepsilon], \quad z(\tau, 0) = z_0(\tau, c_r),$$

of the boundary-value problem

$$\frac{dz}{d\tau} = Az + f + \varepsilon\{\beta(Az + f) + (1 + \varepsilon\beta)Z(z, \varepsilon)\}, \tag{5.93}$$

$$\ell z(a + (\cdot - a)(1 + \varepsilon\beta(\varepsilon))) = (1 + \varepsilon\beta(\varepsilon))$$
$$\times (\alpha + \varepsilon J(z(a + (\cdot - a)(1 + \varepsilon\beta(\varepsilon)), \varepsilon)) : C[a, b^*] \to R^m. \tag{5.94}$$

Thus, the required solution $z = z(\cdot, \varepsilon)\colon [a, b^*] \to R^n$ of the boundary-value problem (5.93), (5.94) is defined in the interval of fixed length $[a, b^*]$.

Indeed, in order to understand the meaning of the change of variables (5.92) in the boundary condition (5.86), we assume, e.g., that condition (5.86) is specified by the Riemann–Stieltjes integral

$$\ell z(\cdot, \varepsilon) = \int_a^{b(\varepsilon)} d\Omega(t) z(t, \varepsilon),$$

where $\Omega(t)$ is an $m \times n$ matrix whose components are functions of bounded variation on $[a, b(\varepsilon)]$. It is known [52] that the indicated representation of a linear vector functional is most general in the space $C[a, b(\varepsilon)]$. As a result of the change of variables (5.92), we arrive at the boundary condition (5.94).

From the geometric point of view, the change of variables (5.91), (5.92) is a linear mapping of the interval $[a, b(\varepsilon)]$ whose length depends on ε onto the interval $[a, b^*]$ of constant length. Moreover, for $\beta > 0$, the interval $[a, b(\varepsilon)]$ suffers linear contraction and, for $\beta < 0$, it is expanded. It is easy to see that, for a periodic boundary-value problem with $a = 0$ and $b^* = T$, the change of variables (5.91), (5.92) coincides with the change of variables used in [65, 101, 149].

In problem (5.93), (5.94), we now set

$$z(\tau, \varepsilon) = z_0(\tau, c_r) + x(\tau, \varepsilon).$$

This yields the following boundary-value problem:

$$\frac{dx}{d\tau} = Ax + \varepsilon\{\beta[A(z_0 + x) + f] + (1 + \varepsilon\beta)Z(z_0 + x, \varepsilon)\}, \tag{5.95}$$

$$\ell x(\cdot, \varepsilon) = \varepsilon\alpha\beta(\varepsilon) + \varepsilon(1 + \varepsilon\beta(\varepsilon))J(z_0(\cdot, c_r^*) + x(\cdot, \varepsilon), \varepsilon) \tag{5.96}$$

in the fixed interval $[a, b^*]$. We seek a solution $x(\tau, \varepsilon)$ of this problem such that $x(\cdot, \varepsilon) \in C^1[\tau]$, $\tau \in [a, b^*]$, $x(\tau, \cdot) \in C[\varepsilon]$, and $x(\tau, 0) = 0$. By applying Theorem 5.4 to problem (5.95), (5.96), we arrive at the following necessary condition for the existence of the required solution of problem (5.85), (5.86):

$$P_{Q_d^*}\left\{\alpha\beta^* + J(z_0(\cdot, c_r^*), 0) - l\int_a^{b^*} K(\cdot, s)\{\beta^*[Az_0(s, c_r^*) + f] + Z(z_0(s, c_r^*), 0)\}ds\right\} = 0.$$

Denote

$$f_0(\tau, c^*) := \beta^*[Az_0(s, c_r^*) + f] + Z(z_0(s, c_r^*), 0),$$
$$\varphi_0(c^*) := \alpha\beta^* + J(z_0(\cdot, c_r^*), 0)$$
$$c^* = \operatorname{col}(c_r^*, \beta^*) \in R^{r+1}.$$

As a result, we get the following equation for the vector c^*:

$$F(c^*) = P_{Q_d^*}\left\{\varphi_0(c^*) - l\int_a^{b^*} K(\cdot, s)f_0(s, c^*)ds\right\} = 0. \tag{5.97}$$

The existence of a real solution of this equation is a necessary condition for the existence of the required solution of problem (5.95), (5.96) and, hence, problems (5.93), (5.94) and (5.85), (5.86).

Theorem 5.7. *Assume that the autonomous boundary-value problem (5.85), (5.86) possesses a solution* $z(\tau,\varepsilon)$ *such that* $z(\cdot,\varepsilon)\in C^1[\tau]$ *and* $z(\tau,\cdot)\in C[\varepsilon]$, *which turns into a generating solution* $z_0(\tau,c_r^*)=z(\tau,0)$, $\tau\in[0,b^*]$, *of problem (5.87), (5.88) for* $\varepsilon=0$. *Then the vector* $c^*=\operatorname{col}(c_r^*,\beta^*)$ *is a solution of equation (5.97).*

Equation (5.97) is the equation for generating constants of the autonomous boundary-value problem (5.85), (5.86). The vector c^* obtained from equation (5.97) specifies the generating solution $z_0(\tau,c_r^*)$ that can be associated with the solution $z(\tau,\varepsilon)$: $z(\cdot,\varepsilon)\in C^1[\tau]$, $z(\tau,\cdot)\in C[\varepsilon]$, $z(\tau,0)=z_0(\tau,c_r^*)$ of problem (5.93), (5.94). The last component β^* of the vector c^* is the first correction to the length of the interval $[a,b^*]$, where the solution of problem (5.85), (5.86) is constructed. If equation (5.97) has no real solutions, then problem (5.85), (5.86) does not have the required solution.

It is easy to see that, for $m=n$, $J=0$, $\alpha=0$, $f=0$, and $lz(\cdot,\varepsilon)=z(0,\varepsilon)-z(T_1(\varepsilon),\varepsilon)=0$, equation (5.97) turns into the equation for generating amplitudes of the autonomous periodic problem. Moreover, condition (5.89) turns into an identity and, unlike the general case, the last component of the vector c_r in representation (5.90) can be made equal to zero by the proper choice of the reference point of the independent variable t [25, 26, 36].

Example 1. As an illustration to Theorem 5.7, we now find the solution $z(\cdot,\varepsilon)\in C^1[t]$, $z(t,\cdot)\in C[\varepsilon]$ of a scalar periodic problem

$$\begin{gathered}\frac{dz}{dt}=\varepsilon\cosh z,\\ lz(\cdot,\varepsilon)=z(0)-z(b(\varepsilon))=0\end{gathered}\tag{5.98}$$

which turns, for $\varepsilon=0$, into the solution $z_0(\cdot)\in C^1[0,b(0)]$ of the generating boundary-value problem

$$\frac{dz_0}{dt}=0,\quad lz_0(\cdot)=z_0(0)-z_0(1)=0,\quad b(0)=1.$$

It is easy to see that the generating problem has a one-parameter family of solutions $z_0(t,c)=c$, $c\in R^1$. We now deduce the equation for generating constants of the boundary-value problem (5.98). By using the accepted notation, we can write

$$\begin{gathered}X(t)=1,\quad Q=lX(\cdot)=0,\quad P_{Q_d^*}=1,\\ K(t,s)=1/2\operatorname{sign}(t-s),\\ f_0(t,c^*)=1(0z_0+0+\cosh c^*)=\cosh c^*,\end{gathered}$$

$$F(c^*)=1\left[0-1/2\,l\int_0^1\operatorname{sign}(\cdot-s)\cosh c^*ds\right]=1/2\cosh c^*l\int_0^1\operatorname{sign}(\cdot-s)ds.$$

The equation for generating constants

$$F(c^*) = -\cosh c^* = 0$$

of problem (5.98) has no real solutions and, hence, the analyzed problem does not have a solution $z(\cdot, \varepsilon) \in C^1[0, b(\varepsilon)]$, $z(t, \cdot) \in C[0, \varepsilon_0]$ which turns into a generating solution for $\varepsilon = 0$. Indeed, integrating the differential system (5.98), we obtain its solution in the form

$$z(t, \varepsilon) = \ln[\tan(\varepsilon t/2 + c)], \quad (\tan(\varepsilon t/2 + c) \geq 0).$$

This solution satisfies the boundary condition

$$z(0) - z(1 + \varepsilon\beta(\varepsilon)) = \ln\left[\frac{\tan c}{\tan(\varepsilon(1 + \varepsilon\beta(\varepsilon))/2 + c)}\right] = 0$$

for

$$\tan c - \tan(\varepsilon(1 + \varepsilon\beta(\varepsilon))/2 + c) = 0,$$

and, therefore, $\sin(\varepsilon(1+\varepsilon\beta(\varepsilon))) = 0$. This equality does not hold for sufficiently small ε $(0 < \varepsilon < \varepsilon_0)$ and, hence, the boundary-value problem (5.98) does not have the required solution.

Sufficient Condition. Assume that the necessary condition for the existence of solutions of the analyzed problem is satisfied. In this case, the solution of problem (5.95), (5.96) admits the following representation:

$$x(\tau, \varepsilon) = X_r(\tau)c_r + x^{(1)}(\tau, \varepsilon),$$

where

$$x^{(1)}(\tau, \varepsilon) = \varepsilon X(\tau)Q^+ \{\alpha\beta(\varepsilon) + (1 + \varepsilon\beta(\varepsilon))J(z_0(\cdot, c_r^*) + x(\cdot, \varepsilon), \varepsilon)\}$$
$$+ \varepsilon G\{\beta[A(z_0 + x) + f] + (1 + \varepsilon\beta)Z(z_0 + x, \varepsilon)\}(\tau).$$

The vector function $Z(z, \varepsilon)$ and vector functional $J(z(\cdot, \varepsilon), \varepsilon)$ are continuously differentiable with respect to z in the vicinity of the generating solution $z_0(\tau, c_r^*)$ and continuous in ε in the vicinity of the point $\varepsilon = 0$. Hence, we can select terms linear in x and terms of order zero with respect to ε in the vector function $Z(z_0 + x, \varepsilon)$ and vector functional $J(z_0 + x, \varepsilon)$. As a result, we obtain

$$Z(z_0(\tau, c_r^*) + x(\tau, \varepsilon), \varepsilon) = Z(z_0(\tau, c_r^*), 0) + A_1(\tau)x + \varphi_1(x, \varepsilon), \tag{5.99}$$

where

$$A_1(\tau) = \frac{\partial Z(z, 0)}{\partial z}\Big|_{z=z_0(\tau, c_r^*)}, \quad \varphi_1(0, 0) = 0, \quad \frac{\partial \varphi_1(0, 0)}{\partial x} = 0,$$

and

$$\{\alpha\beta(\varepsilon)+(1+\varepsilon\beta(\varepsilon))J(z_0(\cdot,c_r^*)+x(\cdot,\varepsilon),\varepsilon)\}$$
$$=\alpha\beta^*+J(z_0(\cdot,c_r^*,0))+\alpha\bar\beta(\varepsilon)+\ell_1x(\cdot,\varepsilon)$$
$$+J_1(z_0(\cdot,c_r^*)+x(\cdot,\varepsilon),\varepsilon)+\varepsilon\beta(\varepsilon)J(z_0(\cdot,c_r^*)+x(\cdot,\varepsilon),\varepsilon). \quad (5.100)$$

Here, $l_1x(\cdot,\varepsilon)$ is a linear (as a function of x) part of the functional

$$J(z_0(\cdot,c_r^*)+x(\cdot,\varepsilon),\varepsilon),$$

$J_1(x(\cdot,\varepsilon),\varepsilon)$ is its nonlinear part, $J_1(0,0)=0$, and $\partial J_1(0,0)/\partial x=0$.

In view of the indicated representation of the solution and expressions (5.99) and (5.100), we can transform the right-hand sides of system (5.95) and the boundary condition (5.96). Thus, we get

$$\beta[A(z_0+x)+f]+(1+\varepsilon\beta)Z(z_0+x,\varepsilon)$$
$$=\beta A[X_r(\tau)c_r^*+\bar z_0(\tau,f)]+\beta Ax+\beta f$$
$$+Z(z_0,0)+A_1(\tau)x+\varphi_1(x,\varepsilon)+\varepsilon\beta Z(z,\varepsilon)$$
$$=f_0(\tau,c^*)+\bar\beta Az_0(\tau,c_r^*)+\beta f+\beta^*Ax$$
$$+\bar\beta Ax+A_1(\tau)x+\varphi_1(x,\varepsilon)+\varepsilon\beta Z(z,\varepsilon)$$
$$=f_0(\tau,c^*)+[\beta^*A+A_1(\tau)]x+[Az_0(\tau,c_r^*)+f]\bar\beta+R(x,\varepsilon)$$
$$=f_0(\tau,c^*)+[\beta^*A+A_1(\tau)]X_r(\tau)c_r$$
$$+[\beta^*A+A_1(\tau)]x^{(1)}+[Az_0(\tau,c_r^*)+f]\bar\beta+R(x,\varepsilon)$$
$$=f_0(\tau,c^*)+\bar A(\tau)c+[\beta^*A+A_1(\tau)]x^{(1)}(\tau,\varepsilon)+R(x,\varepsilon),$$

$$\{\alpha\beta(\varepsilon)+(1+\varepsilon\beta(\varepsilon))J(z_0(\cdot,c_r^*)+x(\cdot,\varepsilon),\varepsilon)\}$$
$$=\varphi_0(c^*)+A_2c(\varepsilon)+\ell_1x^{(1)}(\cdot,\varepsilon)+R_1(x(\cdot,\varepsilon),\varepsilon),$$

where

$$\bar A(\tau):=\{[\beta^*A+A_1(\tau)]X_r(\tau),Az_0(\tau,c_r^*)+f\},\quad A_2:=[\ell_1X_r;\alpha]$$

are $n\times(r+1)$ and $m\times(r+1)$ matrices,

$$c=\mathrm{col}\,(c_r,\bar\beta)\in R^{r+1},\quad c(\cdot)\in C[0,\varepsilon_0],\quad \bar\beta=\beta-\beta^*\in R^1,$$

$$R(x,\varepsilon)=\bar\beta Ax+\varepsilon\beta Z(z_0+x,\varepsilon)+\varphi_1(x,\varepsilon),$$
$$R_1(x(\cdot,\varepsilon),\varepsilon)=J_1(z_0(\cdot,c_r^*)+x(\cdot,\varepsilon),\varepsilon)+\varepsilon\beta(\varepsilon)J(z_0(\cdot,c_r^*)+x(\cdot,\varepsilon),\varepsilon).$$

As a result, we arrive at the problem of construction of a solution $x(\tau, \varepsilon)$ such that

$$x(\cdot, \varepsilon) \in C^1[\tau], \quad \tau \in [a, b^*], \quad x(\tau, \cdot) \in C[\varepsilon], \quad \varepsilon \in [0, \varepsilon_0], \quad x(\tau, 0) = 0,$$

for the differential system

$$\frac{dx}{d\tau} = Ax + \varepsilon\{f_0(\tau, c^*) + \bar{A}(\tau)c + [\beta^* A + A_1(\tau)]x^{(1)}(\tau, \varepsilon) + R(x, \varepsilon)\} \quad (5.101)$$

with the boundary condition

$$\ell x(\cdot, \varepsilon) = \varepsilon\{\varphi_0(c^*) + A_2 c(\varepsilon) + \ell_1 x^{(1)}(\cdot, \varepsilon) + R_1(x(\cdot, \varepsilon), \varepsilon)\}. \quad (5.102)$$

In this notation, the condition of solvability of the boundary-value problem (5.95), (5.96) and, hence, of problem (5.101), (5.102) takes the form

$$P_{Q_d^*}\Big\{\varphi_0(c^*) + A_2 c(\varepsilon) + \ell_1 x^{(1)}(\cdot, \varepsilon) + R_1(x(\cdot, \varepsilon), \varepsilon)$$
$$- \ell \int_a^{b^*} K(\cdot, s)\{f_0(s, c^*) + \bar{A}(s)c$$
$$+ [\beta^* A + A_1(s)]x^{(1)}(s, \varepsilon) + R(x, \varepsilon)\}ds\Big\} = 0.$$

By using the equation for generating constants (5.97), we obtain

$$P_{Q_d^*}\Big\{A_2 c(\varepsilon) + \ell_1 x^{(1)}(\cdot, \varepsilon) + R_1(x(\cdot, \varepsilon), \varepsilon)$$
$$- \ell \int_a^{b^*} K(\cdot, s)\{\bar{A}(s)c(\varepsilon) + [\beta^* A + A_1(s)]x^{(1)}(s, \varepsilon) + R(x, \varepsilon)\}ds\Big\} = 0,$$

whence

$$P_{Q_d^*}\Big\{A_2 - \ell \int_a^{b^*} K(\cdot, s)\bar{A}(s)ds\Big\}c(\varepsilon)$$
$$= -P_{Q_d^*}\Big\{l_1 x^{(1)}(\cdot, \varepsilon) + R_1(x(\cdot, \varepsilon), \varepsilon)$$
$$- l \int_a^{b^*} K(\cdot, s)\{[\beta^* A + A_1(s)]x^{(1)} + R(x, \varepsilon)\}ds\Big\}.$$

If we now introduce a $d \times (r+1)$ matrix with constant entries

$$B_0 := P_{Q_d^*}\Big\{A_2 - \ell \int_a^{b^*} K(\cdot, s)\bar{A}(s)ds\Big\},$$

then we get the following equation for the vector $c \in R^{r+1}$, $c = c(\cdot) \in C[\varepsilon]$:

$$B_0 c = -P_{Q_d^*}\Big\{l_1 x^{(1)}(\cdot,\varepsilon) + R_1(x(\cdot,\varepsilon),\varepsilon) - l\int_a^{b^*} K(\cdot,s)\{[\beta^* A + A_1(s)]x^{(1)} + R(x,\varepsilon)\}ds\Big\}. \tag{5.103}$$

In order that this equation be solvable, it is necessary and sufficient that

$$P_{B_0^*} P_{Q_d^*}\Big\{l_1 x^{(1)}(\cdot,\varepsilon) + R_1(x(\cdot,\varepsilon),\varepsilon) - l\int_a^{b^*} K(\cdot,s)\{[\beta^* A + A_1(s)]x^{(1)} + R(x,\varepsilon)\}ds\Big\} = 0, \tag{5.104}$$

where $P_{B_0^*}$ is a $d \times d$ matrix orthoprojector ($P_{B_0^*} : R^d \to N(B_0^*)$). Assume that $P_{B_0^*} = 0$, which means that the matrix B_0 is of full rank, i.e., $\operatorname{rank} B_0 = d$. Then condition (5.104) is satisfied. In this case, system (5.103) is solvable and possesses at least one solution

$$c = -B_0^+ P_{Q_d^*}\Big\{l_1 x^{(1)}(\cdot,\varepsilon) + R_1(x(\cdot,\varepsilon),\varepsilon) - l\int_a^{b^*} K(\cdot,s)\{[\beta^* A + A_1(s)]x^{(1)} + R(x,\varepsilon)\}ds\Big\}.$$

If I_1 is an $r \times (r+1)$ matrix with constant entries:

$$I_1 = \begin{bmatrix} 1 & 0 & \cdots & 0 & 0 \\ 0 & 1 & \cdots & 0 & 0 \\ \cdots & \cdots & \cdots & \cdots & \cdots \\ 0 & 0 & \cdots & 1 & 0 \end{bmatrix},$$

then, under the assumption that $P_{B_0^*} = 0$, one of the solutions $x(\tau,\varepsilon)$: $x(\cdot,\varepsilon) \in C^1[\tau]$, $x(\tau,\cdot) \in C[\varepsilon]$, $x(\tau,0) = 0$ of the boundary-value problem (5.101), (5.102) can be found from the following equivalent operator system:

$$x(\tau,\varepsilon) = X_r(\tau) I_1 c + x^{(1)}(\tau,\varepsilon), \tag{5.105}$$

$$c = -B_0^+ P_{Q_d^*}\Big\{l_1 x^{(1)}(\cdot,\varepsilon) + R_1(x(\cdot,\varepsilon),\varepsilon) - l\int_a^{b^*} K(\cdot,s)\{[\beta^* A + A_1(s)]x^{(1)} + R(x,\varepsilon)\}ds\Big\},$$

$$x^{(1)}(\tau,\varepsilon) = \varepsilon X(\tau)Q^{+}\{\varphi_0(c^*) + A_2 c(\varepsilon) + \ell_1 x^{(1)}(\cdot,\varepsilon) + R_1(x(\cdot,\varepsilon),\varepsilon)\}$$
$$+ \varepsilon G\{f_0(s,c^*) + \bar{A}(s)c + [\beta^* A + A_1(s)]x^{(1)}(s,\varepsilon) + R(x,\varepsilon)\}(\tau).$$

It is known [19] that the operator system (5.105) belongs to the class of systems whose solution can be found by the method of simple iterations convergent for all $\varepsilon \in [0,\varepsilon_*]$. The lower bound of the quantity ε_* can be established by using the Lyapunov majorizing equations.

Thus, assume that the matrix B_0 is of full rank, i.e., $\operatorname{rank} B_0 = d$. It is natural to seek the first approximation $x_1(\tau,\varepsilon)$ to the solution $x(\tau,\varepsilon)$ of system (5.105) in the form of a solution of the boundary-value problem

$$\frac{dx_1}{d\tau} = Ax_1 + \varepsilon f_0(\tau,c^*),$$

$$lx_1(\cdot,\varepsilon) = \varepsilon\varphi_0(c^*),$$

where $c^* = \operatorname{col}(c_r^*,\beta^*) \in R^{r+1}$ satisfies equation (5.97).

The required solution has the form

$$x_1(\tau,\varepsilon) = X_r(\tau)I_1 c_0 + x_1^{(1)}(\tau,\varepsilon), \quad c_0 = 0,$$

$$x_1^{(1)}(\tau,\varepsilon) = \varepsilon X(\tau)Q^{+}\varphi_0(c^*) + \varepsilon G f_0(s,c^*)(\tau).$$

The second approximation $x_2(\tau,\varepsilon)$ is sought as a solution of the problem

$$\frac{dx_2}{d\tau} = Ax_2 + \varepsilon\{f_0(\tau,c^*) + \bar{A}(\tau)c_1$$
$$+ [\beta^* A + A_1(\tau)]x_1^{(1)}(\tau,\varepsilon) + R(x_1^{(1)},\varepsilon)\},$$
$$lx_2(\cdot,\varepsilon) = \varepsilon\{\varphi_0(c^*) + \ell_1 x_1(\cdot,\varepsilon)\}$$

in the form

$$x_2(\tau,\varepsilon) = X_r(\tau)I_1 c_1 + x_2^{(1)}(\tau,\varepsilon),$$

$$x_2^{(1)}(\tau,\varepsilon) = \varepsilon X(\tau)Q^{+}\{\varphi_0(c^*) + A_2 c_1 + \ell_1 x_1^{(1)}(\cdot,\varepsilon) + R_1(x_1^{(1)}(\cdot,\varepsilon),\varepsilon)\}$$
$$+ \varepsilon G\{f_0(s,c^*) + \bar{A}_1(s)c_1 + [\beta^* A + A_1(s)]x_1^{(1)}(s,\varepsilon) + R(x_1^{(1)},\varepsilon)\}(\tau).$$

The condition of solvability of the system deduced for the second approximation yields the following equation for the vector c_1:

$$B_0 c_1 = -P_{Q_d^*}\Bigg\{l_1 x_1^{(1)}(\cdot,\varepsilon) + R_1(x_1^{(1)}(\cdot,\varepsilon),\varepsilon)$$
$$- l\int_a^{b^*} K(\cdot,s)\{[\beta^* A + A_1(s)]x_1^{(1)} + R(x_1^{(1)},\varepsilon)\}ds\Bigg\}.$$

The solution of this equation has the form

$$c_1 = -B_0^+ P_{Q_d}^* \Big\{ l_1 x_1^{(1)}(\cdot,\varepsilon) + R_1(x_1^{(1)}(\cdot,\varepsilon),\varepsilon) - l \int_a^{b^*} K(\cdot,s)\{[\beta^* A + A_1(s)]x_1^{(1)} + R(x_1^{(1)},\varepsilon)\}ds \Big\} + P_\rho c_\rho,$$

where P_ρ is an $(r+1)\times\rho$ matrix whose columns form a complete system of ρ linearly independent columns of the orthoprojector P_{B_0},

$$\rho = \operatorname{rank} P_{B_0} = n - m + 1,$$

and $c_\rho = c_\rho(\cdot) \in C[\varepsilon]$, $c_\rho \in R^\rho$, is a vector such that $c_\rho(0) = 0$.

We continue the process of calculations and, as a result, conclude that one of the solutions of the operator system (5.105) can be obtained with the help of the following iterative procedure:

$$c_k = -B_0^+ P_{Q_d^*} \Big\{ l_1 x_k^{(1)}(\cdot,\varepsilon) + R_1(x_k(\cdot,\varepsilon),\varepsilon) - l \int_a^{b^*} K(\cdot,s)\{[\beta^* A + A_1(s)]x_k^{(1)}(s,\varepsilon) + R(x_k(s,\varepsilon),\varepsilon)\}ds \Big\},$$

$$\begin{aligned} x_{k+1}^{(1)}(\tau,\varepsilon) &= \varepsilon X(\tau) Q^+ \{\varphi_0(c^*) + A_2 c_k + \ell_1 x_k^{(1)}(\cdot,\varepsilon) + R_1(x_k^{(1)}(\cdot,\varepsilon),\varepsilon)\} \\ &\quad + \varepsilon G\{ f_0(s,c^*) + \bar{A}(s) c_k \\ &\qquad + [\beta^* A + A_1(s)]x_k^{(1)}(s,\varepsilon) + R(x_k(s,\varepsilon),\varepsilon)\}(\tau), \\ x_{k+1}(\tau,\varepsilon) &= X_r(\tau) I_1 c_k + x_{k+1}^{(1)}(\tau,\varepsilon), \\ x_0(\tau,\varepsilon) &= x_0^{(1)}(\tau,\varepsilon) = 0, \quad k = 0,1,2,3,\ldots. \end{aligned} \tag{5.106}$$

Assume that the homogeneous generating problem (5.87), (5.88) has an r-parameter ($r = n - n_1$, $\operatorname{rank} Q = n_1$) family of nontrivial solutions and condition (5.89) is satisfied. Then the inhomogeneous generating problem is solvable and possesses a solution of the form (5.90). The sufficient conditions for the existence of solutions of the original weakly nonlinear problem (5.85), (5.86) are established by the following assertion:

Theorem 5.8. *Assume that the matrix B_0 is of full rank ($P_{B_0^*} = 0$). Then, for each root $c^* = \operatorname{col}(c_r^*, \beta^*) \in R^{r+1}$ of the equation for generating constants (5.97), the boundary-value problem (5.95), (5.96) possesses at least one solution $x(\tau,\varepsilon)$ such that $x(\cdot,\varepsilon) \in C^1[a,b^*]$, $x(\tau,\cdot) \in C[0,\varepsilon_*]$, and $x(\tau,0) = 0$. This solution can be obtained from the operator system (5.105) as a result of the iterative procedure (5.106) convergent for sufficiently small $\varepsilon \in [0,\varepsilon_*]$.*

In this case, the boundary-value problem (5.93), (5.94) has at least one solution $z(\tau,\varepsilon)$ such that $z(\cdot,\varepsilon) \in C^1[0,b^*]$, $z(\tau,\cdot) \in C[0,\varepsilon_0]$, and $z(\tau,0) = z_0(\tau,c_r^*)$. This solution can be found with the help of the iterative process (5.106) and the following formula:

$$z_k(\tau,\varepsilon) = z_0(\tau,c_r^*) + x_k(\tau,\varepsilon), \quad k = 0,1,2,\ldots.$$

In view of the change of the independent variable (5.92), Theorem 5.8 specifies the solutions $z(\cdot,\varepsilon) \in C^1[a,b(\varepsilon)]$, $z(t,\cdot) \in C[0,\varepsilon_0]$, $z(t,0) = z(t,c^*)$, of the boundary-value problem (5.85), (5.86) depending on an arbitrary constant $c_\rho \in R^\rho$ $(\rho = \operatorname{rank} P_{B_0} = n - m + 1)$. Furthermore, the $(r+1)$th component $\bar\beta = \bar\beta(\varepsilon) = \beta(\varepsilon) - \beta^*$ of the vector constant $c = c(\varepsilon) \in R^{r+1}$ is a correction to the length of the interval used for the construction of the required solution $z(t,\varepsilon)$ of the original problem.

In the case $m = n$ (most frequently encountered in the theory of oscillations), the equality $P_{B_0^*} = 0$ means that $\det B_0 \neq 0$.

In view of the fact that the last component of the vector c_r can be made zero by the proper choice of the reference point for the independent variable t, the periodic boundary-value problems $\big(m = n,\ J(z(\cdot,\varepsilon),\varepsilon) = 0,\ \alpha = 0,\ f = 0,\ lz(\cdot,\varepsilon) = z(0,\varepsilon) - z(T_1(\varepsilon),\varepsilon)\big)$ can be studied with the help of the following theorem obtained as a consequence of Theorem 5.8 [36]:

Theorem 5.9. *Assume that the boundary-value problem (5.85), (5.86) satisfies the conditions presented above. Then, for any simple* $(\det B_0 \neq 0)$ *root* $c^* = \operatorname{col}(c_{r-1}^*,\beta^*) \in R^r$ *of the equation for generating constants*

$$F(c^*) = P_{Q_r^*} l \int_0^T K(\cdot,s) f_0(s,c^*)\,ds = 0,$$

the boundary-value problem possesses a unique $T_1(\varepsilon)$*-periodic solution* $z(t,\varepsilon)$ *such that* $z(\cdot,\varepsilon) \in C^1[0,T_1(\varepsilon)]$ *and* $z(t,\cdot) \in C[\varepsilon]$ *that turns into the generating solution* $z(t,c_{r-1}^*)$ *for* $\varepsilon = 0$. *The indicated solution can be obtained as a result of the iterative process*

$$c_k = -B_0^{-1} P_{Q_d^*} l \int_0^T K(\cdot,s)\big\{[\beta^* A + A_1(s)]x_k^{(1)}(s,\varepsilon) + R(x_k(s,\varepsilon),\varepsilon)\big\}ds,$$

$$x_{k+1}^{(1)}(\tau,\varepsilon) = \varepsilon \int_0^T G(\tau,s)\big\{f_0(s,c^*) + \bar A(s)c_k$$

$$+ [\beta^* A + A_1(s)]x_k^{(1)}(s,\varepsilon) + R(x_k(s,\varepsilon),\varepsilon)\big\}ds,$$

$$x_{k+1}(\tau,\varepsilon) = X_{r-1}(\tau) I_1 c_k + x_{k+1}^{(1)}(\tau,\varepsilon),$$

$$z_{k+1}(\tau,\varepsilon)=z_0(\tau,c_{r-1}^*)+x_{k+1}(\tau,\varepsilon),$$

$$x_0(\tau,\varepsilon)=x_0^{(1)}(\tau,\varepsilon)\equiv 0,\quad k=0,1,2,\dots,$$

convergent for $\varepsilon\in[0,\varepsilon_*]\subset[0,\varepsilon_0]$.

5.4 General Scheme of Investigation of the Boundary-Value Problems

Let us now outline the general scheme used for the analysis of weakly perturbed linear and nonlinear boundary-value problems.

1. For a given weakly nonlinear boundary-value problem, we consider the corresponding generating ($\varepsilon=0$) boundary-value problem. To do this, for the known normal fundamental matrix $X(t)$ of the homogeneous differential system (5.3), we construct the $m\times n$ matrix $Q=lX(\cdot)$ and the orthoprojectors P_Q and P_{Q^*}. Further, it is necessary to determine whether the problem is critical ($\operatorname{rank} Q<m\sim P_{Q^*}\neq 0$) or noncritical ($\operatorname{rank} Q=m\sim P_{Q^*}=0$). Then we check the criterion of solvability (5.9) of the generating boundary-value problem and find the family of generating solutions $z_0(t,c_r)$ of the form (5.11).

2. In the noncritical case ($P_{Q^*}=0$), the criterion of solvability (5.15) is satisfied for the generating boundary-value problem. It has at least one generating solution $z_0(t,c_r)$ given by (5.11) ($r=n-m$). Then, for any nonlinearity satisfying (5.36), the original weakly nonlinear noncritical boundary-value problem (5.32) possesses at least one solution $z(t,\cdot)\in C[\varepsilon]$ which turns into the generating solution for $\varepsilon=0$. This solution can be obtained from the n-dimensional operator system (5.34), (5.37) as a result of the iterative process (5.39), (5.40) convergent for $\varepsilon\in[0,\varepsilon_*]$. The range of convergence of the iterative process and the accuracy of approximate solutions can be estimated by the method of finite majorizing Lyapunov equations [19, 65].

3. In the critical case ($\operatorname{rank} Q=n_1<m$), the investigation of boundary-value problems of the form (5.43) becomes much more complicated. Assume that the criterion of solvability (5.9) is satisfied for the generating boundary-value problem and that it has an r-parameter ($r=n-n_1$) family of generating solutions $z_0(t,c_r)$ of the form (5.11). Then, by using the equation for generating constants (5.46), we determine constants $c_r=c_0^*\in R^r$ specifying the generating solution associated with the solution $z(t,\cdot)\in C[\varepsilon]$, $z(t,0)=z_0(t,c_0^*)$ of the original boundary-value problem (5.43).

Further, we construct the $d\times r$ matrix B_0. If $\operatorname{rank} B_0=d\sim P_{B_0^*}=0$, then the critical boundary-value problem of the first order (5.43) possesses at least one (unique for $n=m$) solution $z(t,\cdot)\in C[\varepsilon]$ which turns into the generating solution $z_0(t, c_0^*)$ for $\varepsilon = 0$. This solution can be obtained from the $(2n + r)$-dimensional operator system (5.47), (5.52) as a result of the iterative process (5.57) convergent for $\varepsilon \in [0, \varepsilon_*]$.

If $\operatorname{rank} B_0 < d \sim P_{B_0^*} \neq 0$, then, in order to establish conditions required for the existence of solutions of the original problem, it is necessary to construct a $d \times r$ matrix B_1 given by relation (5.62) and its orthoprojectors P_{B_1} and $P_{B_1^*}$. If $P_{B_1^*} P_{B_0^*} = 0$, then, for any nonlinearity satisfying relation (5.73), the critical boundary-value problem of the second order (5.43) possesses at least one (unique for $P_{B_0} P_{B_1} = 0$) solution $z(t, \cdot) \in C[\varepsilon]$ which turns into the generating solution $z(t, 0) = z_0(t, c_0^*)$ for $\varepsilon = 0$. The indicated solution can be found from the $2(n + r)$-dimensional operator system (5.66) by using the iterative process (5.71) convergent for $\varepsilon \in [0, \varepsilon_*]$.

4. If the generating ($\varepsilon = 0$) inhomogeneous linear boundary-value problem is unsolvable (condition (5.9) is not satisfied), then, as indicated above, one can either construct a quasisolution of the problem or make it solvable by adding small linear perturbations [17]. In this case, the requirements guaranteeing the appearance of solutions of weakly perturbed linear boundary-value problems are specified by introducing $d \times r$ matrices B_0, B_1, etc. [19, 98]. The solutions of perturbed boundary-value problems can be constructed with the help of the Vishik–Lyusternik algorithm [150] in the form of convergent series in powers of a small parameter (see Sections 5.6 and 6.4).

Thus, the data required to establish the conditions of solvability of weakly perturbed linear and nonlinear boundary-value problems are contained in the $m \times n$ matrix Q and in the chain of $d \times r$ matrices $B_0, B_1, \ldots$, constructed according to the coefficients of the original differential system.

5.5 Periodic Solutions of the Mathieu, Riccati, and Van der Pol Equations

The results obtained in the previous sections remain true and can be applied [16–18, 99] to the investigation of periodic boundary-value problems frequently encountered in applications and fairly well studied in [65, 93, 94, 97, 122, 143]. The examples presented below illustrate some possibilities of the proposed procedures.

Example 1. Consider a periodic boundary-value problem for the scalar Riccati equation

$$\begin{gathered}\dot{z} = a_0(t)z + \varphi(t) + \varepsilon(a_1(t)z + a_2(t)z^2),\\ lz = z(0) - z(\omega) = 0,\\ \varphi(t + \omega) = \varphi(t) \in C[0, \omega],\\ a_i(t + \omega) = a_i(t) \in C[0, \omega], \quad i = 0, 1, 2. \end{gathered} \tag{5.107}$$

This equation plays an important role in the analysis of oscillations, absence of oscillations, reducibility, and other features of the qualitative behavior of second-order linear systems.

We first consider the noncritical case where

$$X(t) = \exp \int_0^t a_0(t)dt, \qquad Q = lX = 1 - \exp \int_0^\omega a_0(t)dt,$$

$$\int_0^\omega a_0(t)dt \neq 0, \qquad \text{and} \qquad \operatorname{rank} Q = n \equiv 1.$$

This means that the generating problem corresponding to problem (5.107), i.e.,

$$\dot{z} = a_0(t)z + \varphi(t), \qquad lz = 0, \tag{5.108}$$

possesses a unique ω-periodic solution

$$z_0(t) = \int_0^\omega G(t,\tau)\varphi(\tau)d\tau, \tag{5.109}$$

where $G(t,\tau)$ is the Green function of problem (5.108) given by the formula

$$G(t,\tau) = g(t,\tau) = \frac{1}{2}\left\{ \operatorname{sign}(t-\tau) + \frac{1+\exp\left(\int_0^\omega a_0(s)ds\right)}{1-\exp\left(\int_0^\omega a_0(s)ds\right)} \right\} \exp\left(\int_\tau^t a_0(s)ds\right).$$

Our aim is to establish conditions for the existence and develop an algorithm for the construction of an ω-periodic solution of problem (5.108) $z = z(t,\varepsilon) \in C[\varepsilon]$, $\varepsilon \in [0, \varepsilon_*]$, which turns into the ω-periodic solution $z_0(t)$ of the generating problem (5.107) for $\varepsilon = 0$. For this purpose, we perform the change of variables $z(t,\varepsilon) = z_0(t) + x(t,\varepsilon)$ in relations (5.107) and, as a result, arrive at the equivalent problem of finding the ω-periodic solution $x(t,\cdot) \in C[\varepsilon]$, $\varepsilon \in [0,\varepsilon_*]$, $x(t,0) = 0$, of the following boundary-value problem:

$$\dot{x} = a_0(t)x + \varepsilon\{a_1(t)(z_0 + x) + a_2(t)(z_0 + x)^2\}, \quad lx = 0. \tag{5.110}$$

According to Theorem 5.3, an ω-periodic solution always exists and is unique in the analyzed noncritical case $\left(\int_0^\omega a_0(t)dt \neq 0\right)$. It can be found from the scalar operator equation

$$x(t,\varepsilon) = \varepsilon \int_0^\omega g(t,\tau)\{a_1(\tau)(z_0(\tau) + x(\tau,\varepsilon)) + a_2(\tau)(z_0(\tau) + x(\tau,\varepsilon))^2\}d\tau$$

as a result of the iterative process

$$x_{k+1}(t,\varepsilon) = \varepsilon \int_0^\omega g(t,\tau)\{a_1(\tau)(z_0(\tau) + x_k(\tau,\varepsilon)) + a_2(\tau)(z_0(\tau) + x_k(\tau,\varepsilon))^2\}d\tau,$$

$$k = 0, 1, 2, \ldots, \qquad x_0(t,\varepsilon) = 0,$$

convergent on $[0, \varepsilon_*]$.

In order to determine the unique ω-periodic solution $z(t,\cdot) \in C[\varepsilon]$, $\varepsilon \in [0, \varepsilon_*]$, $z(t,0) = z_0(t)$, of the periodic boundary-value problem (5.107) for the Riccati equation, it is necessary to use the formula $z_k(t,\varepsilon) = z_0(t) + x_k(t,\varepsilon)$, $k = 0, 1, 2, \ldots$.

We now consider the critical case where

$$Q = lX = 0 \sim \int_0^\omega a_0(s)ds = 0, \quad \operatorname{rank} Q = n_1 = 0, \quad \text{and} \quad r = n - n_1 = 1.$$

It is also assumed that the generating boundary-value problem (5.108) possesses a one-parameter $(r = 1)$ family of solutions

$$z_0(t, c_r) = X_r(t)c_r + \int_0^\omega G(t,\tau)\varphi(\tau)d\tau$$

provided that

$$\int_0^\omega H(t)\varphi(t)dt = 0, \quad H(t) = h(t) = \exp\left(-\int_0^t a_0(s)ds\right),$$

where $X_r(t) = X(t) = \exp \int_0^t a_0(s)ds$ and $G(t,\tau)$ is the generalized Green function given by relation (5.14). After simple transformations, we obtain

$$P_Q = P_{Q^*} = 1,$$

$$G(t,\tau) = g(t,\tau) = K(t,\tau) = \frac{1}{2}\exp\left(\int_\tau^t a_0(s)ds\right)\operatorname{sign}(t-\tau).$$

Let us now establish conditions for the existence and develop an algorithm for the construction of ω-periodic solutions $z(t,\cdot) \in C[\varepsilon]$, $\varepsilon \in [0, \varepsilon_0]$, $z(t,0) = z_0(t, c_r)$, of the boundary-value problem (5.43). Theorem 5.4 gives a necessary condition for the existence of solutions of this sort. According to this theorem, the constant $c_r \in R^r$ in the generating solution can be found from the equation for generating constants (5.46). In the analyzed case, the equation

$$F(c_r) = \int_0^\omega h(t)\{a_1(t)z_0(t, c_r) + a_2(t)z_0^2(t, c_r)\}dt = 0$$

can be rewritten in the form of a quadratic algebraic equation

$$F(c_r) = ac_r^2 + bc_r + c = 0, \tag{5.111}$$

where

$$a = \int_0^\omega h(t)\{a_1(t)X_r(t) + 2a_2(t)X_r(t)\int_0^\omega g(t,\tau)\varphi(\tau)d\tau\}dt,$$

$$b = \int_0^\omega h(t)\{a_1(t)X_r(t) + 2a_2(t)X_r(t)\int_0^\omega g(t,\tau)\varphi(\tau)d\tau\}dt,$$

$$c = \int_0^\omega h(t)\Big\{a_1(t)\int_0^\omega g(t,\tau)\varphi(\tau)d\tau + a_2(\tau)\Big(\int_0^\omega g(t,\tau)\varphi(\tau)d\tau\Big)^2\Big\}dt.$$

We assume that equation (5.111) has simple roots: $c_r = c_0^* = c_{0i}^*$ $(i=1,2)$, i.e.,

$$B_0 = \frac{\partial F(c_r)}{\partial c_r}\Big|_{c_r=c_0^*} = b_0 = \int_0^\omega h(t)p(t)X_r(t)dt \neq 0$$

and

$$p(t) = a_1(t) + 2a_2(t)z_0(t, c_0^*).$$

For the deviation $x(t,\varepsilon)$ from the generating solution $z_0(t, c_0^*)$, we obtain an ω-periodic solution $x(t,\cdot) \in C[\varepsilon]$, $\varepsilon \in [0, \varepsilon_0]$, $x(t, 0) = 0$, of the boundary-value problem

$$\dot{x} = a_0(t)x + \varepsilon\{f_0(t, c_0^*) + p(t)x + a_2(t)x^2\}, \quad lx = 0, \tag{5.112}$$

with

$$f_0(t, c_0^*) = a_1(t)z_0(t, c_0^*) + a_2(t)z_0^2(t, c_0^*).$$

In the set of ω-periodic solutions, this problem is equivalent to the following three-dimensional operator system:

$$b_0c = -\int_0^\omega h(t)\{p(t)x^{(1)}(t,\varepsilon) + a_2(t)x^2(t,\varepsilon)\}dt,$$

$$x^{(1)}(t,\varepsilon) = \int_0^\omega g(t,\tau)\{f_0(\tau, c_0^*) + p(\tau)[X_r(\tau)c + x^{(1)}] + a_2(\tau)x^2(t,\varepsilon)\}d\tau, \tag{5.113}$$

$$x(t,\varepsilon) = X_r(t)c + x^{(1)}(t,\varepsilon).$$

According to Theorem 5.5, for any simple root of equation (5.111), i.e., for $b_0 = b_0(c_0^*) \neq 0$ $(P_{B_0^*} = P_{B_0} = 0)$, the boundary-value problem (5.112) [the Riccati equation (5.107)] possesses a unique ω-periodic solution $x(t,\cdot) \in C[\varepsilon]$, $\varepsilon \in [0, \varepsilon_0]$, $x(t,0) = 0$, determined from the operator system (5.113) by the method of simple iterations similar to (5.56).

In the case where $b_0 = 0 \sim P_{B_0^*} = 1 \neq 0$, we can pass from the operator system (5.113) to the system

$$\varepsilon b_1 c = -\int_0^\omega h(t)\{p(t)x^{(2)}(t,\varepsilon) + a_2(t)x^2(t,\varepsilon)\}dt,$$

$$x^{(2)}(t,\varepsilon) = \varepsilon \int_0^\omega g(t,\tau)\big\{f_0(\tau, c_0^*) + p(\tau)[\varepsilon g(\tau)c + x^{(2)}(\tau,\varepsilon)] + a_2(\tau)x^2(\tau,\varepsilon)\big\}d\tau, \tag{5.114}$$

$$x(t,\varepsilon) = \varepsilon g_1(t)c + x^{(2)}(t,\varepsilon),$$

where

$$g_1(t) = \int_0^\omega g(t,\tau)p(\tau)X_r(\tau)d\tau, \quad b_1 = \int_0^\omega h(t)p(t)g_1(t)dt.$$

By virtue of Theorem 5.6, this operator system (and, hence, the Riccati equation) possess a unique ω-periodic solution for each $c_r = c_{0i}^*$ satisfying condition (5.111) provided that $b_1 \neq 0$ $(P_{B_1} = P_{B_1^*} = 0)$ and

$$\int_0^\omega h(t)\{p(t)x_1(t,\varepsilon) + a_2(t)x_1^2(t,\varepsilon)\}dt = 0.$$

The indicated solution can be found from system (5.114) by the method of simple iterations.

Example 2. In order to illustrate the applicability of the iterative process (5.71), we consider the well-studied [65, 81, 93, 99] problem of construction of 2π-periodic solutions of the Mathieu equation

$$\ddot{z} + (a + \varepsilon \cos 2t)z = 0, \tag{5.115}$$

which can be represented in the form of a system

$$\dot{y} = \begin{bmatrix} 0 & 1 \\ -k^2 & 0 \end{bmatrix} y + \varepsilon \begin{bmatrix} 0 & 0 \\ -h(\varepsilon) - \cos 2t & 0 \end{bmatrix} y, \tag{5.116}$$

where

$$a = k^2 + \varepsilon h(\varepsilon), \quad k = 0, 1, 2, \ldots, \quad h(\varepsilon) = h_0 + \varepsilon h_1 + \varepsilon^2 h_2 + \ldots,$$

and ε is a small positive parameter. For the sake of definiteness, we consider the case $k = 1$. Our aim is to establish conditions under which system (5.116) has a 2π-periodic solution. The indicated conditions are imposed on the unknown parameters h_i and then used to find these parameters.

The generating system obtained from (5.116) by setting $\varepsilon = 0$ possesses a two-parameter family of 2π-periodic solutions

$$y_0(t, c_0) = X(t)c_0,$$

where

$$X(t) = \begin{bmatrix} \cos t & \sin t \\ -\sin t & \cos t \end{bmatrix}$$

and

$$c_0 = \begin{bmatrix} c_{01} \\ c_{02} \end{bmatrix}$$

is an arbitrary column vector of constants. By the change of variables

$$y(t, \varepsilon) = y_0(t, c_0) + x(t, \varepsilon)$$

in system (5.116), we arrive at the problem of finding 2π-periodic solutions $x(t, \cdot) \in C[\varepsilon]$, $x(t, 0) = 0$, of the following system:

$$\dot{x} = \begin{bmatrix} 0 & 1 \\ -1 & 0 \end{bmatrix} x + \varepsilon \begin{bmatrix} 0 & 0 \\ -h(\varepsilon) - \cos 2t & 0 \end{bmatrix} (y_0 + x). \tag{5.117}$$

The necessary condition for the existence of 2π-periodic solutions of system (5.117) takes the form

$$\int_0^{2\pi} H(t)A_1(t)y_0(t, c_0)dt = B_0 c_0 = 0, \tag{5.118}$$

where

$$H(t) = \begin{bmatrix} \cos t & -\sin t \\ \sin t & \cos t \end{bmatrix}, \quad A_1(t) = \begin{bmatrix} 0 & 0 \\ -h_0 - \cos 2t & 0 \end{bmatrix},$$

$$B_0 = \begin{bmatrix} 0 & h_0 - 1/2 \\ h_0 + 1/2 & 0 \end{bmatrix} \pi.$$

In order that the equation for generating amplitudes (5.118) have a nontrivial solution $c_0 \neq 0$, it is necessary and sufficient that $\det B_0 = h_0^2 - 1/4 = 0$. This equality is used to find h_0. Thus, let $h_0 = -1/2$. Then

$$B_0 = \pi \begin{bmatrix} 0 & -1 \\ 0 & 0 \end{bmatrix}, \quad B_0^+ = \frac{1}{\pi} \begin{bmatrix} 0 & 0 \\ -1 & 0 \end{bmatrix}, \quad P_{B_0} = \begin{bmatrix} 1 & 0 \\ 0 & 0 \end{bmatrix},$$

$$P_{B_0^*} = \begin{bmatrix} 0 & 0 \\ 0 & 1 \end{bmatrix}, \quad \text{and} \quad c_0^* = P_{B_0} c_0.$$

By using the generalized Green matrix,

$$G(t,s) = g(t,s)X(t)X^{-1}(s), \quad g(t,s) = \begin{cases} \dfrac{s}{2\pi} & \text{for} \quad 0 \le s < t, \\ \dfrac{s}{2\pi} - 1 & \text{for} \quad t \le s \le 2\pi, \end{cases}$$

we obtain

$$x_1^{(2)}(t,\varepsilon) = \varepsilon G_1(t) c_0^* = \frac{\varepsilon}{16} \begin{bmatrix} \cos 3t \\ -3\sin 3t \end{bmatrix}, \quad c_0^* = \begin{bmatrix} 1 \\ 0 \end{bmatrix}.$$

Further,

$$B_1 = \frac{\pi}{32} \begin{bmatrix} 0 & 0 \\ -1 & 0 \end{bmatrix}, \quad B_1^+ = \frac{32}{\pi} \begin{bmatrix} 0 & -1 \\ 0 & 0 \end{bmatrix}, \quad P_{B_1} = \begin{bmatrix} 0 & 0 \\ 0 & 1 \end{bmatrix}, \quad P_{B_1^*} = \begin{bmatrix} 1 & 0 \\ 0 & 0 \end{bmatrix}.$$

The conditions $P_{B_0} P_{B_1} = 0$ and $P_{B_1^*} P_{B_0^*} = 0$ are satisfied and, hence, for the applicability of the iterative process (5.71) to the construction of the 2π-periodic solution $x(t,\cdot) \in C[\varepsilon]$, $x(t,0) = 0$, of system (5.117), it is necessary that conditions (5.73) be satisfied with

$$R(x_1^{(1)}(t,\varepsilon), t, \varepsilon) = \varepsilon \begin{bmatrix} 0 & 0 \\ -h_1 - \varepsilon h_2 - \ldots & 0 \end{bmatrix} (y_0(t, c_0^*) + x_1^{(1)}(t,\varepsilon)),$$

$$x_1^{(1)}(t,\varepsilon) = x_1^{(2)}(t,\varepsilon).$$

Note that conditions (5.73) are satisfied for $h_1 = -1/32$ and $h_i = 0$, $i = 2, 3, \ldots$. Thus, a 2π-periodic solution $x(t,\cdot) \in C[\varepsilon]$ of the system

$$\dot{x} = \begin{bmatrix} 0 & 1 \\ -1 & 0 \end{bmatrix} x + \varepsilon \begin{bmatrix} 0 & 0 \\ 1/2 - \cos 2t & 0 \end{bmatrix} (y_0(t, c_0^*) + x) + \frac{\varepsilon}{32} \begin{bmatrix} 0 & 0 \\ 1 & 0 \end{bmatrix} (y_0 + x)$$

can be obtained by using the iterative procedure (5.71). After necessary transformations, this procedure takes the form

$$\begin{aligned} \varepsilon c_{k-1}^{(1)} &= \frac{32}{\pi} \int_0^{2\pi} \begin{bmatrix} -1/2\cos 3t & 0 \\ 0 & 0 \end{bmatrix} x_k^{(2)}(t,\varepsilon)\,dt \\ &\quad + \frac{\varepsilon}{\pi} \int_0^{2\pi} \begin{bmatrix} \cos t & 0 \\ 0 & 0 \end{bmatrix} \left\{ \begin{bmatrix} \cos t \\ -\sin t \end{bmatrix} + x_k(t,\varepsilon) \right\} dt, \\ c_k^{(0)} &= \frac{1}{\pi} \int_0^{2\pi} \begin{bmatrix} 0 & 0 \\ -\sin t + 1/2\sin 3t & 0 \end{bmatrix} x_k^{(2)}(t,\varepsilon)\,dt \\ &\quad + \frac{\varepsilon}{32\pi} \int_0^{2\pi} \begin{bmatrix} 0 & 0 \\ -\sin t & 0 \end{bmatrix} x_k(t,\varepsilon)\,dt, \end{aligned} \tag{5.119}$$

$$x_{k+1}^{(2)}(t,\varepsilon) = x_1^{(2)}(t,\varepsilon) + \varepsilon X(t)\left\{\frac{1}{2\pi}\int_0^{2\pi} sX^{-1}(s)\{\ldots\}ds - \int_t^{2\pi} X^{-1}(s)\{\ldots\}ds\right\},$$

$$x_{k+1}(t,\varepsilon) = \begin{bmatrix} 0 & \sin t \\ 0 & \cos t \end{bmatrix} c_k^{(0)} + \frac{\varepsilon}{16}\begin{bmatrix} \cos 3t & 0 \\ -3\sin 3t & 0 \end{bmatrix} c_{k-1}^{(1)} + x_{k+1}^{(2)}(t,\varepsilon),$$

$$k = 1, 2, \ldots, \quad x_0(t,\varepsilon) = x_0^{(2)}(t,\varepsilon) = 0,$$

where

$$\{\ldots\} = \begin{bmatrix} 0 & 0 \\ 0 & \sin s - 1/2\sin 3s \end{bmatrix} c_k^{(0)} + \frac{\varepsilon}{32}\begin{bmatrix} 0 & 0 \\ -\cos s + \cos 3s - \cos 5s & 0 \end{bmatrix} c_{k-1}^{(1)}$$
$$+ \begin{bmatrix} 0 & 0 \\ 1/2 - \cos 2s & 0 \end{bmatrix} x_k^{(2)}(s,\varepsilon) + \frac{\varepsilon}{32}\left\{\begin{bmatrix} 0 \\ \cos s \end{bmatrix} + \begin{bmatrix} 0 & 0 \\ 1 & 0 \end{bmatrix} x_k(s,\varepsilon)\right\}.$$

Hence, for $k = 0$, we obtain

$$\varepsilon c_{-1}^{(1)} = 0, \quad c_0^{(0)} = 0, \quad x_1^{(2)}(t,\varepsilon) = \frac{\varepsilon}{16}\begin{bmatrix} \cos 3t \\ -3\sin 3t \end{bmatrix}, \quad x_1^{(2)}(t,\varepsilon) = x_1(t,\varepsilon),$$

and, for $k = 1$, we get

$$\varepsilon c_0^{(1)} = 0, \quad c_{21}^{(0)} = 0,$$

$$x_1^{(2)}(t,\varepsilon) = \begin{bmatrix} (q - q^2 - 2q^3)\cos 3t + (q^2/3 + q^3)\cos 5t \\ -3(q - q^2 - 2q^3)\sin 3t - 5(q^2/3 + q^3)\sin 5t \end{bmatrix},$$

$$x_2(t,\varepsilon) = x_2^{(2)}(t,\varepsilon), \quad q = \frac{\varepsilon}{16}.$$

By using the relation for $c_k^{(0)}$ and the fact that the first components in the expressions for $x_k^{(2)}(t,\varepsilon)$ and $x_k(t,\varepsilon)$, $k = 1, 2, \ldots$, are cosine polynomials, we conclude that $c_k^{(0)} = 0$ for $k = 0, 1, 2, \ldots$ and the iterative process (5.119) becomes somewhat simpler.

To within the terms of the same order as q^2 and higher, the obtained approximation $x_2(t,\varepsilon)$ to the exact 2π-periodic solution of system (5.117) coincides with the well-known results established for the Mathieu equation by different methods [81].

It is known that 2π-periodic solutions of the Mathieu equation exist if and only if the point (a,ε) in the plane of parameters lies on the boundary of the regions of stability [81]. It has already been shown that, for $a = a(\varepsilon) = 1 - \varepsilon/2 - \varepsilon^2/32 - \ldots$, the 2π-periodic solution of system (5.117) and, hence, of the Mathieu equation (5.93) exists at least for $\varepsilon \in [0, \varepsilon_*]$. Therefore, for $\varepsilon \in [0, \varepsilon_*]$, the expression $a = 1 - \varepsilon/2 - \varepsilon^2/32 - \ldots$ specifies the boundary of the stability region for the Mathieu equation (5.115) in the plane (a, ε) to within the terms of order q^2.

Example 3. In this example, we illustrate the algorithm used for the investigation of autonomous weakly nonlinear boundary-value problems by analyzing the problem of finding periodic solutions $z(t,\varepsilon) = \operatorname{col}(u, v)$ of the well-known Van der Pol equation [101, 121]

$$\frac{dz}{dt} = Az + \varepsilon Z(z),$$

where

$$A = \begin{bmatrix} 0 & 1 \\ -1 & 0 \end{bmatrix} \quad \text{and} \quad Z(z) = \operatorname{col}(0, (1-u^2)v),$$

which turns into a solution of the generating problem

$$\frac{dz_0}{dt} = Az_0$$

for $\varepsilon = 0$. According to the notation introduced above, we have $r = n = 2$,

$$X_r(\tau) = \begin{bmatrix} \cos\tau & \sin\tau \\ -\sin\tau & \cos\tau \end{bmatrix} = X(\tau),$$

$$P_{Q_r^*} lK(\cdot, \tau) = \begin{bmatrix} \cos\tau & -\sin\tau \\ \sin\tau & \cos\tau \end{bmatrix} = X^{-1}(\tau),$$

$$z_0(t, c_r^*) = X_r(t)c_r^*, \quad \text{and} \quad c^* = \operatorname{col}(c_r^*, \beta^*).$$

It is now possible to deduce the equation for generating amplitudes (5.97) of the periodic problem formulated for the Van der Pol equation:

$$F(c^*) = \int_0^{2\pi} X^{-1}(s)\{\beta^* A X_r(s) c_r^*$$
$$+ \operatorname{col}[0, [1 - (c_1^* \cos s + c_2^* \sin s)^2][-c_1^* \sin s + c_2^* \cos s]]\} ds = 0,$$

$$c^* = \operatorname{col}(c_1^*, c_2^*, \beta^*) \in R^3.$$

As a result of integration, we arrive at the following algebraic system:

$$F(c^*) = \pi \begin{cases} 2\beta^* c_2^* + c_1^* - 1/4\, c_1^{*3} - 1/4\, c_1^* c_2^{*2} = 0, \\ -2\beta^* c_1^* + c_2^* - 1/4\, c_1^{*2} c_2^* - 1/4\, c_2^{*3} = 0. \end{cases}$$

The vector $c_1^* = 2,\ c_2^* = 0,\ \beta^* = 0$ is one of the roots of this system.

For this root, we construct the matrices

$$A_1(\tau) = \begin{bmatrix} 0 & 0 \\ 8\sin\tau\cos\tau & 1 - 4\cos^2\tau \end{bmatrix}$$

and

$$B_0 = \int_0^{2\pi} X^{-1}(s)\bar{A}_1(s)ds = -2\pi \begin{bmatrix} 1 & 0 & 0 \\ 0 & 0 & 2 \end{bmatrix}.$$

The matrix B_0 is of full rank and, thus, we have the critical case of the first order. In this case, by virtue of Theorem 5.9, the analyzed equation possesses a one-parameter family of periodic solutions $T_1(\varepsilon) = 2\pi(1+\varepsilon\beta(\varepsilon))$ with $\beta(0) = \beta^* = 0$. The first approximation $z_1(\tau,\varepsilon)$ to the solution of this equation is given by the formula

$$x_1^{(1)}(\tau,\varepsilon) = \varepsilon\int_0^{2\pi} G(\tau,s)f_0(s,c^*)ds = \varepsilon\int_0^{2\pi} K(\tau,s)f_0(s,c^*)ds$$

$$= \varepsilon/2X(\tau)\Big\{\int_0^{\tau} X^{-1}(s)Z(z_0,0)ds - \int_{\tau}^{2\pi} X^{-1}(s)Z(z_0,0)ds\Big\}.$$

After elementary transformations, we get the first approximation $x_1^{(1)}(\tau,\varepsilon)$ to the unknown vector $x^{(1)}(\tau,\varepsilon)$:

$$x_1^{(1)}(\tau,\varepsilon) = \varepsilon/4\,\mathrm{col}\,(3\sin\tau - \sin 3\tau, 3\cos\tau - 3\cos 3\tau).$$

We now find the pseudoinverse matrix B_0^+ and the corresponding orthoprojectors:

$$B_0^+ = -1/4\pi \begin{bmatrix} 2 & 0 \\ 0 & 0 \\ 0 & 1 \end{bmatrix}, \quad P_{B_0} = \begin{bmatrix} 0 & 0 & 0 \\ 0 & 1 & 0 \\ 0 & 0 & 0 \end{bmatrix}, \quad P_{B_0^*} = \begin{bmatrix} 0 & 0 \\ 0 & 0 \end{bmatrix}.$$

According to relation (5.106), the first approximation c_1 to the constant c has the form

$$c_1 = -B_0^+ \int_0^{2\pi} X^{-1}(s)\{Z(z_0(s,c_\tau^*) + x_1^{(1)}(s,\varepsilon),\varepsilon) - Z(z_0(s,c_\tau^*),0)\}ds + P_\rho c_\rho,$$

where $P_\rho = \mathrm{col}\,(0,1,0)$ is the matrix formed by the sole ($\rho = 1$) nonzero column of the orthoprojector P_{B_0}, $c_\rho = c_\rho(\varepsilon) \in C[\varepsilon]$, and $c_\rho(0) = 0 \in R^1$. Integrating this relation, we obtain

$$c_1(\varepsilon) = -\varepsilon\pi(-1/4) \begin{bmatrix} 2 & 0 \\ 0 & 0 \\ 0 & 1 \end{bmatrix} \begin{bmatrix} 0 \\ 1 \end{bmatrix} + \begin{bmatrix} 0 \\ 1 \\ 0 \end{bmatrix} c_\rho(\varepsilon)$$

$$= \varepsilon\pi/4 \begin{bmatrix} 0 \\ 0 \\ 1 \end{bmatrix} + \begin{bmatrix} 0 \\ c_\rho(\varepsilon) \\ 0 \end{bmatrix}.$$

Thus, the first approximation $z_1(\tau, \varepsilon) = \operatorname{col}(z_{a1}, z_{b1})$ to the exact solution $z(\tau, \varepsilon)$ has the form

$$z_{a1}(\tau, \varepsilon) = 2\cos\tau + c_\rho(\varepsilon)\sin\tau + \varepsilon\sin^3\tau,$$

$$z_{b1}(\tau, \varepsilon) = -2\sin\tau + c_\rho(\varepsilon)\cos\tau + 3\varepsilon\sin^2\tau\cos\tau.$$

Note that the first approximation $\beta_1 = \beta_1(\varepsilon) = \varepsilon/4$ to the vector $\beta(\varepsilon)$ is nonzero and, hence, the first approximation to the required period $T_1(\varepsilon)$ is a function of ε, namely,

$$T_1(\varepsilon) = 2\pi(1 + \varepsilon^2/4) = \pi(4 + \varepsilon^2)/2.$$

The proposed calculations carried out for the analyzed periodic boundary-value problem can be simplified if we fix the initial time. In this case, B_0 turns into a 2×2 matrix, i.e.,

$$B_0 = -2\pi \begin{bmatrix} 1 & 0 \\ 0 & 2 \end{bmatrix},$$

and we arrive at the well-known first approximation to the unique periodic solution of the original Van der Pol equation [101].

5.6 Differential Systems with Delay

The proposed approach to the analysis of boundary-value problems for systems of ordinary differential equations can also be applied (with proper modifications) to systems with delayed argument. It should be emphasized that the linear parts of the boundary-value problems for systems of functional differential equations [71] (systems with delay can be regarded as a special case of systems of this sort) are not necessarily Fredholm. The presence of the Fredholm property depends on the statement of the initial-value problem and the spaces in which this problem is analyzed.

Initial-Value Problems. Consider a linear equation with concentrated delay

$$\dot{z}(t) - \sum_{i=1}^{k} A_i(t) z(h_i(t)) = g(t), \quad t \in [a, b], \quad z(s) = \psi(s) \text{ for } s < a, \quad (5.120)$$

where $A_i(t)$ are $n \times n$ matrices and $h_i(t) \le t$ are functions measurable for $t \in [a, b]$.

Most often (see [71, 110]), the solution of the differential equation (5.120) with delay is constructed in the space of continuously differentiable functions as a continuous extension of a function $\psi(s)$ defined on the initial set to the interval $[a, b]$. According to this definition, the initial function $\psi(s)$ and the solution $z(s)$ must be continuously conjugated at the point $s = a$, i.e., $\psi(a) = z(a)$.

This enables us to introduce the notion of infinite-dimensional fundamental matrix (used for the investigation of the initial-value problem (5.120)) whose dimension coincides with the dimension of a basis in the space of initial functions.

Following [8], we now present some basic facts about the initial-value problem (5.120) for differential systems with delay and finite-dimensional fundamental matrices.

Let $h_i\colon [a,b] \to R^1$ and $\psi\colon R^1 \setminus [a,b] \to R^n$ be given functions and let

$$(S_{h_i}z)(t) = \begin{cases} z(h_i(t)) & \text{for } h_i(t) \in [a,b], \\ 0 & \text{for } h_i(t) \notin [a,b], \end{cases} \tag{5.121}$$

where S_{h_i} is the operator of so-called inner composition (see [8, p. 10]). We also set

$$\psi^{h_i}(t) = \begin{cases} 0 & \text{for } h_i(t) \in [a,b], \\ \psi(h_i(t)) & \text{for } h_i(t) \notin [a,b]. \end{cases} \tag{5.122}$$

Further, in view of relations (5.121) and (5.122), equation (5.120) can be rewritten in the form

$$(Lz)(t) := \dot{z}(t) - \sum_{i=1}^{k} A_i(t)(S_{h_i}z)(t) = \varphi(t), \tag{5.123}$$

where

$$\varphi(t) = g(t) + \sum_{i=1}^{k} A_i(t)\psi^{h_i}(t). \tag{5.124}$$

Transformations (5.121) and (5.122) enable us to conjugate the initial function $\psi(s)$, $s<a$, in problem (5.120) with the absolute term and apply the well-developed methods of linear functional analysis to equation (5.123). We study equation (5.123) under the assumption that the operator L bounded on $[a,b]$ maps the Banach space $D_p^n[a,b]$ of absolutely continuous functions $z\colon [a,b] \to R^n$ with the norm

$$||z||_{D_p^n} = ||\dot{z}||_{L_p^n} + ||z(a)||_{R^n}$$

into the Banach space $L_p^n[a,b]$ $(1 < p < \infty)$ of integrable vector functions $\varphi\colon [a,b] \to R^n$ equipped with the norm standard for this space.

According to [8, p. 13], a vector function $z(t) \in D_p^n[a,b]$ absolutely continuous on $[a,b]$ and such that $\dot{z}(t) \in L_p^n[a,b]$ is called a *solution of the differential system (5.123) with delay* if it satisfies system (5.123) almost everywhere on $[a,b]$.

In what follows, we consider equation (5.123) rewritten in the form

$$\dot{z}(t) = A(t)(S_h z)(t) + \varphi(t),$$

where $A(t) = (A_1(t), \ldots, A_k(t))$ is an $n \times N$ matrix $(N = nk)$ formed by $n \times n$

matrices $A_i(t)$, $(S_h z)(t) = \operatorname{col}[(S_{h_1}z)(t), \dots, (S_{h_k}z)(t)]$ is an N-dimensional column vector, and $\varphi(t)$ is an n-dimensional column vector given by relation (5.124). The operator of inner composition S_h maps the space D_p^n into the space $L_p^N = \underbrace{L_p^n \times \ldots \times L_p^n}_{k \text{ times}}$, i.e., $S_h\colon D_p^n \to L_p^N$. The operator $S_{h_i}\colon D_p^n \to L_p^n$ admits the following representation:

$$(S_{h_i}z)(t) = \int_a^b \chi_{h_i}(t,s)\dot{z}(s)ds + \chi_{h_i}(t,a)z(a), \tag{5.125}$$

where $\chi_{h_i}(t,s)$ is the characteristic function of the set

$$\Omega = \{(t,s) \in [a,b] \times [a,b]\colon a \le s \le h_i(t) \le b\}.$$

This means that (see [8, p. 17] or [72])

$$\chi_{h_i}(t,s) = \begin{cases} 1 & \text{for } (t,s) \in \Omega, \\ 0 & \text{for } (t,s) \notin \Omega. \end{cases}$$

It is well known that the inhomogeneous operator equation (5.123) with delay is solvable for any right-hand side $\varphi(t) \in L_p^n[a,b]$ and possesses an n-parameter family of solutions of the form

$$z(t) = X(t)c + \int_a^b K(t,\tau)\varphi(\tau)\,d\tau, \tag{5.126}$$

where $K(t,\tau)$ is an $n \times n$ *Cauchy matrix.* For any fixed τ, this matrix is a solution of the following matrix Cauchy problem:

$$\frac{\partial K(t,\tau)}{\partial t} = A(t)(S_h K(\cdot,\tau))(t), \quad K(\tau,\tau) = I.$$

In what follows, we assume that the matrix $K(t,\tau)$ is defined in the square $[a,b]\times[a,b]$, where $K(t,\tau) \equiv 0$ for $a \le t < \tau \le b$. The finite-dimensional fundamental $n \times n$ matrix of the homogeneous ($\varphi(t) \equiv 0$) equation with delay corresponding to equation (5.123) has the form $X(t) = K(t,a)$. By $(S_h K(\cdot,\tau))(t)$ we denote the $N \times n$ matrix whose columns are obtained by applying the operator of inner composition S_h to the corresponding columns of the $n \times n$ matrix $K(t,\tau)$.

Fredholm Boundary-Value Problems. Consider the following linear inhomogeneous boundary-value problem

$$(Lz)(t) := \dot{z}(t) - A(t)(S_h z)(t) = \varphi(t), \quad t \in [a,b], \tag{5.127}$$

$$lz = \alpha, \tag{5.128}$$

where $L\colon D_p^n[a,b] \to L_p^n[a,b]$ is a bounded linear differential operator with delay and $l = \operatorname{col}[l_1, \dots, l_m]$ is an m-dimensional bounded vector functional (the number m of components of this functional is, generally speaking, not equal to the dimension n of the differential system). The functionals l_i map the space $D_p^n[a,b]$ into the space R and, moreover, $l\colon D_p^n[a,b] \to R^m$, $\alpha \in R^m$. Furthermore, the rows of the matrices $A_i(t)$ and the column vector $\varphi(t)$ belong to the space $L_p^n[a,b]$, i.e., $A_i(t), \varphi(t) \in L_p^n[a,b]$. It is well known (see [8, p. 33] or [88, p. 86]), that the boundary-value problem posed above specifies a Fredholm operator that maps the space $D_p^n[a,b]$ into the space $L_p^n[a,b] \times R^m$.

Our aim is to establish necessary and sufficient conditions for the solvability of the problem posed above and construct a representation of its solution $z(t) \in D_p^n[a,b]$.

The general solution of equation (5.127) has the form (5.126). Thus, substituting relation (5.126) in the boundary conditions (5.128), we arrive at the following algebraic system of equations for $c \in R^n$:

$$Qc = \alpha - l \int_a^b K(\cdot, \tau)\varphi(\tau)\, d\tau \tag{5.129}$$

where $Q = lX(\cdot)$ is an $m \times n$ constant matrix such that $\operatorname{rank} Q = n_1$. System (5.129) enables us to find a constant $c \in R^n$ for which the solution (5.126) of system (5.127) is, at the same time, a solution of the boundary-value problem (5.127), (5.128).

By using the theory of pseudoinverse matrices and orthoprojectors (see, e.g., [124] or Theorem 3.9 in [38, p. 92]), we establish necessary and sufficient conditions for the solvability of the algebraic system (5.129) and the existence of solutions of the boundary-value problem (5.127), (5.128).

Let

$$P_Q\colon R^n \to N(Q) = \ker Q \quad \text{and} \quad P_{Q^*}\colon R^m \to N(Q^*) = \ker Q^* = \operatorname{coker} Q$$

be, respectively, the $n \times n$ and $m \times m$ matrix orthoprojectors to the kernel and cokernel of the matrix Q such that $P_Q^2 = P_Q = P_Q^*$ and $P_{Q^*}^2 = P_{Q^*} = P_{Q^*}^*$, where the symbol $*$ denotes the operation of transposition. Also let Q^+ be the $n \times m$ matrix pseudoinverse to Q in the Moore–Penrose sense.

The algebraic system (5.129) is solvable if and only if its right-hand side belongs to the orthogonal complement $N^{\perp}(Q^*) = R(Q)$ to the subspace $N(Q^*)$. This means that

$$P_{Q^*}\left\{\alpha - l \int_a^b K(\cdot, \tau)\varphi(\tau)\, d\tau\right\} = 0.$$

Since

$$\operatorname{rank} P_Q = n - \operatorname{rank} Q = n - n_1 = r$$

and

$$\operatorname{rank} P_{Q^*} = m - \operatorname{rank} Q^* = m - n_1 = d,$$

we denote by $P_{Q_d^*}$ the $d \times m$ matrix whose rows form a complete system of d linearly independent rows of the $m \times m$ matrix P_{Q^*}. Further, let P_{Q_r} be an $n \times r$ matrix whose columns form a complete system of r linearly independent columns of the $n \times n$ matrix P_Q. In this case, the last condition admits the following representation:

$$P_{Q_d^*}\left\{\alpha - l\int_a^b K(\cdot,\tau)\varphi(\tau)\,d\tau\right\} = 0. \tag{5.130}$$

If condition (5.130) is satisfied, then

$$c = Q^+\left(\alpha - l\int_a^b K(\cdot,\tau)\varphi(\tau)d\tau\right) + P_{Q_r}c_r, \quad P_{Q_r}c_r \in N(Q) \quad \forall c_r \in R^r$$

is a solution of the algebraic system (5.129). Substituting the value of c obtained as a result in (5.126), we arrive at the general solution of the boundary-value problem (5.127), (5.128):

$$\begin{aligned} z(t,c) &= X(t)P_{Q_r}c_r + X(t)Q^+\alpha \\ &\quad + \int_a^b K(t,\tau)\varphi(\tau)d\tau - X(t)Q^+ l\int_a^b K(\cdot,\tau)\varphi(\tau)d\tau. \end{aligned}$$

This solution can be rewritten in the form

$$z(t,c_r) = X_r(t)c_r + (G\varphi)(t) + X(t)Q^+\alpha, \tag{5.131}$$

where $X_r(t) = X(t)P_{Q_r}$ is the fundamental matrix of the homogeneous boundary-value problem

$$\dot z(t) = A(t)(S_h z)(t), \quad lz = 0. \tag{5.132}$$

The operator $(G\varphi)(t)$ is defined as follows:

$$(G\varphi)(t) = \int_a^b K(t,\tau)\varphi(\tau)d\tau - X(t)Q^+ l\int_a^b K(\cdot,\tau)\varphi(\tau)d\tau.$$

It is called the *generalized Green operator* of the boundary-value problem (5.127), (5.128) (see [38, p. 134]).

Thus, we have proved the following statement:

Theorem 5.10. *The following assertions are true for the boundary-value problem (5.127), (5.128).:*

(i) the operator $\Lambda_0\colon D_p^n[a,b] \to L_p^n[a,b] \times R^m$ *given by the formula*

$$(\Lambda_0 z)(t) \stackrel{\text{def}}{=} \operatorname{col}\left[\dot z(t) - A(t)(S_h z)(t),\ lz\right] \tag{5.133}$$

is a Fredholm operator with

$$\operatorname{ind}\Lambda_0 = \dim\ker\Lambda_0 - \dim\ker\Lambda_0^* = \rho = r - d = n - m,$$

where the operator Λ_0^* is adjoint to the operator Λ_0;

(ii) the homogeneous boundary-value problem (5.132) has exactly r linearly independent solutions $X_r(t)c_r$ for any $c_r \in R^r$ ($\dim\ker\Lambda_0 = r = n - \operatorname{rank} Q = n - n_1$);

(iii) the inhomogeneous boundary-value problem (5.127), (5.128) is solvable if and only if $\varphi(t) \in L_p^n[a,b]$ and $\alpha \in R^m$ satisfy condition (5.130) ($\dim\ker\Lambda_0^* = d = m - \operatorname{rank} Q^* = m - n_1$) and possesses the r-parameter family of solutions (5.131).

In what follows, these results are used to establish new conditions required for the existence of solutions of perturbed linear and nonlinear boundary-value problems for equations with delay.

Remark 5.1. If the vector functional l satisfies the relation

$$l\int_a^b K(\cdot,\tau)\varphi(\tau)d\tau = \int_a^b lK(\cdot,\tau)\varphi(\tau)d\tau, \tag{5.134}$$

then the generalized Green operator $(G\varphi)(t)$ takes the form

$$(G\varphi)(t) = \int_a^b G(t,\tau)\varphi(\tau)d\tau.$$

The $n \times n$ matrix $G(t,\tau)$ is the kernel of the integral representation of the operator $(G\varphi)(t)$. This matrix is called the *generalized Green matrix* and has the form

$$G(t,\tau) = K(t,\tau) - X(t)Q^+lK(\cdot,\tau). \tag{5.135}$$

In what follows, without loss of generality, we can assume that condition (5.134) is satisfied.

Indeed, relation (5.134) holds for both periodic ($lz := z(a) - z(b) = 0$) and multipoint $\left(lz = \sum_{i=1}^k M_i z(t_i)\right)$ boundary conditions. Moreover, it is also true under conditions formulated in the form of Riemann–Stieltjes integrals, namely,

$$lz = \int_a^b d\Phi(t)z(t),$$

where $\Phi(t)$ is an $m \times n$ matrix whose components are functions of bounded variation on $[a, b]$. In the last case, we have

$$lK(\cdot, \tau) = \int_{\tau}^{b} d\Phi(t) K(t, \tau)$$

because $K(t, \tau) \equiv 0$ for $t < \tau$.

Remark 5.2. The condition of solvability (5.130) of problem (5.127), (5.128) is specified for a properly chosen initial function ψ. In fact, by using relation (5.122), we can represent condition (5.130) in the form

$$P_{Q_d^*}\left\{\alpha - l\int_a^b K(\cdot, \tau)\left[g(\tau) + \sum_{i=1}^{k} A_i(\tau)\psi^{h_i}(\tau)\right] d\tau\right\} = 0.$$

This enables us to guarantee the solvability of problem (5.127), (5.128) by varying the function ψ. At the same time, for arbitrary inhomogeneities $g(t) \in L_p^n[a, b]$ and $\alpha \in R^m$ and initial vector function $\psi\colon R^1 \setminus [a, b] \to R^n$, the condition of solvability (5.130) of problem (5.127), (5.128) is not satisfied. Hence, it is necessary to be able to regularize boundary-value problems if they are solvable not everywhere.

Perturbed Boundary-Value Problems.[1] In Banach spaces, we consider the problem of existence and construction of solutions $z\colon [a, b] \to R^n$ for systems of ordinary differential equations with small parameter ε and finitely many measurable delays of the argument of the form

$$\dot{z}(t) = \sum_{i=1}^{k} A_i(t) z(h_i(t)) + \varepsilon \sum_{i=1}^{k} B_i(t) z(h_i(t)) + g(t), \tag{5.136}$$

$$t \in [a, b], \quad h_i(t) \le t.$$

These systems are equipped with initial conditions

$$z(s) = \psi(s), \quad s < a < b,$$

and boundary conditions

$$lz = \alpha, \quad \alpha \in R^m. \tag{5.137}$$

It is also assumed that the unperturbed problem ($\varepsilon = 0$) is unsolvable for arbitrary inhomogeneities $g(t)$ from a space introduced in what follows, $\alpha \in R^m$, and initial functions $\psi\colon R^1 \setminus [a, b] \to R^n$.

---

[1] A part of this subsection was completed together with Prof. M. Grammatikopoulos [27, 28] during the visit of A. Boichuk to the University of Ioannina (Greece) under the Grant of NATO Science Fellowship No. D00850.

Our aim is to establish conditions that should be imposed on the perturbed coefficients $B_i(t)$ and delays $h_i(t)$ to guarantee that the boundary-value problem (5.136), (5.137) possesses a family of solutions or a single solution. We also propose an algorithm for the construction of the required solutions.

Thus, we consider the perturbed inhomogeneous linear boundary-value problem (5.136), (5.137). In view of relations (5.121) and (5.122), this problem can be rewritten in the form

$$\dot{z}(t) = A(t)(S_h z)(t) + \varepsilon B(t)(S_h z)(t) + \varphi(t), \quad lz = \alpha, \quad t \in [a, b], \qquad (5.138)$$

where, as earlier, $A(t) = (A_1(t), \ldots, A_k(t))$ and $B(t) = (B_1(t), \ldots, B_k(t))$ are $n \times N$ matrices ($N = nk$) formed of the $n \times n$ matrices $A_i(t) \in L_p^n[a, b]$ and $B_i(t) \in L_p^n[a, b]$, respectively.

Assume that the generating boundary-value problem

$$\dot{z}(t) = A(t)(S_h z)(t) + \varphi(t), \quad lz = \alpha, \qquad (5.139)$$

which follows from problem (5.138) for $\varepsilon = 0$, is unsolvable for arbitrary inhomogeneities $\varphi(t) \in L_p^n[a, b]$ and $\alpha \in R^m$. In this case, Theorem 5.10 implies that the criterion of solvability (5.130) is not true for problem (5.139) because the inhomogeneities arc arbitrary. Thus, it is necessary to answer the following questions:

Is it possible to make problem (5.139) solvable by introducing linear perturbations? If the answer to the first question is positive, then what kind of perturbations $B_i(t)$ and delays $h_i(t)$ should be used to make the boundary-value problem (5.138) everywhere solvable?

We can answer these questions with the help of the $d \times r$ matrix

$$B_0 = \int_a^b H(\tau) B(\tau)(S_h X_r)(\tau)\, d\tau, \quad H(\tau) = P_{Q_d^*} l K(\cdot, \tau),$$

constructed by using the coefficients of problem (5.138). The method developed in [150] enables us to establish conditions under which the solutions of the boundary-value problem (5.138) can be represented in the form of series (in powers of the small parameter ε) containing finitely many terms with negative powers of ε.

In what follows, we prove an assertion that enables us to solve the problem posed above. In order to formulate this result, we recall that P_{B_0} is the $r \times r$ matrix orthoprojector projecting the space R^r onto the null space $N(B_0)$ of the $d \times r$ matrix B_0 and $P_{B_0^*}$ is the $d \times d$ matrix orthoprojector projecting the space R^d onto the null space $N(B_0^*)$ of the $r \times d$ matrix $B_0^* = B_0^t$. We can now formulate the following statement:

Lemma 5.1. *Consider the boundary-value problem (5.138) and assume that the corresponding generating boundary-value problem (5.139) is unsolvable for arbitrary inhomogeneities* $\varphi(t) \in L_p^n[a,b]$ *and* $\alpha \in R^m$.

If the equivalent relations

$$P_{B_0^*} = 0 \iff \operatorname{rank} B_0 = d \tag{5.140}$$

are true, then, for any $\varphi(t) \in L_p^n[a,b]$ *and* $\alpha \in R^m$, *the boundary-value problem (5.138) has at least one solution in the form of a series*

$$z(t,\varepsilon) = \sum_{i=-1}^{\infty} \varepsilon^i z_i(t) \tag{5.141}$$

convergent for fixed $\varepsilon \in (0, \varepsilon_*]$, *where* ε_* *is a proper constant characterizing the range of convergence of series (5.141).*

Proof. We substitute relation (5.141) in problem (5.138) and equate the coefficients of the same powers of ε. As a result, for ε^{-1}, we arrive at the following homogeneous boundary-value problem for the quantity $z_{-1}(t)$:

$$\dot{z}_{-1} = A(t)(S_h z_{-1})(t), \quad l z_{-1} = 0. \tag{5.142}$$

According to Theorem 5.10, the homogeneous boundary-value problem (5.142) has the r-parameter $(r = n - n_1)$ family of solutions $z_{-1}(t, c_{-1}) = X_r(t)c_{-1}$, where $c_{-1} \in R^r$ is an r-dimensional column vector determined from the condition of solvability of the problem for $z_0(t)$.

For ε^0, we get the following boundary-value problem for the quantity $z_0(t)$:

$$\dot{z}_0 = A(t)z_0 + B(t)(S_h z_{-1})(t) + \varphi(t), \quad l z_0 = \alpha. \tag{5.143}$$

As follows from Theorem 5.10, the criterion of solvability of problem (5.143) has the form

$$P_{Q_d^*}\alpha - \int_a^b H(\tau)\{\varphi(\tau) + B(\tau)(S_h X_r)(\tau)c_{-1}\}d\tau = 0.$$

This condition yields the following algebraic system for $c_{-1} \in R^r$:

$$B_0 c_{-1} = P_{Q_d^*}\alpha - \int_a^b H(\tau)\varphi(\tau)d\tau, \tag{5.144}$$

where

$$B_0 = \int_a^b H(\tau)B(\tau)(S_h X_r)(\tau)\,d\tau, \quad H(\tau) = P_{Q_d^*} l K(\cdot,\tau).$$

The last system is solvable for any $\varphi(t) \in L_p^n[a,b]$ and $\alpha \in R^m$ if and only if $P_{B_0^*} = 0$. In this case, system (5.144) is solvable with respect to $c_{-1} \in R^r$ to within an arbitrary constant vector $P_{B_0}c$ ($\forall c \in R^r$) from the null space of the matrix B_0 and

$$c_{-1} = B_0^+ \left\{ P_{Q_d^*}\alpha - \int_a^b H(\tau)\varphi(\tau)\, d\tau \right\} + P_{B_0}c.$$

This solution can be rewritten in the form

$$c_{-1} = \bar{c}_{-1} + P_{B_\rho}c_\rho \quad \forall\, c_\rho \in R^\rho, \tag{5.145}$$

where

$$\bar{c}_{-1} = B_0^+ \left\{ P_{Q_d^*}\alpha - \int_a^b H(\tau)\varphi(\tau)\, d\tau \right\}, \tag{5.146}$$

P_{B_ρ} is an $r \times \rho$ matrix whose columns form a complete system of ρ linearly independent columns of the $r \times r$ matrix P_{B_0}, and

$$\rho = \operatorname{rank} P_{B_0} = r - \operatorname{rank} B_0 = r - d = n - m.$$

Hence, for the solutions of problem (5.142), we get

$$z_{-1}(t, c_\rho) = \bar{z}_{-1}(t, \bar{c}_{-1}) + X_r(t)P_{B_\rho}c_\rho \quad \forall\, c_\rho \in R^\rho$$

and

$$\bar{z}_{-1}(t, \bar{c}_{-1}) = X_r(t)\bar{c}_{-1}.$$

In the case where relations (5.140) are true, the boundary-value problem (5.143) possesses the r-parameter family of solutions

$$z_0(t, c_0) = X_r(t)c_0 + X(t)Q^+\alpha + \int_a^b G(t,\tau)\big[\varphi(\tau) + B(\tau)S_h(\bar{z}_{-1}(\cdot,\bar{c}_{-1}) + X_r(\cdot)P_{B_\rho}c_\rho)(\tau)\big]d\tau, \tag{5.147}$$

where c_0 is an r-dimensional constant vector determined in the next stage of the process from the condition of solvability of the boundary-value problem for $z_1(t)$.

For ε^1, we get the following boundary-value problem for the quantity $z_1(t)$:

$$\dot{z}_1 = A(t)z_1 + B(t)(S_h z_0)(t), \quad l z_1 = 0. \tag{5.148}$$

The criterion of solvability of problem (5.148) takes the form

$$\int_a^b H(\tau)B(\tau)S_h\Big\{ X_r(\star)c_0 + X(\star)Q^+\alpha + \int_a^b G(\star, s)\big[\varphi(s) + B(s)S_h(\bar{z}_{-1}(\cdot,\bar{c}_{-1}) + X_r(\cdot)P_{B_\rho}c_\rho)(s)\big]ds \Big\}(\tau)d\tau = 0,$$

or, equivalently,

$$B_0 c_0 = -\int_a^b H(\tau)B(\tau)S_h\Big\{X(\star)Q^+\alpha + \int_a^b G(\star, s)\Big[\varphi(s) + B(s)S_h(\bar{z}_{-1}(\cdot, \bar{c}_{-1}) + X_r(\cdot)P_{B_\rho}c_\rho)(s)\Big]ds\Big\}(\tau)d\tau. \quad (5.149)$$

The algebraic system (5.149) has the following family of solutions:

$$\begin{aligned} c_0 = &-B_0^+ \int_a^b H(\tau)B(\tau)S_h\Big\{X(\star)Q^+\alpha \\ &+ \int_a^b G(\star, s)[\varphi(s) + B(s)(S_h\bar{z}_{-1}(\cdot, \bar{c}_{-1}))(s)]ds\Big\}(\tau)d\tau \\ &+ \Big[I_r - B_0^+ \int_a^b H(\tau)B(\tau)S_h\Big\{\int_a^b G(\star, s)B(s)(S_h X_r(\cdot))(s)ds\Big\}(\tau)d\tau\Big]P_{B_\rho}c_\rho \\ = &\ \bar{c}_0 + [\cdot, \cdot, \cdot]P_{B_\rho}c_\rho, \end{aligned} \quad (5.150)$$

where

$$\bar{c}_0 = -B_0^+ \int_a^b H(\tau)B(\tau)S_h\Big\{X(\star)Q^+\alpha + \int_a^b G(\star, s)[\varphi(s) + B(s)(S_h\bar{z}_{-1}(\cdot, \bar{c}_{-1}))(s)]ds\Big\}(\tau)d\tau, \quad (5.151)$$

and

$$[\cdot, \cdot, \cdot] = I_r - B_0^+ \int_a^b H(\tau)B(\tau)S_h\Big\{\int_a^b G(\star, s)B(s)(S_h X_r(\cdot))(s)ds\Big\}(\tau)d\tau.$$

Thus, for the ρ-parameter family of solutions of problem (5.142), we can write

$$z_0(t, c_\rho) = \bar{z}_0(t, \bar{c}_0) + \bar{X}_0(t)P_{B_\rho}c_\rho \quad \forall\ c_\rho \in R^\rho,$$

where

$$\bar{z}_0(t, \bar{c}_0) = X_r(t)\bar{c}_0 + X(t)Q^+\alpha + \int_a^b G(t, \tau)[\varphi(\tau) + B(\tau)(S_h\bar{z}_{-1}(\cdot, \bar{c}_{-1}))(\tau)]d\tau \quad (5.152)$$

and

$$\bar{X}_0(t,\bar{c}_0) = X_r(t)\Bigg[I_r - B_0^+ \int_a^b H(\tau)B(\tau) \times S_h\Big\{\int_a^b G(\star,s)B(s)(S_hX_r(\cdot))(s)ds\Big\}(\tau)d\tau\Bigg] + \int_a^b G(t,\tau)B(\tau)(S_hX_r(\cdot))(\tau)d\tau.$$

Further, in the case where relations (5.140) are satisfied, the boundary-value problem (5.148) has the following r-parameter family of solutions:

$$z_1(t,c_1) = X_r(t)c_1 + \int_a^b G(t,\tau)B(\tau)S_h(\bar{z}_0(\cdot,\bar{c}_0) + \bar{X}_0(\cdot)P_{B_\rho}c_\rho)(\tau)d\tau, \quad (5.153)$$

where c_1 is an r-dimensional constant vector determined in the next stage of the process from the condition of solvability of the boundary-value problem for $z_2(t)$, namely,

$$\dot{z}_2 = A(t)z_2 + B(t)(S_hz_1)(t), \quad lz_2 = 0. \quad (5.154)$$

The criterion of solvability of problem (5.154) has the form

$$\int_a^b H(\tau)B(\tau)S_h\Big\{X_r(\star)c_1 + \int_a^b G(\star,s)B(s)S_h(\bar{z}_0(\cdot,\bar{c}_0) + \bar{X}_0(\cdot)P_{B_\rho}c_\rho)(s)ds\Big\}(\tau)d\tau = 0,$$

or

$$B_0c_1 = -\int_a^b H(\tau)B(\tau) \times S_h\Big\{\int_a^b G(\star,s)B(s)S_h(\bar{z}_0(\cdot,\bar{c}_0) + \bar{X}_0(\cdot)P_{B_\rho}c_\rho)(s)ds\Big\}(\tau)d\tau.$$

Under condition (5.140), the last equation possesses the ρ-parameter family of solutions

$$c_1 = \bar{c}_1 + \{\cdot,\cdot,\cdot\},$$

where

$$\bar{c}_1 = -B_0^+ \int_a^b H(\tau)B(\tau)S_h\Big\{ \int_a^b G(\star, s)B(s)(S_h\bar{z}_0(\cdot, \bar{c}_0))(s)ds \Big\}(\tau)d\tau$$

and

$$\{\cdot,\cdot,\cdot\} = \Big\{ I_r - B_0^+ \int_a^b H(\tau)B(\tau) \\ \times S_h\Big\{ \int_a^b G(\star, s)B(s)(S_h\bar{X}_0(\cdot))(s)ds \Big\}(\tau)d\tau \Big\} P_{B_\rho} c_\rho .$$

Hence, for the quantity $z_1(t, c_1) = z_1(t, c_\rho)$, we obtain

$$z_1(t, c_\rho) = \bar{z}_1(t, \bar{c}_1) + \bar{X}_1(t)P_{B_\rho}c_\rho \quad \forall\ c_\rho \in R^\rho,$$

where

$$\bar{z}_1(t, \bar{c}_1) = X_r(t)\bar{c}_1 + \int_a^b G(t, \tau)B(\tau)(S_h\bar{z}_0(\cdot, \bar{c}_0))(\tau)d\tau \tag{5.155}$$

and

$$\bar{X}_1(t) = X_r(t)\Big[I_r - B_0^+ \int_a^b H(\tau)B(\tau) \\ \times S_h\Big\{ \int_a^b G(\star, s)B(s)(S_h\bar{X}_0(\cdot))(s)ds \Big\}(\tau)d\tau \Big] \\ + \int_a^b G(t, \tau)B(\tau)(S_h\bar{X}_0(\cdot))(\tau)d\tau.$$

If we continue this process and assume that relations (5.140) are true, then, by induction, we conclude that the coefficients $z_i(t, c_i) = z_i(t, c_\rho)$ of series (5.141) can be found from the corresponding boundary-value problems as follows:

$$z_i(t, c_\rho) = \bar{z}_i(t, \bar{c}_i) + \bar{X}_i(t)P_{B_\rho}c_\rho \quad \forall\ c_\rho \in R^\rho, \tag{5.156}$$

where

$$\bar{z}_i(t, \bar{c}_i) = X_r(t)\bar{c}_1 + \int_a^b G(t, \tau)B(\tau)S_h\bar{z}_{i-1}(\cdot, \bar{c}_{i-1})(\tau)d\tau,$$

$$\bar{c}_i = -B_0^+ \int\limits_a^b H(\tau)B(\tau)S_h\left\{\int\limits_a^b G(\star, s)B(s)S_h\bar{z}_{i-1}(\cdot, \bar{c}_{i-1})(s)ds\right\}(\tau)d\tau$$

$$(i = 1, 2, \ldots),$$

$$\bar{X}_i(t) = X_r(t)\left[I_r - B_0^+ \int\limits_a^b H(\tau)B(\tau)\right.$$

$$\left.\times S_h\left\{\int\limits_a^b G(\star, s)B(s)(S_h\bar{X}_{i-1}(\cdot))(s)ds\right\}(\tau)d\tau\right]$$

$$+ \int\limits_a^b G(t,\tau)B(\tau)(S_h\bar{X}_{i-1}(\cdot))(\tau)d\tau \qquad (i = 0, 1, 2, \ldots),$$

and

$$\bar{X}_{-1}(t) = X_r(t).$$

Since the fact that series (5.141) is convergent can be proved by using traditional methods of majorization, the proof of Lemma 5.1 is completed.

Lemma 5.1 enables us to make the following conclusions:

The boundary-value problem (5.138) specifies the following bounded operator:

$$(\Lambda_\varepsilon z)(t) \stackrel{\text{def}}{=} \mathrm{col}\,[\dot{z}(t) - A(t)(S_h z)(t) - \varepsilon B(t)(S_h z)(t),\ lz] \tag{5.157}$$

acting from the space $D_p^n[a, b]$ into the space $L_p^n[a, b] \times R^m$, $1 < p < \infty$. Under condition (5.140), problem (5.138) is always solvable in the analyzed Banach spaces. This means that the image of the operator Λ_ε coincides with the entire space $L_p^n[a, b] \times R^m$, i.e., $\mathrm{Im}\,\Lambda_\varepsilon = L_p^n[a, b] \times R^m$. Therefore, Λ_ε is a normally solvable operator (see [80, 88]), and the boundary-value problem conjugate to the homogeneous problem

$$\dot{z}(t) = A(t)(S_h z)(t) + \varepsilon B(t)(S_h z)(t), \quad lz = 0 \in R^m \tag{5.158}$$

possesses only trivial solutions, i.e., $\dim\ker\Lambda_\varepsilon^* = 0$, $\varepsilon \neq 0$, where Λ_ε^* is the operator adjoint to the operator Λ_ε in the corresponding spaces. Note that, in our problem, it is not necessary to construct the conjugate problem. For the unperturbed boundary-value problem (5.127), (5.128), the corresponding construction can be found in [8, p. 36].

As shown in the proof of Lemma 5.1, $\dim\ker\Lambda_\varepsilon = \rho = r - d$. This fact, together with the already established equality $\dim\ker\Lambda_\varepsilon^* = 0$, means that the normally resolvable operator Λ_ε is a Fredholm operator. Thus, it is easy to see that the well-known fact from the theory of operators (see [88, p. 86] or [80]) concerning preservation of the index of the Fredholm operator Λ_0 (5.133) under small perturbations is true for the differential operator (5.157) with delayed arguments. Indeed, by virtue of Theorem 5.10, we have

$$\dim\ker\Lambda_0 = r, \quad \dim\ker\Lambda_0^* = d.$$

Further, in view of Lemma 5.1, we get

$$\dim\ker\Lambda_\varepsilon = r - d \qquad \text{and} \qquad \dim\ker\Lambda_\varepsilon^* = 0, \quad \varepsilon \neq 0,$$

and, therefore,

$$\operatorname{ind}\Lambda_0 = \operatorname{ind}\Lambda_\varepsilon.$$

This means that we have proved the following statement:

Theorem 5.11. *Consider the boundary-value problem*

$$\dot{z}(t) = A(t)(S_h z)(t) + \varepsilon B(t)(S_h z)(t) + \varphi(t), \quad lz = \alpha \in R^m. \tag{5.159}$$

Assume that the corresponding generating boundary-value problem (5.139) is unsolvable for arbitrary inhomogeneities $\varphi(t) \in L_p^n[a,b]$ *and* $\alpha \in R^m$.

If the condition

$$\operatorname{rank}\left[B_0 = \int_a^b H(\tau)B(\tau)(S_h X_r)(\tau)\,d\tau\right] = d = m - n_1 \tag{5.160}$$

$$(\operatorname{rank} Q = n_1)$$

is satisfied, then the following assertions are true for sufficiently small $\varepsilon \in (0, \varepsilon_*]$:

(i) the operator $\Lambda_\varepsilon \colon D_p^n[a,b] \to L_p^n[a,b] \times R^m$ $(1 < p < \infty)$ *given by relation (5.157) is a Fredholm operator with*

$$\operatorname{ind}\Lambda_\varepsilon = \dim\ker\Lambda_\varepsilon - \dim\ker\Lambda_\varepsilon^* = \rho = r - d = n - m,$$

$$\operatorname{ind}\Lambda_0 = \dim\ker\Lambda_0 - \dim\ker\Lambda_0^* = \rho = r - d = n - m,$$

where Λ_ε^* *is the operator adjoint to the operator* Λ_ε $(\dim\ker\Lambda_0 = r$ *and* $\dim\ker\Lambda_0^* = d)$;

(ii) the homogeneous boundary-value problem (5.158) possesses the ρ*-parameter family of solutions*

$$z_0(t,\varepsilon,c_\rho) = \sum_{i=-1}^{\infty} \varepsilon^i \bar{X}_i(t) P_{B_\rho} c_\rho \quad \forall\, c_\rho \in R^\rho \tag{5.161}$$

$$(\rho = \dim\ker\Lambda_\varepsilon)$$

such that

$$z_0(\cdot,\varepsilon,c_\rho) \in D_p^n[a,b], \quad \dot{z}_0(\cdot,\varepsilon,c_\rho) \in L_p^n[a,b], \quad z_0(t,\cdot,c_\rho) \in C(0,\varepsilon_*];$$

(iii) the boundary-value problem conjugate to (5.158) has only trivial solutions $(\dim\ker\Lambda_\varepsilon^* = 0,\ \varepsilon \neq 0)$;

(iv) for any $\varphi(t) \in L_p^n[a,b]$ *and* $\alpha \in R^m$, *the boundary-value problem (5.159) possesses a* ρ*-parameter set of solutions* $z(t,\varepsilon) = z(t,\varepsilon,c_\rho)$ *such that*

$$z(\cdot,\varepsilon,c_\rho) \in D_p^n[a,b], \quad \dot{z}(\cdot,\varepsilon,c_\rho) \in L_p^n[a,b], \quad z(t,\cdot,c_\rho) \in C(0,\varepsilon_*],$$

given by a series

$$z(t,\varepsilon,c_\rho) = \sum_{i=-1}^{\infty} \varepsilon^i [\bar{z}_i(t,\bar{c}_i) + \bar{X}_i(t) P_{B_\rho} c_\rho] \quad \forall\, c_\rho \in R^\rho \tag{5.162}$$

convergent for fixed $\varepsilon \in (0,\varepsilon_*]$, *where* ε_* *is the same constant as in Lemma 5.1 and the coefficients* $\bar{z}_i(t,\bar{c}_i)$, $\bar{c}_i$, *and* $\bar{X}_i(t)$ *can be found from relations (5.156).*

In the case where the number m of boundary conditions is equal to the dimension n of the differential system (5.159), condition (5.160) yields

$$\operatorname{rank}\left[B_0 = \int_a^b H(\tau)B(\tau)(S_h X_r)(\tau)\,d\tau\right] = r = d$$

and we arrive at the following corollary of Theorem 5.11:

Corollary 5.4. *For a boundary-value problem*

$$\dot{z}(t) = A(t)(S_h z)(t) + \varepsilon B(t)(S_h z)(t) + \varphi(t), \quad lz = \alpha \in R^n, \tag{5.163}$$

assume that the corresponding generating boundary-value problem (5.139) is unsolvable for arbitrary inhomogeneities $\varphi(t) \in L_p^n[a,b]$ *and* $\alpha \in R^n$.

If the condition

$$\det B_0 \neq 0 \tag{5.164}$$

is satisfied, then the following assertions are true:

(i) the operator $\Lambda_\varepsilon \colon D_p^n[a,b] \to L_p^n[a,b] \times R^n$ *given by the formula*

$$(\Lambda_\varepsilon z)(t) \overset{\text{def}}{=} \operatorname{col}[\dot{z}(t) - A(t)(S_h z)(t) - \varepsilon B(t)(S_h z)(t),\ lz] \tag{5.165}$$

is a Fredholm operator of index zero with

$$\operatorname{ind}\Lambda_\varepsilon = \dim\ker\Lambda_\varepsilon - \dim\ker\Lambda_\varepsilon^* = 0,$$

$$\operatorname{ind}\Lambda_0 = \dim\ker\Lambda_0 - \dim\ker\Lambda_0^* = 0$$

$(\dim\ker\Lambda_0 = \dim\ker\Lambda_0^* = r = d)$;

(ii) the homogeneous boundary-value problem

$$\dot{z}(t) = A(t)(S_h z)(t) + \varepsilon B(t)(S_h z)(t), \quad lz = 0 \in R^n \tag{5.166}$$

has only trivial solutions $(\dim\ker\Lambda_\varepsilon = 0,\ \varepsilon \neq 0)$;

(iii) the boundary-value problem conjugate to (5.166) has only trivial solutions $(\dim\ker\Lambda_\varepsilon^* = 0,\ \varepsilon \neq 0)$;

(iv) for any $\varphi(t) \in L_p^n[a,b]$ *and* $\alpha \in R^n$, *the boundary-value problem (5.163) possesses a unique solution* $z(t,\varepsilon)$ *such that*

$$z(\cdot,\varepsilon) \in D_p^n[a,b], \quad \dot z(\cdot,\varepsilon) \in L_p^n[a,b], \quad z(t,\cdot) \in C(0,\varepsilon_*],$$

given by a series

$$z(t,\varepsilon) = \sum_{i=-1}^{\infty} \varepsilon^i \bar z_i(t,\bar c_i) \tag{5.167}$$

convergent for fixed $\varepsilon \in (0,\varepsilon_*]$, *where* ε_* *is the same constant as in Lemma 5.1 and the coefficients* $\bar z_i(t,\bar c_i)$ *and* $\bar c_i$ *can be found from (5.156).*

Remark 5.3. In the case where the effect of delay is absent $(h_i(t) = t,\ i = 1,\dots,k)$, and $m = n$, problem (5.136), (5.137) was studied in [38, p. 252]. Moreover, in the case where the effect of delay is absent $(h_i(t) = t,\ i = 1,\dots,k)$ and, in addition, $A_i(t) = 0$, the periodic $(lz := z(a) - z(b) = 0)$ boundary-value problem (5.136), (5.137) was studied in [82].

Remark 5.4. If condition (5.160) is not satisfied, then, in order to obtain sufficient conditions for the existence of solutions of the boundary-value problem (5.159) for arbitrary inhomogeneities $\varphi(t) \in L_p^n[a,b]$ and $\alpha \in R^m$, the solution $z(t,\varepsilon)$ of problem (5.159) is constructed in the form of series (5.141) with $i \leq -2$.

Remark 5.5. If

$$\operatorname{rank}\left[B_0 = \int_a^b H(\tau)B(\tau)(S_h X_r)(\tau)\,d\tau\right] = d,$$

then the nonlinear boundary-value problem with measurable delays $h_i(t)$

$$\dot z(t) = \sum_{i=1}^{k} A_i(t)z(h_i(t)) + g(t) + \varepsilon\sum_{i=1}^{k} B_i(t)z(h_i(t)) + \varepsilon\sum_{i=1}^{k} R_i(z(h_i(t)),t,\varepsilon),$$

$$z(s) = \psi(s) \quad \text{for} \quad s < a, \quad lz = \alpha \in R^m, \quad t \in [a,b],$$

has at least one solution $z(t,\varepsilon)$ such that

$$z(\cdot,\varepsilon) \in D_p^n[a,b], \quad \dot z(\cdot,\varepsilon) \in L_p^n[a,b],$$

where

$$A_i(t),\ B_i(t),\ g(t),\ R_i(z,\cdot,\varepsilon) \in L_p^n[a,b], \quad R_i(z,t,\varepsilon) = o(z^2).$$

Applications. Example 1. Consider a linear boundary-value problem for a differential equation with delay

$$\dot{z}(t) = \varepsilon \sum_{i=1}^{k} B_i(t) z(h_i(t)) + g(t), \quad t \in [0, T], \tag{5.168}$$

$$z(s) = \psi(s) \quad \text{for} \quad s < 0, \quad z(0) = z(T),$$

where $B_i(t)$ are $n \times n$ matrices, $B_i(t), g(t) \in L_p^n[0, T]$, $\psi \colon R^1 \setminus [a, b] \to R^n$, and $h_i(t)$ are measurable functions. By using the symbols S_{h_i} and ψ^{h_i} (see (5.121), (5.122)), we arrive at the following operator system:

$$\dot{z}(t) = \varepsilon B(t)(S_h z)(t) + \varphi(t), \quad lz = z(0) - z(T) = 0,$$

where $B(t) = (B_1(t), \ldots, B_k(t))$ is an $n \times N$ matrix $(N = nk)$ and

$$\varphi(t) = g(t) + \sum_{i=1}^{k} B_i(t) \psi^{h_i}(t) \in L_p^n[0, T].$$

It is easy to see that

$$X(t) = E, \quad \dot{z}(t) = 0, \quad lX(\cdot) = Q = 0,$$

$$P_Q = P_{Q^*} = E \quad (r = n, \quad d = m = n),$$

$$K(t, \tau) = \begin{cases} E, & 0 \le \tau \le t \le T, \\ 0, & \tau > t, \end{cases}$$

$$lK(\cdot, \tau) = K(0, \tau) - K(T, \tau) = -E,$$

$$H(\tau) = P_{Q^*} lK(\cdot, \tau) = -E.$$

By using representation (5.125), we obtain

$$(S_{h_i} E)(t) = \chi_{h_i}(t, 0) E = E \begin{cases} 1 & \text{for} \quad 0 \le h_i(t) \le T, \\ 0 & \text{for} \quad h_i(t) < 0, \end{cases}$$

and

$$B_0 = \int_0^T H(\tau) B(\tau) (S_h E)(\tau) d\tau$$

$$= -\int_0^T \sum_{i=1}^{k} B_i(\tau) (S_{h_i} E)(\tau) d\tau = -\sum_{i=1}^{k} \int_0^T B_i(\tau) \chi_{h_i}(\tau, 0) d\tau,$$

where B_0 is an $n \times n$ matrix.

If $\det B_0 \neq 0$, then problem (5.168) possesses a unique solution $z(t,\varepsilon)$ such that

$$z(\cdot,\varepsilon) \in D_p^n[0,T], \quad \dot{z}(\cdot,\varepsilon) \in L_p^n[0,T], \quad \text{and} \quad z(t,\cdot) \in C(0,\varepsilon_*]$$

for any $g(t) \in L_p^n[0,T]$ and $\varphi(t) \in L_p^n[0,T]$ and any measurable delays $h_i(t)$.

If, e.g., $h_i(t) = t - \Delta_i$, where $0 < \Delta_i = \text{const} < T$ $(i = 1,\dots,k)$, then

$$\chi_{h_i}(t,0) = \begin{cases} 1 & \text{for} \quad 0 \le h_i(t) = t - \Delta_i \le T, \\ 0 & \text{for} \quad h_i(t) = t - \Delta_i < 0, \end{cases}$$
$$= \begin{cases} 1 & \text{for} \quad \Delta_i \le t \le T + \Delta_i, \\ 0 & \text{for} \quad t < \Delta_i. \end{cases}$$

Hence, the $n \times n$ matrix B_0 can be rewritten as

$$B_0 = \int_0^T H(\tau) \sum_{i=1}^k B_i(\tau)\chi_{h_i}(\tau,0)d\tau$$
$$= -\sum_{i=1}^k \int_0^T B_i(\tau)\chi_{h_i}(\tau,0)d\tau = -\sum_{i=1}^k \int_{\Delta_i}^T B_i(\tau)d\tau,$$

and the condition of solvability of the boundary-value problem (5.168) takes the form

$$\det\Big[B_0 = -\sum_{i=1}^k \int_{\Delta_i}^T B_i(\tau)d\tau\Big] \neq 0.$$

In the case where the effect of delay is absent $(\Delta_i = 0,\ i = 1,\dots,k)$, this condition coincides with the condition presented in [19, 38, 82].

Example 2. Consider a linear boundary-value problem for the following differential equation with delay:

$$\dot{z}(t) = \varepsilon \sum_{i=1}^k B_i(t) z(h_i(t)) + g(t), \quad t \in [0,T],$$
$$z(s) = \psi(s) \quad \text{for} \quad s < 0, \tag{5.169}$$
$$lz := [1, \underbrace{0 \cdots 0}_{(n-1)\,\text{times}}]z(0) - [1, \underbrace{0 \cdots 0}_{(n-1)\,\text{times}}]z(T) = \alpha \in R \quad (m = 1).$$

By using the symbols S_{h_i} and ψ^{h_i}, we arrive at the following boundary-value problem for the corresponding operator system:

$$\dot{z}(t) = \varepsilon B(t)(S_h z)(t) + \varphi(t),$$
$$lz := [1, \underbrace{0 \cdots 0}_{(n-1)\,\text{times}}]z(0) - [1, \underbrace{0 \cdots 0}_{(n-1)\,\text{times}}]z(T) = \alpha \in R \quad (m = 1),$$

where $B(t) = (B_1(t), \dots, B_k(t))$ is an $n \times N$ matrix $(N = nk)$ and

$$\varphi(t) = g(t) + \sum_{i=1}^{k} B_i(t)\psi^{h_i}(t) \in L_p^n[0,T].$$

It is easy to see that

$$X(t) = E, \quad \dot{z}(t) = 0, \quad lX(\cdot) = Q = [\underbrace{0 \cdots 0}_{n \text{ times}}],$$

$$P_Q = E, \quad P_{Q^*} = 1$$

$$(\operatorname{rank} Q = n_1 = 0, \quad r = n, \quad d = m - n_1 = 1),$$

$$K(t,\tau) = \begin{cases} E, & 0 \le \tau \le t \le T, \\ 0, & \tau > t, \end{cases}$$

$$lK(\cdot,\tau) = [1, \underbrace{0 \cdots 0}_{(n-1) \text{ times}}]K(0,\tau) - [1, \underbrace{0 \cdots 0}_{(n-1) \text{ times}}]K(T,\tau) = -[1, \underbrace{0 \cdots 0}_{(n-1) \text{ times}}],$$

$$H(\tau) = P_{Q^*} lK(\cdot,\tau) = -[1, \underbrace{0 \cdots 0}_{(n-1) \text{ times}}].$$

By using representation (5.125), we obtain

$$(S_{h_i}E)(t) = \chi_{h_i}(t,0)E = E \begin{cases} 1 & \text{for} \quad 0 \le h_i(t) \le T, \\ 0 & \text{for} \quad h_i(t) < 0. \end{cases}$$

In order to establish the conditions of solvability of problem (5.169), it suffices to consider the first row of the matrices

$$B_i(t) = \begin{pmatrix} b_{11}^{(i)}(t) & b_{12}^{(i)}(t) & * & * & * & b_{1n}^{(i)}(t) \\ * & * & * & * & * & * \\ * & * & * & * & * & * \end{pmatrix} \qquad (i = 1, \dots, k).$$

Indeed, the $1 \times n$ matrix has the form

$$\begin{aligned} B_0 &= \int_0^T H(\tau)B(\tau)(S_hE)(\tau)d\tau \\ &= \int_0^T H(\tau)\sum_{i=1}^{k} B_i(\tau)(S_{h_i}E)(\tau)d\tau = \int_0^T H(\tau)\sum_{i=1}^{k} B_i(\tau)\chi_{h_i}(\tau,0)d\tau \\ &= -\Bigg[\sum_{i=1}^{k}\int_0^T b_{11}^{(i)}(\tau)\chi_{h_i}(\tau,0)d\tau, \\ &\qquad \sum_{i=1}^{k}\int_0^T b_{12}^{(i)}(\tau)\chi_{h_i}(\tau,0)d\tau, \ \dots, \sum_{i=1}^{k}\int_0^T b_{1n}^{(i)}(\tau)\chi_{h_i}(\tau,0)d\tau\Bigg]. \end{aligned}$$

If one of the inequalities

$$\sum_{i=1}^{k} \int_{0}^{T} b_{1j}^{(i)}(\tau)\chi_{h_i}(\tau, 0)d\tau \neq 0 \quad (j = 1, \dots, n)$$

is true, then $\operatorname{rank} B_0 = d = 1$ and, for any $\varphi(t) \in L_p^n[a, b]$ and $\alpha \in R$ and any measurable delays $h_i(t)$, the boundary-value problem (5.169) possesses a $\rho = (n-1)$-parameter set of solutions $z(t, c_\rho, \varepsilon)$ such that

$$z(\cdot, c_\rho, \varepsilon) \in D_p^n[0, T], \quad \dot{z}(\cdot, c_\rho, \varepsilon) \in L_p^n[0, T], \quad z(t, c_\rho, \cdot) \in C(0, \varepsilon_*],$$

given by series (5.162).

If, e.g., $h_i(t) = t - \Delta_i$, where $0 < \Delta_i = \text{const} < T$ $(i = 1, \dots, k)$, then

$$\chi_{h_i}(t, 0) = \begin{cases} 1 & \text{for} \quad \Delta_i \le t \le T + \Delta_i, \\ 0 & \text{for} \quad t < \Delta_i. \end{cases}$$

Hence, the $1 \times n$ matrix B_0 can be rewritten as

$$\begin{aligned} B_0 &= \int_{0}^{T} H(\tau) \sum_{i=1}^{k} B_i(\tau)\chi_{h_i}(\tau, 0)d\tau \\ &= -\left[\sum_{i=1}^{k} \int_{\Delta_i}^{T} b_{11}^{(i)}(\tau)d\tau, \sum_{i=1}^{k} \int_{\Delta_i}^{T} b_{12}^{(i)}(\tau)d\tau, \dots, \sum_{i=1}^{k} \int_{\Delta_i}^{T} b_{1n}^{(i)}(\tau)d\tau\right], \end{aligned}$$

and one of the conditions of solvability of problem (5.169) takes the form

$$\sum_{i=1}^{k} \int_{\Delta_i}^{T} b_{1j}^{(i)}(\tau)d\tau \neq 0 \quad (j = 1, \dots, n).$$

5.7 Fredholm Boundary-Value Problems for Differential Systems with Single Delay[2]

It is well known that a disadvantage of the proposed approach to the investigation of problem (5.127), (5.128) is connected with the necessity of finding the Cauchy matrix $K(t, s)$ (see (5.126)). This matrix exists but, as a rule, can be found only numerically. Therefore, it is important to find systems of differential equations with delay for which this problem can be directly solved. In what follows, we consider the case of a system with single delay. In this

[2] This research was partially supported by the Grants 1/3238/06 and 1/0771/08 of the Scientific Grant Agency of the Ministry of Education of Slovak Republic and Project APVV-0700-07 of the Slovak Research and Development Agency [164].

case the problem of how to construct the Cauchy matrix is successfully solved *analytically* by using a delayed matrix exponential introduced in [185]. Assume that $a = 0$.

Delayed Exponential. For the investigation of the Cauchy problem for a linear inhomogeneous differential system with constant coefficients and single delay

$$\dot{z}(t) = Az(t-\tau) + g(t), \qquad z(s) = \psi(s) \text{ for } s \in [-\tau, 0], \tag{5.170}$$

we consider a matrix function e_τ^{At} called a delayed matrix exponential [185] and defined as follows:

$$e_\tau^{At} = \begin{cases} \Theta & \text{for } -\infty < t < -\tau, \\ I & \text{for } -\tau \le t < 0, \\ I + A\dfrac{t}{1!} & \text{for } 0 \le t < \tau, \\ \cdots & \\ I + A\dfrac{t}{1!} + A^2\dfrac{(t-\tau)^2}{2!} & \\ \quad + \cdots + A^k\dfrac{(t-(k-1)\tau)^k}{k!} & \text{for } (k-1)\tau \le t < k\tau, \\ \cdots & \end{cases} \tag{5.171}$$

where Θ is the null matrix, A is an $n \times n$ constant matrix, $\tau > 0$, $g(t)$ is an n-dimensional column vector with components in the space L_p $(1 < p < \infty)$ of functions summable on $[0, b]$: $g(t) \in L_p[0, b]$, and $\psi\colon \mathbb{R}^1 \setminus [0, b] \to \mathbb{R}^n$ is a given vector function from the space $D_p[0, b]$. This definition can be briefly formulated as follows:

$$e_\tau^{At} = \sum_{n=0}^{[t/\tau]} A^n \frac{(t-(n-1)\tau)^n}{n!}$$

where $[\cdot]$ is the greatest integer function. By direct verification, one can easily prove that the delayed matrix exponential has the following properties [164, 185]:

1. For any matrix A, the differential rule

$$\frac{d}{dt} e_\tau^{At} = A e_\tau^{A(t-\tau)}$$

is true, i.e., the delayed matrix exponential solves a homogeneous system satisfying the following unit initial conditions:

$$e_\tau^{At} = I \text{ for } -\tau \le t \le 0.$$

2. Integrating the delayed matrix exponential, we get

$$\int_0^t e_\tau^{As} ds = I\frac{t}{1!} + A\frac{(t-\tau)^2}{2!} + \dots + A^k\frac{(t-(k-1)\tau)^{k+1}}{(k+1)!}$$

where $k = [t/\tau] + 1$. If, in addition, the matrix A is regular, then

$$\int_0^t e_\tau^{As} ds = A^{-1}[e_\tau^{A(t-\tau)} - e_\tau^{A\tau}].$$

Here, we recall that the integral of a matrix is a matrix whose entries are integrals of the corresponding entries of the initial matrix.

3. The delayed matrix exponential e_τ^{At}, $t > 0$, is a function continuously differentiable infinitely many times. At the nodes $k\tau$, $k = 1, 2, \dots$ the derivative has discontinuities of order $(k+1)$:

$$\{e_\tau^{At}\}_{t=k\tau-0}^{k+1} = 0, \quad \{e_\tau^{At}\}_{t=k\tau+0}^{k+1} = A^{k+1}.$$

4. The general solution of the inhomogeneous system (5.170) with single delay satisfying a constant initial condition $z(0) = c \in \mathbb{R}^n$ has the form

$$z(t) = e_\tau^{A(t-\tau)}c + \int_0^t e_\tau^{A(t-\tau-s)}g(s)\, ds, \qquad \forall c \in \mathbb{R}^n.$$

Main Results. System (5.170) can be transformed into an equation of the form (5.123):

$$\dot{z}(t) - A(S_h z)(t) = \varphi(t) \tag{5.172}$$

where, in view of (5.121) and (5.122),

$$(S_h z)(t) = \begin{cases} z(t-\tau) & \text{for } t-\tau \in [0,b], \\ 0 & \text{for } t-\tau \notin [0,b], \end{cases}$$

$$\varphi(t) = g(t) + A\psi^h(t) \in L_p[0,b],$$

$$\psi^h(t) = \begin{cases} 0 & \text{for } t-\tau \in [0,b], \\ \psi(t-\tau) & \text{for } t-\tau \notin [0,b]. \end{cases}$$

The general solution of problem (5.172) for an inhomogeneous system with single delay and trivial initial data has the form

$$z(t) = X(t)c + \int_0^b K(t,s)\varphi(s)\, ds, \quad \forall c \in \mathbb{R}^n$$

where, as can easily be verified (by using the delayed matrix exponential defined above),

$$X(t) = e_{\tau}^{A(t-\tau)}, \quad X(0) = e_{\tau}^{-A\tau} = I$$

is a normal fundamental matrix of the homogeneous system (5.172) (or (7.11)) with the unit initial data for $t = 0$: $X(0) = I$, and the Cauchy matrix $K(t,s)$ has the form

$$K(t,s) = e_{\tau}^{A(t-\tau-s)} \text{ and } K(t,s) \equiv 0 \text{ for } 0 \le t < s \le b, \; K(t,0) = X(t).$$

Therefore, system (5.172) with single delay the trivial initial data possesses an n-parameter family of linearly independent solutions

$$z(t) = e_{\tau}^{A(t-\tau)} c + \int_0^t e_{\tau}^{A(t-\tau-s)} \varphi(s)\, ds, \qquad \forall c \in \mathbb{R}^n. \tag{5.173}$$

We now consider two classes of boundary-value problems: periodic problems and general boundary-value problems.

Periodic Problem. We set $b = T$. For system (5.172), we consider a periodic boundary-value with boundary condition

$$z(0) = z(T). \tag{5.174}$$

Then equality (5.173) is a solution of the boundary-value problem (5.172), (5.174) if and only if there exists a constant vector $c \in \mathbb{R}^n$ solving the algebraic system

$$Qc = \int_0^T e_{\tau}^{A(T-\tau-s)} \varphi(s)\, ds, \tag{5.175}$$

where

$$Q := I - e_{\tau}^{A(T-\tau)}, \quad e_{\tau}^{-A\tau} = I.$$

We now consider the following two possible cases:

I. *Noncritical Case.* Let $\det Q \neq 0$. In this case, the algebraic system (5.175) is uniquely solvable and

$$c = Q^{-1} \int_0^T e_{\tau}^{A(T-\tau-s)} \varphi(s)\, ds.$$

By using this quantity, we get a unique solution of the periodic boundary-value problem (5.172), (5.174) with single delay in the form

$$z(t) = e_{\tau}^{A(t-\tau)} Q^{-1} \int_0^T e_{\tau}^{A(T-\tau-s)} \varphi(s)\, ds + \int_0^t e_{\tau}^{A(t-\tau-s)} \varphi(s)\, ds. \tag{5.176}$$

Theorem 5.12. *If* $\det Q \neq 0$, *then the periodic boundary-value problem (5.172), (5.174) with single delay is uniquely solvable in the space* $D_p[0,T]$ *of vector functions absolutely continuous on* $[0,T]$ *for any* $\varphi(t) \in L_p[0,T]$. *The unique solution of this problem is given by the formula*

$$z(t) = \int_0^T G(t,s)\varphi(s)\,ds$$

where

$$G(t,s) := e_\tau^{A(t-\tau)} Q^{-1} K(T,s) + K(t,s). \tag{5.177}$$

Remark 5.6. Note that the matrix $G(t,s)$ given by (5.177) is a Green matrix of the periodic boundary-value problem (5.172), (5.174). With the help of the delayed matrix exponential (5.171), this matrix can be constructed explicitly. In the space $C^1[0,T]$ of functions continuously differentiable on $[0,T]$, the periodic boundary-value problem (5.170), (5.174) is not always solvable even if $\det Q \neq 0$ because the transformation of system (5.170) into system (5.172) is performed ignoring the condition for the solution z and its initial function ψ is continuous at $t = 0$.

II. *Critical Case.* Let $\det Q = 0$. Then the algebraic system (5.175) is not always solvable. Moreover, if it is solvable, then its solution is not unique. In what follows, preserving the notation used above, we denote by Q^+ the unique pseudoinverse matrix (in the Moore–Penrose sense) with respect to a given matrix Q [174]. By $P_Q := I_n - Q^+Q$ we denote an $n \times n$-dimensional matrix (orthogonal projection) projecting the space $\mathbb{R}^n$ onto the kernel of the matrix Q. At the same time, by $P_{Q^*} := I_n - QQ^+$ we denote an $n \times n$-dimensional matrix (orthogonal projection) projecting the space $\mathbb{R}^n$ onto the kernel of the transposed matrix $Q^* := Q^T$. Let $\operatorname{rank} Q := n_1 < n$. In view of the property

$$\operatorname{rank} P_Q = \operatorname{rank} P_{Q^*} = n - \operatorname{rank} Q = n - \operatorname{rank} Q^* = r = n - n_1,$$

we denote by $P_{Q_r^*}$ an $r \times n$-dimensional matrix constructed from r linearly independent rows of the matrix P_{Q^*}. By P_{Q_r} we denote an $n \times r$-dimensional matrix constructed from r linearly independent columns of the matrix P_Q. According to [174, p.79], we can state that the algebraic system (5.175) is solvable if and only if the right-hand side satisfies r linearly independent conditions

$$P_{Q_r^*} \int_0^T e_\tau^{A(T-\tau-s)} \varphi(s)\,ds = 0, \tag{5.178}$$

where $r = n - \operatorname{rank} Q = n - n_1$, and has an r-parameter family of linearly independent solutions

$$c = Q^+ \int_0^T e_\tau^{A(T-\tau-s)} \varphi(s)\, ds + P_{Q_r} c_r\,, \quad \forall c_r \in \mathbb{R}^r.$$

We use this vector constant and, finally, arrive at a family of solutions of the periodic boundary-value problem (5.172), (5.174) for the system with single delay

$$z(t) = e_\tau^{A(t-\tau)} P_{Q_r} c_r + e_\tau^{A(t-\tau)} Q^+ \int_0^T e_\tau^{A(T-\tau-s)} \varphi(s)\, ds + \int_0^t e_\tau^{A(t-\tau-s)} \varphi(s)\, ds.$$

Theorem 5.13. *If* $\operatorname{rank} Q = n_1 < n$, *then the periodic boundary-value problem (5.172), (5.174) for a system with single delay is solvable in the space* $D_p[0,T]$ *if and only if*

$$\varphi(t) = g(t) + A\psi^h(t) \in L_p[0,T]$$

satisfies r *linearly independent conditions (5.178). Then there exists an* r*-parameter family of linearly independent solutions:*

$$z(t, c_r) = e_\tau^{A(t-\tau)} P_{Q_r} c_r + \int_0^T G(t,s)\varphi(s)\, ds, \qquad \forall\, c_r \in \mathbb{R}^r,$$

where

$$G(t,s) := e_\tau^{A(t-\tau)}\, Q^+ K(T,s) + K(t,s)$$

is a generalized Cauchy matrix for the periodic boundary-value problem (5.172), (5.174).

General Boundary-Value Problem. By using the results from [27, 174], we can easily obtain similar results for the general boundary-value problem if the number m of boundary conditions does not coincide with the number n of unknowns in the differential system with single delay. As in the periodic case, we get these results in the *explicit analytic* form.

Consider a boundary-value problem

$$\dot z(t) - A(S_h z)(t) = \varphi(t), \ t \in [0, b]\,, \tag{5.179}$$

$$l z(\cdot) = \alpha \in \mathbb{R}^m \tag{5.180}$$

where α is an m-dimensional constant vector column, l is an m-dimensional linear vector functional defined in the space $D_p[0,b]$ of n-dimensional vector functions $l = \operatorname{col}(l_1, \ldots, l_m)\colon D_p[0,b] \to \mathbb{R}^m$, $l_i\colon D_p[0,b] \to \mathbb{R}$ absolutely continuous on $[0, b]$. As above, we state that, in the considered spaces, this problem is equivalent to problem (5.170), (5.180). It is well known that the indicated problems for functional-differential equations are of Fredholm type (see, e.g., [1, 80, 174]).

We now establish necessary and sufficient conditions and a representation of the solutions $z \in D_p[0, b]$ of the boundary-value problem (5.172), (5.180). Since

$$K(t,s) = e_\tau^{A(t-\tau-s)} \text{ if } 0 \le s \le t \le b, \quad K(t,s) \equiv 0 \text{ if } 0 \le t < s \le b,$$
$$K(t,0) = X(t) = e_\tau^{A(t-\tau)},$$

in view of (5.126), the general solution of system (5.172) takes the form

$$z(t) = e_\tau^{A(t-\tau)} c + \int_0^t e_\tau^{A(t-\tau-s)} \varphi(s)\, ds, \quad \forall c \in \mathbb{R}^n. \tag{5.181}$$

In the algebraic system

$$Qc = \alpha - l \int_0^b K(\,\cdot\,, s)\varphi(s)\, ds, \tag{5.182}$$

deduced by substituting the general solution (5.181) of equation (5.172) in the boundary condition (5.180), the constant matrix $Q = lX = le_\tau^{A(\,\cdot\,-\tau)}$ is $m \times n$-dimensional.

Preserving the same notation as above, we conclude that the condition

$$P_{Q_d^*}\left\{\alpha - l\int_0^b K(\cdot, s)\varphi(s)\, ds\right\} = 0, \tag{5.183}$$

is necessary and sufficient for the algebraic system (5.182) to be solvable and if this condition is satisfied, then system (5.182) has a solution

$$c = P_{Q_r} c_r + Q^+\left\{\alpha - l\int_0^b K(\cdot, s)\varphi(s)\, ds\right\} \tag{5.184}$$

where Q^+ is an $n \times m$-dimensional matrix pseudoinverse with respect to the $m \times n$-dimensional matrix Q.

Substituting the constant $c \in \mathbb{R}^n$ defined by (5.184) in (5.181), we get a formula for the general solution of problem (5.172), (5.180):

$$z(t, c_r) = X_r(t)c_r + (G\varphi)(t) + X(t)Q^+\alpha, \quad \forall c_r \in \mathbb{R}^r. \tag{5.185}$$

where $X_r(t) = X(t)P_{Q_r}$,

$$(G\varphi)(t) := \int_0^b G(t,s)\varphi(s)\,ds$$

is a generalized Green operator, and

$$G(t,s) := e_\tau^{A(t-\tau)}Q^+lK(\,\cdot\,,s) + K(t,s)$$

is a generalized Green matrix corresponding to the boundary-value problem (5.172), (5.180). Hence, the following assertion is true:

Theorem 5.14. *If* $\operatorname{rank} Q = n_1 \le \min(m,n)$, *then the homogeneous problem corresponding to problem (5.172), (5.180) (with* $f(t) = 0, \alpha = 0$*) has exactly* r *(where* $r = n - n_1$*) linearly independent solutions in the space* $D_p[0,b]$. *The inhomogeneous problem (5.172), (5.180) is solvable in the space* $D_p[0,b]$ *if and only if* $\varphi \in L_p[0,b]$ *and* $\alpha \in \mathbb{R}^m$ *satisfy* d *linearly independent conditions (5.183). Then it has an* r*-dimensional family of linearly independent solutions represented in the explicit form (5.185).*

In the case $\operatorname{rank} Q = n$, we get the inequality $m \ge n$, i.e., the boundary-value problem is overdetermined and the number of boundary conditions is not smaller than the number of unknowns. Thus, Theorem 5.14 has the following corollary:

Corollary 5.5. *If* $\operatorname{rank} Q = n$, *then the homogeneous problem has only the trivial solution. The inhomogeneous problem (5.172), (5.180) is solvable if and only if*

$$P_{Q_d^*}\left\{\alpha - l\int_0^b K(\,\cdot\,,s)\varphi(s)\,ds\right\} = 0,$$

where $d = m - n$. *Then the unique solution of the problem can be represented as*

$$z(t) = (G\varphi)(t) + X(t)Q^+\alpha.$$

The case $\operatorname{rank} Q = m$ is also of interest. In this case, we have $m \le n$, i.e., the boundary-value problem is underdetermined. In this case, Theorem 5.14 has the following corollary:

Corollary 5.6. *If* $\operatorname{rank} Q = m$, *then the boundary-value problem possesses an* r*-dimensional (*$r = n - m$*) family of solutions. The inhomogeneous problem (5.172), (5.180) is solvable for arbitrary* $\varphi(t) \in L_p[0,b]$ *and* $\alpha \in R^m$ *and has an* r*-parameter family of solutions*

$$z(t,c_r) = X_r(t)c_r + (G\varphi)(t) + X(t)Q^+\alpha.$$

Finally, combining both special cases mentioned above, we get:

Corollary 5.7. *If* rank $Q = n = m$, *then the homogeneous problem has only the trivial solution. The inhomogeneous boundary-value problem (5.172), (5.180) is solvable for any* $\varphi(t) \in L_p[0, b]$ *and* $\alpha \in R^n$ *and possesses a unique solution*

$$z(t) = (G\varphi)(t) + X(t)Q^{-1}\alpha.$$

Corollary 5.8. *Similar conditions for the existence of solutions of Fredholm boundary-value problems for systems of ordinary differential equations with constant coefficients and several delays specified by pairwise permutable matrices were established in [169].*

5.8 Degenerate Systems of Ordinary Differential Equations

It is known that numerous problems of the control theory, radiophysics, mathematical economics, and linear programming are modeled by systems of ordinary differential equations with degenerate matrix at the derivative. In the literature, systems of this kind are usually called degenerate [191], algebraic-differential [181, 189], or singular [179, 180] systems of differential equations. The authors of the present book call these systems degenerate. An extensive literature devoted to these systems and the corresponding terminology, which is not completely finalized and generally accepted, can be found in [179, 181, 189]. The notion of central canonical form is very important for the theory of degenerate systems. For the first time, this notion was introduced in 1983 by S. L. Campbell and L.R. Petzold [180]. The existence of solutions mainly of the initial problems and some types of boundary-value problems for these systems was studied in [181, 191].

In the present section, we consider perturbed degenerate Fredholm boundary-value problems of nonzero index of the general form for which boundary conditions are defined by a linear vector functional of the most general form under the assumption that the unperturbed differential system is reduced to a central canonical form. As shown in what follows [177], the Fredholm property of the operator specifying the analyzed boundary-value problem in the corresponding space is equivalent to the requirement that the number of boundary conditions does not coincide with the dimension of the original differential system minus the number of nilpotent Jordan blocks in the central canonical form. Thus, we consider most complicated (and insufficiently well studied) undetermined and overdetermined boundary-valued problems for systems of ordinary differential equations with degenerate matrix at the derivative and illustrate the efficiency of the proposed approach to the investigation of these systems.

Statement of the Problem. Consider the following degenerate linear inhomogeneous boundary-value problem with small parameter:

$$B(t)\,\frac{dx}{dt} = A(t)\,x + \varepsilon A_1(t)\,x + f(t), \qquad t \in [a,b], \tag{5.186}$$

$$l\,x(\cdot) = \alpha + \varepsilon l_1 x, \qquad \alpha \in \mathbb{R}^m, \tag{5.187}$$

where $A(t)$, $B(t)$, and $A_1(t)$ are $(n \times n)$ matrices and $f(t)$ is an n-dimensional column vector whose components are real functions continuously differentiable on $[a,b]$ sufficiently many times (in what follows, we present the relationship with the rank of the matrix $B(t)$), $\det B(t) = 0 \ \forall t \in [a;b]$, α is an m-dimensional column vector of constants, l and l_1 are linear vector functionals defined in the space of n-dimensional vector functions continuous on $[a,b]$: $l = \operatorname{col}(l_1, ..., l_m) : C[a,b] \to \mathbb{R}^m$, $l_1 = \operatorname{col}(l_1^1, ..., l_m^1) : C[a,b] \to \mathbb{R}^m$, and l_i, $l_i^1 : C[a,b] \to \mathbb{R}$, $i = \overline{1,m}$.

Assume that the generating boundary-value problem

$$B(t)\,\frac{dx}{dt} = A(t)\,x + f(t), \quad l\,x(\cdot) = \alpha \in \mathbb{R}^m, \qquad t \in [a;b], \tag{5.188}$$

obtained from (5.186) and (5.187) for $\varepsilon = 0$ does not have solutions for arbitrary inhomogeneities $f(t)$ from the corresponding space and any $\alpha \in \mathbb{R}^m$.

The following question arises:

Is it possible to make the boundary-value problem (5.188) solvable with the help of linear perturbations? If this is true, then for what perturbation terms [the matrix $A_1(t)$ in the differential system (5.186) and the vector functional l_1 in the boundary condition (5.187)], the boundary-value problem (5.186), (5.187) is solvable for any inhomogeneities $f(t)$ from the corresponding space and any $\alpha \in \mathbb{R}^m$?

Unperturbed Problem. To answer this question, we first study the solvability of the generating boundary-value problem (5.188). To this end, we use the theorem on reducibility of a degenerate linear system to the central canonical form proved in [192].

Assume that the following conditions are satisfied:

(i) $\operatorname{rank} B(t) = \text{const} = n - r_1 \quad \forall t \in [a;b]$;

(ii) the matrix $B(t)$ on the segment $[a;b]$ has a complete Jordan collection of vectors $\varphi_i^{(j)}(t)$, $j = \overline{1,s_i}$, $i = \overline{1,r_1}$ with respect to the operator

$$L(t) = A(t) - B(t)\,\frac{d}{dt},$$

which are determined from the relations

$$B(t)\,\varphi_i^{(1)}(t) = 0, \quad i = \overline{1,r_1},$$

$$B(t)\,\varphi_i^{(j)}(t) = L(t)\,\varphi_i^{(j-1)}(t), \quad j = \overline{2, s_i}, \quad i = \overline{1, r_1},$$
$$B(t)\,x \neq L(t)\,\varphi_i^{s_i}(t);$$

(iii) $A(t)$, $B(t) \in C^{3q-2}[a;b]$; $f(t) \in C^{q-1}[a;b]$, where $q = \max\limits_i s_i$.

It follows from condition (ii) [191] that the conjugate matrix $B^*(t)$ has a similar Jordan collection with respect to the operator

$$L^*(t) = A^*(t) + \frac{d}{dt} B^*(t)$$

formed by the vectors $\psi_i^{(j)}(t)$, $j = \overline{1, s_i}$, $i = \overline{1, r_1}$. The vectors used to form these collections are linearly independent for all $t \in [a;b]$ and the condition of completeness is that the determinant formed by the scalar products of the vectors $L\,\varphi_i^{(s_i)}(t)$ with basis elements of the null space of the matrix $B^*(t)$ is not equal to zero:

$$\det \left\| \left(L(t)\,\varphi_i^{(s_i)}(t)\,, \psi_i^{(1)}(t)\right) \right\|_{i=\overline{1,r_1}} \neq 0 \quad \forall t \in [a;b]\,.$$

As shown in [192], under these conditions, there exist nondegenerate matrices $P(t)$, $Q_1(t) \in C^{q-1}[a;b]$ for all $t \in [a;b]$ such that, as a result of multiplication by $P(t)$ and the change $x = Q_1(t)\,y$, system (5.188) is reduced to the central canonical form:

$$\begin{pmatrix} E_{n-s} & 0 \\ 0 & I \end{pmatrix} \frac{dy}{dt} = \begin{pmatrix} M(t) & 0 \\ 0 & E_s \end{pmatrix} y + P(t) f(t),$$

where $s = s_1 + s_2 + \ldots + s_{r_1}$, E_s and E_{n-s} are the identity matrices of orders s and $n - s$, respectively, and $I = \operatorname{diag}\{I_1, \ldots, I_{r_1}\}$ is a quasidiagonal matrix composed of nilpotent Jordan cells I_i of dimension s_i.

By using this result, in [191, 192], it is proved that, under conditions (i)–(iii), the general solution of system (5.188) takes the form

$$x(t) = X_{n-s}(t)\,c + \int\limits_a^t X_{n-s}(t)\,Y_{n-s}^*(\tau)\,f(\tau)\,d\tau$$
$$- \Phi(t) \sum_{k=0}^{q-1} \mathrm{I}^k \frac{d^k}{dt^k} \left\{ \left[\Psi^*(t)\,L(t)\Phi(t) \right]^{-1} \Psi^*(t)\,f(t) \right\} \qquad (5.189)$$
$$\forall c \in \mathbb{R}^{n-s}.$$

Here, $X_{n-s}(t)$ is an $(n \times (n-s))$ matrix formed by $(n-s)$ linearly independent solutions of the homogeneous system

$$B(t) \frac{dx}{dt} = A(t) x, \qquad (5.190)$$

$Y_{n-s}(t)$ is the matrix of the same order formed by $(n-s)$ linearly independent solutions of the system

$$\frac{d}{dt}(B^*(t)y) = -A^*(t)y \tag{5.191}$$

adjoint to system (5.190), c is an arbitrary $(n-s)$-dimensional vector of constants, $\Phi(t)$ and $\Psi(t)$ are $(n \times s)$ matrices composed of the vectors from the Jordan collections considered above:

$$\Phi(t) = \Big[\varphi_1^{(5.186)}(t), \ldots, \varphi_1^{(s_1)}(t);$$

$$\varphi_2^{(5.186)}(t), \ldots, \varphi_2^{(s_2)}(t); \ldots; \varphi_{r_1}^{(1)}(t), \ldots, \varphi_{r_1}^{(s_{r_1})}(t)\Big],$$

$$\Psi(t) = \Big[\psi_1^{(s_1)}(t), \ldots, \psi_1^{(5.186)}(t);$$

$$\psi_2^{(s_2)}(t), \ldots, \psi_2^{(5.186)}(t); \ldots; \psi_{r_1}^{(s_{r_1})}(t), \ldots, \psi_{r_1}^{(1)}(t)\Big].$$

Following [191, p. 63], we say that the matrices $X_{n-s}(t)$ and $Y_{n-s}(t)$ are fundamental. These matrices are constructed to guarantee that

$$Y_{n-s}^*(t)\, B(t) X_{n-s}(t) = E_{n-s},$$

which is always possible [191] because, for any choice of linearly independent solutions of systems (5.190) and (5.191) composing $X_{n-s}(t)$ and $Y_{n-s}(t)$, the product $Y_{n-s}^*(t)\, B(t) X_{n-s}(t)$ is a constant nondegenerate matrix.

By using the general solution (5.189) of the degenerate differential system (5.188), we determine the condition of solvability and the form of the general solution of the linear inhomogeneous boundary-value problem (5.188). Relation (5.189) gives a solution of the boundary-value problem (5.188) if and only if the algebraic system with $m \times (n-s)$ matrix $Q := l X_{n-s}(\cdot)$

$$Qc = \alpha - l\Bigg(\int_a^{\cdot} X_{n-s}(\cdot) Y_{n-s}^*(\tau) f(\tau) d\tau$$

$$- \Phi(\cdot) \sum_{k=0}^{q-1} I^k \frac{d^k}{dt^k}\Big(\Big[\Psi^*(t) L \Phi(t)\Big]^{-1} \Psi^*(t) f(t)\Big)(\cdot)\Bigg) \tag{5.192}$$

is solvable with respect to $c \in \mathbb{R}^{n-s}$.

According to [60], system (5.192) is solvable if and only if its right-hand side belongs to the orthogonal complement $N^{\perp}(Q^*) = R(Q)$ of the kernel $N(Q^*) = \ker Q^*$ of the conjugate matrix Q^*, i.e., the following condition is satisfied:

$$P_{Q_d^*}\Bigg(\alpha - l\Bigg(\int_a^{\cdot} X_{n-s}(\cdot)\, Y_{n-s}^*(\tau)\, f(\tau)\, d\tau$$

$$- \Phi(\cdot) \sum_{k=0}^{q-1} \mathrm{I}^k \frac{d^k}{dt^k} ([\Psi^*(t) L\Phi(t)]^{-1} \Psi^*(t) f(t))(\cdot) \Big) \Big) = 0, \qquad (5.193)$$

$$[d = m - n_1],$$

where $P_{Q^*} = E_m - QQ^+$ is an $(m \times m)$ matrix (orthoprojector) that projects the space $\mathbb{R}^m$ onto the kernel of the matrix $Q^* = Q^T$ $(\ker Q^* = \operatorname{coker} Q)$. Let $\operatorname{rank} Q = n_1 \leq \min(m, n-s)$. Since $\operatorname{rank} P_{Q^*} = d = m - n_1$, we denote a $(d \times m)$ matrix formed by d linearly independent rows of the matrix P_{Q^*} by $P_{Q^*_d}$. As a result, criterion (5.193) consists of d linearly independent conditions and Q^+ is the unique Moore–Penrose pseudoinverse matrix for the matrix Q. Substituting the vector constant $c \in \mathbb{R}^{n-s}$ determined from Eq. (5.192) in the general solution (5.189) of the differential system (5.188), it is possible to prove [175] that the degenerate linear inhomogeneous boundary-value problem (5.188) has an r-parameter family of linearly independent solutions

$$x(t, c_r) = X_r(t) c_r + (Gf)(t) + X_{n-s}(t) Q^+ \alpha, \qquad (5.194)$$

$$[r = (n-s) - n_1] \quad \forall c_r \in \mathbb{R}^r,$$

where $X_r(t) = X_{n-s}(t), P_{Q_r}$ is an $(n \times r)$ matrix, and $P_Q = E_{n-s} - Q^+Q$ is an $((n-s) \times (n-s))$ matrix (orthoprojector) projecting the space $\mathbb{R}^{n-s}$ onto the kernel of the matrix Q, if and only if condition (5.193) is satisfied. Since $\operatorname{rank} P_Q = r = (n-s) - n_1$, we now denote an $((n-s) \times r)$-dimensional matrix formed by r linearly independent columns of the matrix P_Q [38] by P_{Q_r} [174]. This enables us to select the maximum number of linearly independent solutions; $(Gf)(t)$ is a generalized Green operator of the degenerate boundary-value problem (5.188) acting upon the vector function $f(t)$ as follows:

$$\begin{aligned}(Gf)(t) := -X_{n-s}(t) Q^+ l \Bigg(&\int_a^{\cdot} X_{n-s}(\cdot) Y_{n-s}^*(\tau) f(\tau) d\tau \\ &- \Phi(\cdot) \sum_{k=0}^{q-1} \mathrm{I}^k \frac{d^k}{dt^k} \Big([\Psi^*(t) L\Phi(t)]^{-1} \Psi^*(t) f(t) \Big)(\cdot) \Bigg) \\ + \int_a^t X_{n-s}(t) &Y_{n-s}^*(\tau) f(\tau) d\tau \\ - \Phi(t) \sum_{k=0}^{q-1} &\mathrm{I}^k \frac{d^k}{dt^k} \Big([\Psi^*(t) L\Phi(t)]^{-1} \Psi^*(t) f(t) \Big).\end{aligned}$$

Hence, the following statement is true for the unperturbed degenerate boundary-value problem:

Theorem 5.15. *1. The degenerate homogeneous boundary-value problem*

$$B(t)\frac{dx}{dt} = A(t)x, \quad l\,x\,(\cdot) = 0 \in \mathbb{R}^m,$$

has an r-parameter family of linearly independent solutions:

$$x\,(t,\,c_r) = X_r\,(t)\,c_r \quad \forall\, c_r \in \mathbb{R}^r; \quad (r = (n-s) - n_1, \;\; n_1 := \operatorname{rank} Q).$$

2. The unperturbed degenerate inhomogeneous boundary-value problem (5.188) is solvable if and only if the inhomogeneities $f\,(t) \in C^{q-1}\,[a,b]$ and $\alpha \in \mathbb{R}^m$ satisfy $d = m - n_1$ linearly independent conditions (5.193). Moreover, it has an r-parameter family of linearly independent solutions (5.194).

3. The index of the operator L specifying the original degenerate homogenous boundary-value problem

$$Lx := \operatorname{col}\left\{B(t)\frac{dx}{dt} - A(t)x\,,\; lx\right\}$$

is equal to

$$\operatorname{ind} L := \dim\ker L - \dim\operatorname{coker} L = r - d = n - s - m.$$

As an illustration of the efficiency of this theorem, we study the solvability of the Cauchy problem for degenerate systems

$$B\,(t)\,\frac{dx}{dt} = A\,(t)\,x + f\,(t)\,, \quad x\,(a) \;=\; \alpha \in \mathbb{R}^n, \qquad t \in [a;b]\,. \tag{5.195}$$

Unlike nondegenerate systems, it turns out that the Cauchy problem (5.195) for these degenerate systems is not always solvable. In order to apply Theorem 5.15, we use the following functional specifying the initial conditions:

$$l\,x\,(\cdot) \;:= x(a) \in \mathbb{R}^n, (\,m = n\,)$$

as a vector functional l; here, the $m \times (n-s)$ matrix $Q \;:= l\,X_{n-s}\,(\cdot)$ is an $n \times (n-s)$ matrix of the form $Q \;:=\; X_{n-s}\,(a)$ and, hence, its rank n_1 is equal to $n_1 = \operatorname{rank} Q = n-s$. The orthoprojectors P_Q and P_{Q^*} used in Theorem 5.15 have the orders $n \times n$ and, moreover, $d = n - n_1 = s$ and $r = n - s - n_1 = 0$. Hence, the matrices in Theorem 5.15 have the form: $P_{Q^*_d} = P_{Q^*_s}$; these are $s \times n$ matrices and $P_{Q_r} = 0$. As a result, an analog of Theorem 5.15 for the Cauchy problem can be formulated as follows:

Theorem 5.16. *1. The degenerate homogeneous Cauchy problem*

$$B(t)\frac{dx}{dt} = A(t)x, \quad x\,(a) \;=\; 0 \in \mathbb{R}^n,$$

has only the trivial solution.

2. The degenerate inhomogeneous Cauchy problem (5.195) is solvable if and only if the inhomogeneities $f(t) \in C^{q-1}[a,b]$ *and* $\alpha \in \mathbb{R}^n$ *satisfy* s *linearly independent conditions*

$$P_{Q_s^*}\left(\alpha - \Phi(a)\sum_{k=0}^{q-1} \mathrm{I}^k \frac{d^k}{dt^k}\left([\Psi^*(t)\, L\Phi(t)]^{-1}\Psi^*(t)\, f(t)\right)(a)\right) = 0; \qquad (5.196)$$

moreover, the Cauchy problem (5.195) possesses a unique solution

$$x(t) = (Gf)(t) + X_{n-s}(t)\, Q^+\alpha, \qquad (5.197)$$

where $(Gf)(t)$ *is the generalized Green operator of the degenerate Cauchy problem (5.195) acting upon the vector function* $f(t)$ *as follows:*

$$(Gf)(t) := X_{n-s}(t)Q^+\left(\Phi(a)\sum_{k=0}^{q-1} \mathrm{I}^k \frac{d^k}{dt^k}\left([\Psi^*(t)\, L\Phi(t)]^{-1}\Psi^*(t)\, f(t)\right)(a)\right)$$
$$-\Phi(t)\sum_{k=0}^{q-1} \mathrm{I}^k \frac{d^k}{dt^k}\left([\Psi^*(t)\, L\Phi(t)]^{-1}\Psi^*(t)\, f(t)\right).$$

3. The index of the operator L *specifying the original degenerate homogeneous Cauchy problem*

$$Lx := \mathrm{col}\,\{B(t)\frac{dx}{dt} - A(t)x\ , x(a)\}$$

is equal to

$$\mathrm{ind}\, L := \dim\ker L - \dim\mathrm{coker}\, L = -d = -s.$$

Remark 5.7. It is easy to see that, in the case where $r = (n-s) - n_1 = d = m - n_1$, i.e., the number of boundary conditions m is equal to the number n of unknowns in the differential system (5.188) without $s = s_1 + s_2 + \ldots + s_{r_1}$ nilpotent Jordan blocks in the central canonical form of the corresponding order $s_i : m = n - s$, the degenerate linear inhomogeneous boundary-value problem (5.188) in the analyzed space is a Fredholm problem of index zero. Otherwise, it is a Fredholm problem of nonzero index, i.e., according to the classical S. Krein classification [174], the analyzed problem is a Noether problem. It is worth noting that, for the Cauchy problem for a degenerate differential system, this problem is a Fredholm problem of nonzero index. It is well known that the Cauchy problem for nondegenerate systems of ordinary differential equations is a Fredholm problem of index zero. For $\det B(t) \neq 0$, i.e., $s = 0$, we arrive at the boundary-value problem for nondegenerate systems of ordinary differential equations well studied in previous sections.

We now have all necessary information for the solution of the perturbed boundary-value problem (5.186), (5.187) posed above.

Perturbed Problem. We solve the degenerate perturbed boundary-value problem (5.186), (5.187) formulated at the beginning of this section by the Vishik–Lyusternik method [150]. According to our assumptions, the generating problem does not have solutions. Hence, it makes sense to seek the solution of the boundary-value problem (5.186), (5.187) in the form of a part of the series in powers of the small parameter $\varepsilon = 0$:

$$x(t,\varepsilon) = \sum_{i=-1}^{+\infty} \varepsilon^i x_i(t) = \frac{x_{-1}(t)}{\varepsilon} + x_0(t) + \varepsilon x_1(t) + \varepsilon^2 x_2(t) + \ldots . \tag{5.198}$$

We substitute series (5.198) in the boundary-value problem (5.186), (5.187) and equate the coefficients of the same powers of ε. For ε^{-1}, we obtain the following degenerate linear homogeneous boundary-value problem for the determination of the coefficient $x_{-1}(t)$ of series (5.198):

$$B(t)\,\dot{x}_{-1}(t) = A(t)\,x_{-1}(t), \qquad l\,x_{-1}(\cdot) = 0 \tag{5.199}$$

According to Theorem 5.15, the homogeneous degenerate boundary-value problem (5.199) has an r-parameter family of linearly independent solutions

$$x_{-1} = x_{-1}(t, c_{-1}) = X_{n-s}(t)\,P_{Q_r} c_{-1} = X_r(t)\,c_{-1}, \qquad c_{-1} \in \mathbb{R}^r.$$

An arbitrary r-dimensional column vector c_{-1} is determined from the condition of solvability of the degenerate linear inhomogeneous boundary-value problem for the coefficient $x_0(t)$ of series (5.198):

$$\begin{cases} B(t)\,\dot{x}_0(t) = A(t)\,x_0(t) + A_1(t)\,x_{-1}(t, c_{-1}) + f(t), \\ lx_0(\cdot) = \alpha + l_1 x_{-1}(\cdot, c_{-1}). \end{cases} \tag{5.200}$$

The criterion of solvability (5.193) of problem (5.200) takes the form

$$P_{Q_d^*}\Big(\alpha + l_1 x_{-1}(\cdot, c_{-1}) - l\Big(\int_a^{\cdot} X_{n-s}(\cdot) Y_{n-s}^*(\tau)\big[A_1(\tau)x_{-1}(\tau, c_{-1}) + f(\tau)\big]d\tau$$

$$- \Phi(\cdot)\sum_{k=0}^{q-1} \mathrm{I}^k \frac{d^k}{dt^k}\Big(\big[\Psi^*(t)L\Phi(t)\big]^{-1}\Psi^*(t)\big[A_1(t)x_{-1}(t, c_{-1}) + f(t)\big]\Big)(\cdot)\Big)\Big) = 0.$$

In view of the form of $x_{-1}(t, c_{-1})$, we derive the following algebraic system for $c_{-1} \in \mathbb{R}^r$ from this criterion:

$$B_0 c_{-1} = -P_{Q_d^*}\Big(\alpha - l\Big(\int_a^{\cdot} X_{n-s}(\cdot) Y_{n-s}^*(\tau) f(\tau) d\tau$$

$$- \Phi(\cdot)\sum_{k=0}^{q-1} \mathrm{I}^k \frac{d^k}{dt^k}\Big(\big[\Psi^*(t)L\Phi(t)\big]^{-1}\Psi^*(t)f(t)\Big)(\cdot)\Big)\Big), \tag{5.201}$$

where B_0 is the $d \times r$ matrix

$$B_0 := P_{Q_d^*}\Big(l_1 X_r(\cdot) - l\Big(\int_a^{\cdot} X_{n-s}(\cdot)\, Y_{n-s}^*(\tau)\, A_1(\tau) X_r(\tau) d\tau - \Phi(\cdot) \sum_{k=0}^{q-1} \mathrm{I}^k \frac{d^k}{dt^k}\Big(\big[\Psi^*(t)\, L\Phi(t)\big]^{-1} \Psi^*(t)\, A_1(t) X_r(t)\Big)(\cdot)\Big)\Big) \tag{5.202}$$

constructed with regard for the perturbation coefficient $A_1(t)$ of system (5.186) and the vector functional l_1 in the boundary condition (5.187). For the solvability of system (5.201), it is necessary and sufficient that the following condition be satisfied:

$$P_{B_0^*} P_{Q_d^*}\Big(\alpha - l\Big(\int_a^{\cdot} X_{n-s}(\cdot)\, Y_{n-s}^*(\tau)\, f(\tau)\, d\tau - \Phi(\cdot) \sum_{k=0}^{q-1} \mathrm{I}^k \frac{d^k}{dt^k}\Big(\big[\Psi^*(t)\, L\Phi(t)\big]^{-1} \Psi^*(t)\, f(t)\Big)(\cdot)\Big)\Big) = 0.$$

In view of the fact that condition (5.193) is not satisfied because, according to our assumption, the generating problem does not have solutions, we use the last relation to obtain the condition $P_{B_0^*} = 0$ equivalent to the condition [60]

$$\operatorname{rank} B_0 = d. \tag{5.203}$$

Here, $P_{B_0^*}$ is a $(d \times d)$ matrix (orthoprojector) projecting the space $\mathbb{R}^d$ onto the kernel of the matrix $B_0^* = B_0^T$. The set of solutions of the algebraic system (5.201) for $c_{-1} \in \mathbb{R}^r$ has the form

$$c_{-1} = \overline{c}_{-1} + P_{B_0}\overline{c} \quad \forall \overline{c} \in \mathbb{R}^r \qquad \text{or} \qquad c_{-1} = \overline{c}_{-1} + P_\rho c_\rho \quad \forall\, c_\rho \in \mathbb{R}^\rho,$$

where

$$\overline{c}_{-1} = -B_0^+ P_{Q_d^*}\Big(\alpha - l\Big(\int_a^{\cdot} X_{n-s}(\cdot)\, Y_{n-s}^*(\tau)\, f(\tau)\, d\tau - \Phi(\cdot) \sum_{k=0}^{q-1} \mathrm{I}^k \frac{d^k}{dt^k}\Big(\big[\Psi^*(t)\, L\Phi(t)\big]^{-1} \Psi^*(t)\, f(t)\Big)(\cdot)\Big)\Big),$$

B_0^+ is the unique matrix pseudoinverse to B_0, and P_{B_0} is an $(r \times r)$ matrix (orthoprojector) that projects the space $\mathbb{R}^r$ onto the $\ker B_0$. Since $\operatorname{rank} P_{B_0} = r - \operatorname{rank} B_0 = r - d = n - s - m = \rho$, we replace the matrix P_{B_0} by the $(r \times \rho)$ matrix P_ρ formed by ρ linearly independent columns of the matrix P_{B_0}.

In view of the expression for c_{-1}, the homogeneous boundary-value problem (5.199) has a ρ-parameter family of solutions

$$x_{-1}(t, c_\rho) = \overline{x}_{-1}(t, \overline{c}_{-1}) + X_r(t)\, P_\rho c_\rho \quad \forall\, c_\rho \in \mathbb{R}^\rho, \tag{5.204}$$

where

$$\overline{x}_{-1}(t, \overline{c}_{-1}) = X_r(t)\,\overline{c}_{-1}.$$

Under condition (5.203), the general solution of the boundary-value problem (5.200) has the form

$$\begin{aligned} x_0(t,\ c_0) &= X_r(t)\,c_0 + \big(G\,[A_1(\cdot)\,\overline{x}_{-1}(\cdot, \overline{c}_{-1}) + f(\cdot)]\big)(t) \\ &\quad + X_{n-s}(t)\,Q^{+}\,[\alpha + l_1\overline{x}_{-1}(\cdot, \overline{c}_{-1})] \\ &\quad + \Big[\big(G\,[A_1(\cdot)\,X_r(\cdot)]\big)(t) + X_{n-s}(t)\,Q^{+}l_1X_r(\cdot)\Big]P_\rho c_\rho \\ &= X_r(t)\,c_0 + F_{-1}(t) + K_{-1}(t)P_\rho c_\rho, \end{aligned}$$

where

$$\begin{aligned} F_{-1}(t) &= \big(G\,[A_1(\cdot)\,\overline{x}_{-1}(\cdot, \overline{c}_{-1}) + f(\cdot)]\big)(t) \\ &\qquad + X_{n-s}(t)\,Q^{+}\,[\alpha + l_1\overline{x}_{-1}(\cdot, \overline{c}_{-1})]; \\ K_{-1}(t) &= \big(G\,[A_1(\cdot)\,X_r(\cdot)]\big)(t) + X_{n-s}(t)\,Q^{+}l_1X_r(\cdot); \end{aligned}$$

c_0 is an r-dimensional vector of constants determined from the condition of solvability of the boundary-value problem for the coefficient $x_1(t)$ of series (5.198) in the next step. We obtain the following degenerate linear inhomogeneous boundary-value problem for the determination of the coefficient $x_1(t)$ of ε^1 in series (5.198):

$$\begin{cases} B(t)\dot{x}_1(t) = A(t)x_1(t) + A_1(t)x_0(t, c_0), \\ lx_1(\cdot) = l_1x_0(\cdot, c_0). \end{cases} \tag{5.205}$$

By using the expression for $x_0(t,\ c_0)$, we obtain the following relation from the criterion of solvability (5.193) of the boundary-value problem (5.205):

$$\begin{aligned} B_0c_0 = -P_{Q_d^*}\Bigg[& l_1\{F_{-1}(\cdot) + K_{-1}(\cdot)P_\rho c_\rho\} \\ & - l\Bigg(\int_a^{\cdot} X_{n-s}(\cdot)Y_{n-s}^*(\tau)A_1(\tau)\{F_{-1}(\tau) + K_{-1}(\tau)P_\rho c_\rho\}d\tau \\ & - \Phi(\cdot)\sum_{k=0}^{q-1} \mathrm{I}^k \frac{d^k}{dt^k}\Big([\Psi^*(t)L\Phi(t)]^{-1} \\ & \times \Psi^*(t)A_1(t)\{F_{-1}(t) + K_{-1}(t)P_\rho c_\rho\}\Big)(\cdot)\Bigg)\Bigg]. \end{aligned}$$

Under condition (5.203), we get the following relation for $c_0 \in \mathbb{R}^r$ from the last algebraic system:

$$c_0 = -B_0^+ P_{Q_d^*} \Bigg[l_1\{F_{-1}(\cdot) + K_{-1}(\cdot)P_\rho c_\rho\}$$

$$- l\bigg(\int_a^{\cdot} X_{n-s}(\cdot)Y_{n-s}^*(\tau)A_1(\tau)\{F_{-1}(\tau) + K_{-1}(\tau)P_\rho c_\rho\}d\tau$$

$$- \Phi(\cdot)\sum_{k=0}^{q-1} \mathrm{I}^k \frac{d^k}{dt^k}\Big([\Psi^*(t)L\Phi(t)]^{-1}$$

$$\times \Psi^*(t)A_1(t)\{F_{-1}(t) + K_{-1}(t)P_\rho c_\rho\}\Big)(\cdot)\bigg)\Bigg] + P_\rho c_\rho.$$

Thus, we find

$$c_0 = \overline{c}_0 + D_0 P_\rho c_\rho \quad \forall\, c_\rho \in \mathbb{R}^\rho.$$

where

$$\overline{c}_0 = -B_0^+ P_{Q_d^*}\Bigg[l_1 F_{-1}(\cdot) - l\bigg(\int_a^{\cdot} X_{n-s}(\cdot)Y_{n-s}^*(\tau)A_1(\tau)F_{-1}(\tau)d\tau$$

$$- \Phi(\cdot)\sum_{k=0}^{q-1} \mathrm{I}^k \frac{d^k}{dt^k}\Big([\Psi^*(t)L\Phi(t)]^{-1}\Psi^*(t)A_1(t)F_{-1}(t)\Big)(\cdot)\bigg)\Bigg];$$

$$D_0 = I_r - B_0^+ P_{Q_d^*}\Bigg[l_1 K_{-1}(\cdot) - l\bigg(\int_a^{\cdot} X_{n-s}(\cdot)Y_{n-s}^*(\tau)A_1(\tau)K_{-1}(\tau)d\tau$$

$$- \Phi(\cdot)\sum_{k=0}^{q-1} \mathrm{I}^k \frac{d^k}{dt^k}\Big([\Psi^*(t)L\Phi(t)]^{-1}\Psi^*(t)A_1(t)K_{-1}(t)\Big)(\cdot)\bigg)\Bigg].$$

Hence, under condition (5.203), the boundary-value problem (5.200), has a ρ-parameter family of linearly independent solutions

$$x_0\,(t,\ c_\rho) = \overline{x}_0(t, \overline{c}_0) + \overline{X}_0\,(t)\,P_\rho c_\rho \quad \forall\, c_\rho \in \mathbb{R}^\rho,$$

where

$$\overline{x}_0(t, \overline{c}_0) = X_r(t)\overline{c}_0 + F_{-1}\,(t)\,;$$
$$\overline{X}_0\,(t) = X_r(t)D_0 + K_{-1}\,(t)\,.$$

Under condition (5.203), the boundary-value problem (5.205) has a ρ-parameter family of solutions

$$x_1\,(t,\ c_1) = X_r\,(t)\,c_1 + F_0\,(t) + K_0\,(t)\,P_\rho c_\rho,$$

where c_1 is an r-dimensional vector of constants determined from the condition of solvability of the boundary-value problem for the coefficient $x_2\,(t)$ of series (5.198) in the next step.

By induction, we determine the coefficients $x_i(t)$ of series (5.198) from the following boundary-value problem:

$$\begin{cases} B(t)\dot{x}_i(t) = A(t)x_i(t) + A_1(t)x_{i-1}(t, c_{i-1}), \\ lx_i(\cdot) = l_1 x_{i-1}(\cdot, c_{i-1}). \end{cases}$$

According to Theorem 5.15, by analogy with the previous steps of the iterative procedure, we prove that, under condition (5.203), the boundary-value problem has a ρ-parameter family of linearly independent solutions

$$x_i(t, c_i) = \overline{x}_i(t, \overline{c}_i) + \overline{X}_i(t) P_\rho c_\rho \quad \forall c_\rho \in \mathbb{R}^\rho, \quad i = 1, 2, \dots, \tag{5.206}$$

where all terms are determined from the following iterative procedure:

$$\overline{x}_i(t, \overline{c}_i) = X_r(t)\overline{c}_i + F_{i-1}(t), \tag{5.207}$$

$$\overline{c}_i = -B_0^+ P_{Q_d^*}\Bigg(l_1 F_{i-1}(\cdot) - l\Bigg(\int_a^{\cdot} X_{n-s}(\cdot) Y_{n-s}^*(\tau) A_1(\tau) F_{i-1}(\tau) d\tau$$

$$- \Phi(\cdot) \sum_{k=0}^{q-1} \mathrm{I}^k \frac{d^k}{dt^k} \Big([\Psi^*(t) L \Phi(t)]^{-1} \Psi^*(t) A_1(t) F_{i-1}(t) \Big)(\cdot) \Bigg)\Bigg), \tag{5.208}$$

$$i = 1, 2, \dots,$$

$$\overline{X}_i(t) = X_r(t) D_i + K_{i-1}(t), \qquad \overline{X}_{-1}(t) = X_r(t), \tag{5.209}$$

$$D_i = I_r - B_0^+ P_{Q_d^*}\Bigg[l_1 K_{i-1}(\cdot) - l\Bigg(\int_a^{\cdot} X_{n-s}(\cdot) Y_{n-s}^*(\tau) A_1(\tau) K_{i-1}(\tau) d\tau$$

$$- \Phi(\cdot) \sum_{k=0}^{q-1} \mathrm{I}^k \frac{d^k}{dt^k} \Big([\Psi^*(t) L \Phi(t)]^{-1} \Psi^*(t) A_1(t) K_{i-1}(t) \Big)(\cdot) \Bigg)\Bigg], \tag{5.210}$$

$$F_{i-1}(t) = \Big(G\left[A_1(\cdot) \overline{x}_{i-1}(\cdot, \overline{c}_{-1}) \right] \Big)(t) + X_{n-s}(t) Q^+ l_1 \overline{x}_{i-1}(\cdot, \overline{c}_{i-1}), \tag{5.211}$$

$$K_{i-1}(t) = \Big(G\left[A_1(\cdot) \overline{X}_{i-1}(\cdot) \right] \Big)(t) + X_{n-s}(t) Q^+ l_1 \overline{X}_{i-1}(\cdot). \tag{5.212}$$

Hence, the following statement is true:

Theorem 5.17. *Suppose that, for any inhomogeneities $f(t) \in C^{q-1}[a, b]$ and $\alpha \in \mathbb{R}^m$, the degenerate generating boundary-value problem (5.188) does not have solutions with $A(t), B(t) \in C^{3q-2}[a, b]$; $\det B(t) = 0 \;\; \forall t \in [a, b]$, and $l : C[a, b] \to \mathbb{R}^m$. The boundary-value problem (5.186), (5.187), under condition (5.203):*

$$\operatorname{rank} B_0 = d, \qquad (d = m - \operatorname{rank} Q),$$

has a $\rho = (n - s - m)$-parameter family of linearly independent solutions in the *form of a part of the series*

$$x(t, c_\rho) = \sum_{i=-1}^{+\infty} \varepsilon^i \Big[\overline{x}_i\,(t, \overline{c}_i) + \overline{X}_i\,(t)\,P_\rho c_\rho\Big] \quad \forall\, c_\rho \in \mathbb{R}^\rho, \tag{5.213}$$

with coefficients $\overline{x}_i(t, \overline{c}_i)$, $\overline{c}_i$, and $\overline{X}_i(t)$ specified by relations (5.207)–(5.212).

Remark 5.8. 1. By analogy with [150, 166, 173], we can prove that, for every fixed sufficiently small $\varepsilon \in (0; \varepsilon_0]$, series (5.213) converges for any $t \in [a, b]$ and estimate the rate of convergence of the series.

2. Parallel with the boundary-value problem (5.186), (5.187), without noticeable complications, we can consider the problem with the matrix $B\,(t)+\varepsilon B_1\,(t)$ instead of the matrix $B\,(t)$. In this case, all assertions of the theorem remain true by virtue of the following expression for B_0:

$$B_0 = P_{Q_d^*}\Bigg[l_1 X_r(\cdot) - l\bigg(\int_a^{\cdot} X_{n-s}(\cdot)Y_{n-s}^*(\tau)\,\{A_1(\tau)X_r(\tau) - B_1(\tau)\dot{X}_r(\tau)\}\,d\tau$$

$$- \Phi(\cdot)\sum_{k=0}^{q-1} \mathrm{I}^k \frac{d^k}{dt^k}\Big([\Psi^*(t)L\Phi(t)]^{-1}$$

$$\times\, \Psi^*(t)\,\{A_1(t)X_r(t) - B_1(t)\dot{X}_r(t)\}\Big)(\cdot)\bigg)\Bigg].$$

3. If $\det B\,(t) \neq 0\;\forall\, t \in [a, b]$, then we get a nondegenerate Fredholm boundary-value problem for systems of ordinary differential equations, which has been studied in detail in the previous sections [38, 174].

4. Condition (5.203) is sufficient for the existence of solutions of problem (5.186), (5.187). This condition can be satisfied if the boundary-value problem (5.186), (5.187) is such that the number of boundary conditions m does not exceed the number of unknowns n minus the sum s of orders of the Jordan nilpotent blocks of the corresponding order ($m \le n-s$) in the central canonical form of the original equation. If condition (5.203) is not satisfied, then the solution of problem (5.186) in the form of series (5.198) does not exist. However, this solution may exist in the form of a part of series of the form (5.198) in powers of the small parameter ε starting from $-2, -3, \ldots$ etc.

Example. As an illustration of the established theorems, we consider the problem of determination of the solution of a boundary-value problem

$$\begin{pmatrix} \sin 2t - 1 & \cos 2t \\ -\cos 2t & \sin 2t + 1 \end{pmatrix} \frac{dx}{dt} = \begin{pmatrix} 2 & 0 \\ 0 & -2 \end{pmatrix} x + \varepsilon A_1(t)x + f(t), \tag{5.214}$$

$$lx(\cdot) = \begin{pmatrix} 1 & 1 \end{pmatrix} x(0) + \begin{pmatrix} 1 & 1 \end{pmatrix} x(2\pi) = \alpha + \varepsilon l_1 x(\cdot), \tag{5.215}$$

where

$$A_1(t) = \{a_{ij}(t)\}_{i,j=1}^{2}, \qquad l_1 x(\cdot) = \begin{pmatrix} 1 & -1 \end{pmatrix} x(0).$$

The fundamental matrix of the corresponding homogeneous unperturbed ($\varepsilon = 0, f = 0$) system (5.214) has the form

$$X_r(t) = X_{n-s}(t) = \begin{pmatrix} \cos t - \sin t \\ -(\cos t + \sin t) \end{pmatrix} e^{-t}.$$

The matrix Q and the orthoprojectors P_Q and P_{Q^*} onto the kernel and cokernel of the matrix Q take the form

$$Q = 0, \quad P_{Q^*} = 1, \quad P_Q = 1, \quad P_{Q_r} = 1, \quad P_{Q_d^*} = 1;$$

$$n = 2, \quad m = d = r = s = q = 1.$$

The generating boundary-value problem obtained from (5.214), (5.215) for $\varepsilon = 0$ does not have solutions for arbitrary inhomogeneities $f(t) \in C[0, 2\pi]$ and $\alpha \in \mathbb{R}^1$. We establish the conditions under which the boundary-value problem (5.214), (5.215) becomes solvable with the help of linear perturbations. The matrix $Y_{n-s}(t) = \frac{1}{2}\begin{pmatrix} \sin t \\ -\cos t \end{pmatrix} e^{-t}$ is the fundamental matrix of the system adjoint to the homogeneous differential system (5.214),

$$\Phi(t) = \begin{pmatrix} \cos t + \sin t \\ \cos t - \sin t \end{pmatrix}; \qquad \Psi(t) = \begin{pmatrix} \sin t + \cos t \\ \sin t - \cos t \end{pmatrix}.$$

Then, according to relation (5.202), the matrix B_0 is a scalar and has the form

$$B_0 = 2 + \frac{1}{2}\{a_{11}(0) - a_{21}(0) - a_{12}(0) + a_{22}(0)$$
$$+ [a_{11}(2\pi) - a_{21}(2\pi) - a_{12}(2\pi) + a_{22}(2\pi)]e^{-2\pi}\}.$$

The conditions for the components a_{ij} of the perturbation matrix $A_1(t)$ and for the perturbation l_1 in the boundary condition under which $B_0 \neq 0$ are sufficient conditions for the solvability of the boundary-value problem (5.214), (5.215). The problem can be made solvable both with the help of perturbations in the differential system (5.186) and with the help of perturbations in the boundary condition (5.187). Thus, if the perturbation matrix $A_1(t)$ of the differential system is a null matrix, then condition (5.203) is satisfied due to the presence of perturbation in the boundary condition because $B_0 = 2$ and, in this case, the boundary-value problem (5.214), (5.215) has a unique ($\rho = n - s - m = 2 - 1 - 1 = 0$) solution in the form of series (5.213) convergent for sufficiently small fixed $\varepsilon \in (0; \varepsilon_0]$.

Chapter 6

IMPULSIVE BOUNDARY-VALUE PROBLEMS FOR SYSTEMS OF ORDINARY DIFFERENTIAL EQUATIONS[1]

In the mathematical modeling of actual processes with short-term perturbations, it is often possible to neglect the duration of perturbations. These perturbations are regarded as "instantaneous." The indicated idealization leads to the necessity of investigation of dynamical systems with discontinuous trajectories, which are often called dynamical systems with impulsive action.

The problem of investigation of ordinary differential equations with impulsive action is not new. Thus, Krylov and Bogolyubov [90] demonstrated the possibility of investigation of systems of differential equations with impulsive action by the successive application of the asymptotic methods of nonlinear mechanics.

The systematic investigation of the mathematical problems encountered in the theory of differential systems with impulsive action was originated by Myshkis and Samoilenko [111], Samoilenko and Perestyuk [139], and Halanay and Wexler [69]. Later, the ideas proposed in these works were developed and generalized in numerous publications (see [3, 8, 29, 30, 135, 155]). It became clear that the theory of boundary-value problems for differential systems with impulsive action can be developed and new original results can be obtained by using the classical methods of the periodic theory of nonlinear oscillations [14] together with methods based on the theory of generalized inverse operators and generalized Green operators [19]. This direction in the theory of differential systems with impulsive action is studied in the present chapter.

6.1 Linear Boundary-Value Problems. Criterion of Solvability

We consider the problem of necessary and sufficient conditions for the solvability of a system of ordinary differential equations with impulsive action at fixed

[1] The research presented in this chapter was supported by the Foundation for Fundamental Research of the Ukrainian State Committee on Science and Technology (Grant No. 01.07/00109).

points of time (see [69, 139, 155])

$$\dot{z} = A(t)z + f(t), \quad t \neq \tau_i, \quad \Delta z\Big|_{t=\tau_i} = S_i z(\tau_i - 0) + a_i,$$
$$t \in [a, b], \quad \tau_i \in (a, b), \quad i = 1, \dots, p, \tag{6.1}$$

with boundary conditions

$$lz = \alpha, \quad \alpha \in R^m. \tag{6.2}$$

We use the assumptions and notation from [19, 29, 139]:

$$A(t), f(t) \in C([a, b] \setminus \{\tau\}_I)$$

are an $n \times n$ matrix function and an $n \times 1$ vector function, respectively, $C([a, b] \setminus \{\tau_i\}_I)$ is the space of vector functions continuous or piecewise continuous for $t \in [a, b]$ with discontinuities of the first kind for $t = \tau_i$, S_i are $n \times n$ constant matrices such that $E + S_i$ are nonsingular, which means that the solutions of the impulsive system admit unambiguous continuation through the points of discontinuity, a_i is an n-dimensional column vector of constants, $a_i \in R^n$, $-\infty < a < \tau_1 < \dots < \tau_i < \dots < \tau_p < b < \infty$ for $i = 1, \dots, p$, $l = \operatorname{col}(l_1, \dots, l_m)$ is a bounded linear m-dimensional vector functional, $\alpha = \operatorname{col}(\alpha_1, \dots, \alpha_m) \in R^m$, and $\Delta z\big|_{t=\tau_i} = z(\tau_i + 0) - z(\tau_i - 0)$.

The solution $z(t)$ is sought in the space of n-dimensional piecewise continuously differentiable vector functions $z(t) \in C^1([a, b] \setminus \{\tau_i\}_I)$. The norms in the spaces $C([a, b] \setminus \{\tau_i\}_I)$ and $C^1([a, b] \setminus \{\tau_i\}_I)$ are introduced in a standard way (by analogy with [3, 8, 69, 139]).

In a special case ($m = n$, $lz = z(0) - z(T) = \alpha = 0$), this problem turns into the well-known periodic boundary-value problem with impulsive action solved in [139] (in the noncritical case) and [30] (in the critical case). A similar statement of the problem for systems of ordinary differential equations without impulsive action was studied in the previous chapter.

Parallel with the inhomogeneous boundary-value problem (6.1), (6.2), we consider the following homogeneous boundary-value problem:

$$\dot{z} = A(t)z, \quad t \neq \tau_i, \quad \Delta z\Big|_{t=\tau_i} = S_i z(\tau_i - 0), \tag{6.3}$$

$$lz = 0, \quad t \in [a, b], \quad \tau \in (a, b), \quad i = 1, \dots, p. \tag{6.4}$$

We preserve the classification of boundary-value problems introduced above and recall the following definition:

Definition 6.1. The boundary-value problems with impulsive action (6.1), (6.2) for which the problems conjugate to the corresponding linear homogeneous boundary-value problems (6.3), (6.4) have only trivial (nontrivial) solutions are called noncritical (critical).

Let Q be the $m \times n$ matrix obtained by substituting the normal fundamental matrix $X(t) = X(t,a)$, $X(a) = E$, of the system with impulsive action (6.3) in the boundary condition, i.e., $Q = lX(\cdot)$. Then the noncritical case corresponds to the equality $\operatorname{rank} Q = m$ and, in the critical case, we have $\operatorname{rank} Q < m$ (see [19, 65, 93]).

Furthermore, the normal fundamental matrix $X(t,a)$ of the impulsive system (6.3) can be expressed via the normal fundamental matrix $U(t,a)$ of the system without pulses

$$\frac{dU}{dt} = A(t)U, \quad U(a,a) = E,$$

as follows [139]:

$$X(t,a) = U(t,a), \quad a \leq t \leq \tau_1 < b,$$

$$X(t,a) = U(t,\tau_j) \prod_{v=j}^{1} (E + S_v) U(\tau_v, \tau_{v-1}), \tag{6.5}$$

$$a < \tau_j < t \leq \tau_{j+1} < b, \quad j = 1, 2, \ldots, \quad \tau_0 = a.$$

Our aim is to establish necessary and sufficient conditions for the solvability of the boundary-value problem (6.1), (6.2). It is known [135, 139] that any solution $z(t,c) \in C^1([a,b] \setminus \{\tau_i\}_I)$, $z(0,c) = c \in R^n$, of the impulsive system (6.1) has the form

$$z(t,c) = X(t)c + \bar{z}(t), \tag{6.6}$$

where $\bar{z}(t)$ is an arbitrary special solution of the inhomogeneous system (6.1). The required special solution has the form

$$\bar{z} = \int_a^b K(t,\tau) f(\tau) d\tau + \sum_{i=1}^{p} \bar{K}(t,\tau_i) a_i.$$

This fact can be checked by the direct substitution of the proposed expression in (6.1). Here, $K(t,\tau)$ is the Green matrix of the Cauchy problem for the system with impulsive action (6.1). This Green matrix is, as a rule, called a Cauchy matrix, and its choice is ambiguous. Thus, any matrix of the form

$$K(t,\tau) = \begin{cases} X(t)X^{-1}(\tau), & a \leq \tau \leq t \leq b, \\ 0, & a \leq t < \tau \leq b, \end{cases}$$

$$\bar{K}(t,\tau_i) = K(t,\tau_i + 0),$$

or

$$K(t,\tau) = \begin{cases} -X(t)X^{-1}(\tau), & a \leq t \leq \tau \leq b, \\ 0, & a \leq \tau < t \leq b, \end{cases}$$

$$\bar{K}(t,\tau_i) = K(t,\tau_i - 0)(E + S_i)^{-1}$$

can play the role of the Cauchy matrix.

An important role in the study of periodic boundary-value problems for systems with impulsive action is played by conjugate systems with impulsive action [139]. However, in what follows, we show that, in the analysis of the conditions of solvability for the boundary-value problem (6.1), (6.2) with impulsive action, one can use the theory of generalized inverse matrices and projectors instead of the investigation of conjugate systems. In order that relation (6.6) be a solution of the boundary-value problem (6.1), (6.2), it is necessary and sufficient that the boundary condition be satisfied. Substituting relation (6.6) in the boundary condition (6.2), we arrive at the system of algebraic equations for the vector $c \in R^n$:

$$Qc + l\bar{z} = \alpha, \tag{6.7}$$

where Q is an $m \times n$ matrix and

$$l\bar{z} = l\int_a^b K(\cdot,\tau)f(\tau)d\tau + l\sum_{i=1}^{p} \overline{K}(\cdot,\tau_i)a_i.$$

Theorem 3.10 enables us to obtain a condition under which system (6.7) and, hence, the boundary-value problem (6.1), (6.2) are solvable and determine a constant $c \in R^n$ for which the solution $z(t,c)$ of the differential system (6.1) with impulsive action given by relation (6.6) is, at the same time, a solution of the boundary-value problem (6.1), (6.2).

We first consider the critical case where the problem conjugate to the homogeneous system (6.3), (6.4) has nontrivial solutions, which is equivalent to the condition $\operatorname{rank} Q = n_1 < m$ [19, 65, 101].

Let $\operatorname{rank} Q = n_1 < m$, let P_Q be an $n \times n$ matrix orthoprojector $P_Q\colon R^n \to N(Q)$ $(P_Q^2 = P_Q = P_Q^*)$, let P_{Q^*} be an $m \times m$ matrix orthoprojector P_{Q^*} : $R^m \to N(Q^*)$ $(P_{Q^*}^2 = P_{Q^*} = P_{Q^*}^*)$, and let Q^+ be the unique $n \times m$ matrix pseudoinverse to Q in the Moore–Penrose sense [109, 117, 147]. The matrices P_Q, P_{Q^*}, and Q^+ can be found by using the relations presented in Chapter 2.

We now return to system (6.7) and conclude that it is solvable with respect to $c \in R^n$ (and, hence, the boundary-value problem (6.1), (6.2) is solvable) if and only if the term $\alpha - l\bar{z}$ belongs to the orthogonal complement to the kernel of the matrix $Q^* = Q^T$ adjoint to Q, i.e., in the case where

$$P_{Q^*}\{\alpha - l\bar{z}\} = P_{Q^*}\left\{\alpha - l\int_a^b K(\cdot,\tau)f(\tau)d\tau - l\sum_{i=1}^{p} \bar{K}(\cdot,\tau_i)a_i\right\} = 0. \tag{6.8}$$

This condition is necessary and sufficient for the solvability of equation (6.7). The required solution of this equation has the form [19, 29]

$$c = Q^+\left\{\alpha - l\int_a^b K(\cdot,\tau)f(\tau)d\tau - l\sum_{i=1}^{p} \bar{K}(\cdot,\tau_i)a_i\right\} + P_Q c. \tag{6.9}$$

If we now substitute the vector $c \in R^n$ given by relation (6.9) in representation (6.6), then we get the general solution of the boundary-value problem (6.1), (6.2)

$$z(t,c) = X(t)P_Q c + X(t)Q^+\left\{\alpha - l\int_a^b K(\cdot,\tau)f(\tau)d\tau - l\sum_{i=1}^p \bar{K}(\cdot,\tau_i)a_i\right\}$$
$$+\int_a^b K(t,\tau)f(\tau)d\tau + \sum_{i=1}^p \bar{K}(t,\tau_i)a_i. \qquad (6.10)$$

Since $\operatorname{rank} P_{Q^*} = m - \operatorname{rank} Q^* = m - n_1 = d$ $(\operatorname{rank} Q^* = \operatorname{rank} Q = n_1 \le \min(n,m))$, we conclude that relation (6.8) consists of d linearly independent conditions. Therefore, the $m \times m$ matrix P_{Q^*} can be replaced by the $d \times m$ matrix $P_{Q^*_d}$ formed by d linearly independent rows of P_{Q^*}. Further, since $\operatorname{rank} P_Q = n - \operatorname{rank} Q = n - n_1 = r$, the first term in relation (6.10) contains r arbitrary constants. Therefore, it can be replaced by the term $X_r(t)c_r$, where $X_r(t) = X(t)P_{Q_r}$, P_{Q_r} is an $n \times r$ matrix formed by r linearly independent columns of the matrix P_Q, and c_r is an arbitrary column vector from R_r.

Thus, we have proved the following assertion:

Theorem 6.1. *If* $\operatorname{rank} Q = n_1$, *then the homogeneous boundary-value problem (6.3), (6.4) has exactly* $r = n - n_1$ *linearly independent solutions. The inhomogeneous problem (6.1), (6.2) is solvable if and only if* $f(t) \in (C[a,b] \setminus \{\tau_i\}_I)$, $a_i \in R^n$, *and* $\alpha \in R^m$ *satisfy the condition*

$$P_{Q^*_d}\left\{\alpha - l\int_a^b K(\cdot,\tau)f(\tau)d\tau - l\sum_{i=1}^p \bar{K}(\cdot,\tau_i)a_i\right\} = 0, \quad d = m - n_1. \qquad (6.11)$$

In this case, the problem possesses the r*-parameter system of solutions*

$$z(t,c_r) = X_r(t)c_r + \left(G\begin{bmatrix} f \\ a_i \end{bmatrix}\right)(t) + X(t)Q^+\alpha,$$

where the operator $\left(G\begin{bmatrix} f \\ a_i \end{bmatrix}\right)(t)$ *is given by the formula*

$$\left(G\begin{bmatrix} f \\ a_i \end{bmatrix}\right)(t) \stackrel{\text{def}}{=} \Bigg[\int_a^b K(t,\tau) * d\tau - X(t)Q^+ l\int_a^b K(\cdot,\tau) * d\tau,$$
$$\sum_{i=1}^p \bar{K}(t,\tau) * - X(t)Q^+ l\sum_{i=1}^p \bar{K}(\cdot,\tau) * \Bigg]\begin{bmatrix} f(\tau) \\ a_i \end{bmatrix} \qquad (6.12)$$

and is called the generalized Green operator of the original boundary-value problem with impulsive action [29, 30, 135].

Theorem 6.1 stated in the general form yields the following assertion:

Theorem 6.2. *If* rank $Q = n_1 = n$, *then the homogeneous boundary-value problem (6.3), (6.4) has only the trivial solution. The inhomogeneous boundary-value problem (6.1), (6.2) is solvable if and only if* $f(t) \in C([a,b] \setminus \{\tau_i\}_I)$, $a_i \in R^n$, *and* $\alpha \in R^m$ *satisfy the conditions*

$$P_{Q_d^*}\left\{\alpha - l\int_a^b K(\cdot,\tau)f(\tau)d\tau - l\sum_{i=1}^{p}\bar{K}(\cdot,\tau_i)a_i\right\} = 0, \quad d = m - n.$$

In this case, the analyzed problem possesses a unique solution

$$z_0(t) = \left(G\begin{bmatrix} f \\ a_i \end{bmatrix}\right)(t) + X(t)Q^+\alpha.$$

As a special case of problem (6.1), (6.2), we consider the following linear inhomogeneous boundary-value problem with constant coefficients:

$$\dot{z} = Az + f(t), \quad t \neq \tau_i, \ \Delta z\Big|_{t=\tau_i} = Sz(\tau_i - 0) + a_i, \tag{6.13}$$

$$lz = \alpha, \quad \alpha \in R^m. \tag{6.14}$$

The corresponding homogeneous problem has the form

$$\dot{z} = Az, \quad t \neq \tau_i, \quad \Delta z\Big|_{t=\tau_i} = Sz(\tau_i - 0), \tag{6.15}$$

$$lz = 0. \tag{6.16}$$

It is known [139] that the normal fundamental matrix $X(t,a)$ of system (6.15) with impulsive action has the form

$$X(t,a) = e^{A(t-\tau_i)} \prod_{a<\tau_v<t} (E+S)e^{A(\tau_v-\tau_{v-1})}, \quad \tau_0 = a. \tag{6.17}$$

Thus, if the matrices A and S are commuting [139], then the matrix exponent e^{At} commutes with the matrix S, and relation (6.17) can be rewritten as

$$X(t,a) = e^{A(t-a)}(E+S)^{i(t,a)}, \tag{6.18}$$

where $i(t,a)$ is the number of points τ_i in the interval $[a,t]$, i.e., $i(t,a) = k$ for $\tau_k < t < \tau_{k+1}$.

If we know the normal fundamental matrix $X(t,a)$ given by relation (6.17) for the homogenous system (6.15) corresponding to (6.13), then we can readily construct the matrix

$$Q = lX(\cdot,a) = le^{A(\cdot-\tau_i)} \prod_{a<\tau_v<\cdot} (E+S)e^{A(\tau_v-\tau_{v-1})}, \quad \tau_0 = a, \tag{6.19}$$

and the Green matrix of the Cauchy problem for system (6.13).

For the matrix Q given by relation (6.19), we construct the matrices P_Q, P_{Q^*}, and Q^+ and, thus, completely solve the problem of conditions required for the solvability of the boundary-value problem (6.13), (6.14) with impulsive action and the general form of its solution.

The periodic boundary-value problem with impulsive action

$$\dot{z} = Az + f(t), \quad t \neq \tau_i, \quad \Delta z\Big|_{t=\tau_i} = S_i z(\tau_i - 0) + a_i, \tag{6.20}$$

$$lz = z(0) - z(T) = 0, \quad m = n, \tag{6.21}$$

is also a special case of the boundary-value problem with impulsive action (6.1), (6.2). Here, $A(t)$ and $f(t)$ are, respectively, T-periodic $n \times n$ matrix and $n \times 1$ vector functions (either continuous or piecewise continuous with discontinuities of the first kind). S_i are $n \times n$ matrices with constant entries, a_i are n-dimensional vectors, and τ_i are time points such that

$$S_{i+p} = S_i, \quad a_{i+p} = a_i, \quad \tau_{i+p} = \tau_i + T \quad (0 < \tau_0 < \tau_1 < \ldots < \tau_p < T)$$

for all $i \in N$. The matrices $E + S_i$, where E is the $n \times n$ identity matrix, are nonsingular.

The general solution $z(t, c)$, $z(0, c) = c \in R^n$, of system (6.20) in the space of vector functions piecewise continuously differentiable with respect to $t \in [0, T] \setminus \{\tau_i\}$ (with discontinuities of the first kind at the point $t = \tau_i$) has the form

$$z(t, c) = X(t)c + \int_0^T K(t, \tau) f(\tau) d\tau + \sum_{i=1}^{p} \bar{K}(t, \tau_i) a_i, \tag{6.22}$$

where $X(t)$ $(X(0) = E)$ is the fundamental matrix of the system with impulsive action

$$\dot{z} = A(t)z, \quad t \neq \tau_i, \quad \Delta z\Big|_{t=\tau_i} = S_i z(\tau_i - 0) \tag{6.23}$$

corresponding to (6.20) and $K(t, \tau)$ is the Green matrix of the Cauchy problem for the impulsive system (6.20) on the interval $[0, T]$, which can be chosen in the form

$$K(t, \tau) = \begin{cases} X(t) X^{-1}(\tau) & \text{if } 0 \le \tau \le t \le T, \\ 0 & \text{if } 0 \le t < \tau \le T, \end{cases}$$

$$\bar{K}(t, \tau_i) = K(t, \tau_i + 0).$$

Relation (6.22) gives a T-periodic solution of the impulsive boundary-value problem (6.20), (6.21) if and only if $z(0, c) = z(T, c)$. This condition yields the following algebraic system for the vector constant $c \in R^n$:

$$Qc = \int_0^T K(T, \tau) f(\tau) d\tau + \sum_{i=1}^{p} \bar{K}(T, \tau_i) a_i, \tag{6.24}$$

where

$$Q = X(0) - X(T) = E - X(T)$$

is an $n \times n$ matrix. The condition $\det Q \neq 0$ is equivalent to the well-known noncritical case characterized by the fact that the homogeneous impulsive boundary-value problem (6.20), (6.21) ($f(t) = 0,\ a_i = 0$) may have only trivial T-periodic solutions. In this case, system (6.24) is uniquely solvable with respect to $c \in R^n$ and, hence, the inhomogeneous differential system possesses a unique T-periodic solution [139].

The critical case ($\det Q = 0$) was studied in [30]. In this case, the homogeneous impulsive boundary-value problem ($f(t) = 0,\ a_i = 0$) corresponding to problem (6.20), (6.21) possesses an r-parameter family of T-periodic solutions ($r = n - n_1,\ n_1 = \operatorname{rank} Q$). A necessary and sufficient condition for the solvability of the algebraic system (6.24) with respect to $c \in R^n$ has the form

$$P_{Q^*}\left\{\int_0^T K(T,\tau)f(\tau)d\tau + \sum_{i=1}^p \bar{K}(T,\tau_i)a_i\right\} = 0. \tag{6.25}$$

Since $\operatorname{rank} P_{Q^*} = n - n_1 = d = r$, we conclude that relation (6.25) contains r linearly independent conditions. Let $P_{Q_r^*}$ be the $r \times n$ matrix formed by r linearly independent rows of the orthoprojector P_{Q^*}. This enables us to rewrite condition (6.25) in the form

$$P_{Q_r^*}\left\{\int_0^T K(T,\tau)f(\tau)d\tau + \sum_{i=1}^p \bar{K}(T,\tau_i)a_i\right\} = 0. \tag{6.26}$$

Under this condition, system (6.25) possesses the following r-parameter family of solutions:

$$c = Q^+\left\{\int_0^T K(T,\tau)f(\tau)d\tau + \sum_{i=1}^p \bar{K}(T,\tau_i)a_i\right\} + P_Q c, \tag{6.27}$$

where Q^+ is the unique matrix pseudoinverse to the matrix Q in the Moore–Penrose sense and given by relation (3.21). Substituting relation (6.27) in solution (6.22), we get the r-parameter family of T-periodic solutions of the boundary-value problem (6.20), (6.21)

$$z(t,c) = X(t)P_Q c + X(t)Q^+\left\{\int_0^T K(T,\tau)f(\tau)d\tau + \sum_{i=1}^p \bar{K}(T,\tau_i)a_i\right\}$$
$$+ \int_0^T K(t,\tau)f(\tau)d\tau + \sum_{i=1}^p \bar{K}(t,\tau_i)a_i.$$

Since $\operatorname{rank} P_Q = n - \operatorname{rank} Q = r$, the first term contains r arbitrary constants and can be replaced by the term $X_r(t)c_r$, where $X_r(t) = X(t)P_{Q_r}$ is the $n \times r$ matrix whose columns form a complete system of r linearly independent T-periodic solutions of the homogeneous impulsive system (6.23) and c_r is an arbitrary column vector from R^r. Therefore, the general solution of the boundary-value problem (6.20), (6.21) has the form

$$z(t, c_r) = X_r(t)c_r + \int_0^T G_0(t,\tau)f(\tau)d\tau + \sum_{i=1}^{p} \bar{G}_0(t,\tau_i)a_i, \tag{6.28}$$

where $G_0(t,\tau)$ is the generalized Green matrix of the original boundary-value problem (6.20), (6.21) given by the formulas

$$\begin{aligned} G_0(t,\tau) &= K(t,\tau) + X(t)Q^+K(T,\tau), \\ \bar{G}_0(t,\tau_i) &= G_0(t,\tau_i+0). \end{aligned} \tag{6.29}$$

Thus, the following statement is true under the assumptions made above concerning the periodic boundary-value problem (6.20), (6.21):

Theorem 6.3. *If* $\operatorname{rank} Q = n_1 < n$, *then the homogeneous boundary-value problem* $(f(t) = 0,\ a_i = 0)$ *corresponding to problem (6.20), (6.21) has exactly* r *linearly independent* T*-periodic solutions* $(r = n - n_1)$. *The inhomogeneous boundary-value problem (6.20), (6.21) is solvable if and only if* $f(t) \in C([0,T] \setminus \{\tau_i\}_I)$ *and* $a_i \in R^n$ *satisfy conditions (6.26). In this case, it possesses the* r*-parameter family of* T*-periodic solutions given by relation (6.28) from the space of vector functions piecewise continuously differentiable with respect to* $t \in [0,T] \setminus \{\tau_i\}$ *with discontinuities of the first kind at the points* $t = \tau_i$, $i = 1, \ldots, p$.

If $\operatorname{rank} Q = n_1 = n$, then we have the noncritical case ($\det Q \neq 0$, $Q^+ = Q^{-1}$, $P_Q = P_{Q^*} = 0$), condition (6.26) is always satisfied, and the boundary-value problem (6.20), (6.21) possesses a unique solution ($r = 0$) given by the same formula (6.28) ($X_r(t) = 0$) in which the Green matrix $G_0(t,\tau)$ (6.29) has the form ($Q^+ = Q^{-1}$)

$$G_0(t,\tau) = K(t,\tau) + X(t)Q^{-1}K(T,\tau).$$

6.2 Generalized Green Operator for the Semihomogeneous Boundary-Value Problem and Its Properties

In the previous section, we have constructed the generalized Green operator for the boundary-value problem (6.1), (6.2), namely,

$$\left(G\begin{bmatrix} f \\ a_i \end{bmatrix}\right)(t) \stackrel{\text{def}}{=} \Bigg[\int_a^b K(t,\tau) * d\tau - X(t)Q^+ l \int_a^b K(\cdot,\tau) * d\tau, \\ \sum_{i=1}^{p} \bar{K}(t,\tau_i) * - X(t)Q^+ l \sum_{i=1}^{p} \bar{K}(\cdot,\tau_i) * \Bigg] \begin{bmatrix} f(\tau) \\ a_i \end{bmatrix}.$$

This operator resolves the semihomogeneous boundary-value problem

$$\Lambda z = \begin{bmatrix} f \\ a_i \\ 0 \end{bmatrix},$$

where

$$\Lambda \cdot \stackrel{\text{def}}{=} \operatorname{col} \left[\frac{d\cdot}{dt} - A\cdot, \quad \Delta \cdot \Big|_{t=\tau_i} - S_i\cdot, \quad l\, \cdot \right]$$

is a linear operator acting from $C^1([a,b]\setminus\{\tau_i\}_I)$ into $C([a,b]\setminus\{\tau_i\}_I) \times R^n \times R^m$. Moreover, Λ is a Fredholm operator [135] whose index is equal to $n-m$:

$$\begin{aligned} \operatorname{ind} \Lambda &= \dim\ker \Lambda - \dim\ker \Lambda^* = \operatorname{ind} Q \\ &= r - d = n - n_1 - (m - n_1) = n - m < \infty. \end{aligned}$$

For $m = n$ $(r = d)$, Λ is a Fredholm operator of index zero $(\operatorname{ind}\Lambda = 0)$.
If the functional l is such that

$$l \int_a^b K(\cdot,\tau) * d\tau = \int_a^b lK(\cdot,\tau) * d\tau,$$

then

$$\left(G \begin{bmatrix} f \\ a_i \end{bmatrix} \right)(t) = \left[\int_a^b G_0(t,\tau) * d\tau, \ \sum_{i=1}^{p} \bar{G}_0(t,\tau_i) * \right] \begin{bmatrix} f(\tau) \\ a_i \end{bmatrix},$$

where $G_0(t,\tau)$ is the generalized Green matrix of the boundary-value problem (6.1), (6.2) without pulses. This matrix has the form (see Chapter 5)

$$G_0(t,\tau) = K(t,\tau) - X(t)Q^+ lK(\cdot,\tau),$$

$$\bar{G}_0(t,\tau_i) = G_0(t,\tau_i - 0)(E + S_i)^{-1}$$

or

$$\bar{G}_0(t,\tau_i) = G_0(t,\tau_i + 0)$$

and possesses the following basic properties:

(a) for $t \neq \tau$, each column of the matrix $G_0(t,\tau)$ is a continuously differentiable solution of the homogeneous differential system (6.3) without pulses:

$$\frac{\partial G_0(t,\tau)}{\partial t} = A(t) G_0(t,\tau), \quad t \neq \tau;$$

(b) the matrix $G_0(t,\tau)$ has a discontinuity at the point $t = \tau$:

$$G_0(\tau + 0,\tau) - G_0(\tau - 0,\tau) = E;$$

(c) the matrix $G_0(t,\tau)$ satisfies the boundary condition

$$lG_0(\cdot,\tau) = P_{Q^*}lK(\cdot,\tau), \quad P_{Q^*}\colon R^m \to N(Q^*).$$

Indeed, the definitions of the Cauchy matrix $K(t,\tau)$ and the Green matrix $G_0(t,\tau)$ enable us to make the following conclusions:

Since

$$\frac{\partial K(t,\tau)}{\partial t} = \frac{\partial X(t)}{\partial t}X^{-1}(\tau) = A(t)X(t)X^{-1}(\tau) = A(t)K(t,\tau),$$
$$a \le \tau \le t \le b,$$

we get

$$\frac{\partial G_0(t,\tau)}{\partial t} = \frac{\partial K(t,\tau)}{\partial t} - \frac{\partial X(t)}{\partial t}Q^+lK(\cdot,\tau)$$
$$= A(t)K(t,\tau) - A(t)X(t)lK(\cdot,\tau) = A(t)G_0(t,\tau), \quad t \ne \tau.$$

Since

$$K(\tau+0,\tau) - K(\tau-0,\tau) = E,$$

we have

$$G_0(\tau+0,\tau) - G_0(\tau-0,\tau) = K(\tau+0,\tau) - X(\tau+0)Q^+lK(\cdot,\tau)$$
$$- \{K(\tau-0,\tau) - X(\tau-0)Q^+lK(\cdot,\tau)\} = E$$

and, finally,

$$lG_0(\cdot,\tau) = lK(\cdot,\tau) - lX(\cdot)Q^+lK(\cdot,\tau) = (E - QQ^+)lK(\cdot,\tau) = P_{Q^*}lK(\cdot,\tau).$$

Remark 6.1. For the Green matrices

$$G_0(t,\tau) = K(t,\tau) - X(t)Q^{-1}lK(\cdot,\tau)$$

of the noncritical Fredholm $(m = n)$ boundary-value problems of index zero without pulses, property (c) takes the form

$$lG_0(\cdot,\tau) = 0$$

because $P_{Q^*} = 0$.

6.3 Regularization of Linear Impulsive Boundary-Value Problems

In the present section, we illustrate the applicability of the criterion of solvability (6.11) of impulsive boundary-value problems by analyzing the problem of regularization of a boundary-value problem (which is solvable not everywhere) by the action of pulses [8]. Assume that the boundary-value problem (6.1), (6.2) is unsolvable for some $f_0(t) \in C([a,b] \setminus \{\tau_i\}_I)$, $a_i^0 \in R^n$, and $\alpha_0 \in R^m$, i.e., criterion (6.11) is not satisfied. In this problem, we introduce an impulsive action $\left(\text{for } t = \bar{\tau}_1\right)$ as follows:

$$\Delta z\Big|_{t=\bar{\tau}_1} = \bar{S}_1 z + \bar{a}_1, \quad \tau_i \neq \bar{\tau}_1 \in]a,b[, \quad \exists (E + \bar{S}_1)^{-1}. \tag{6.30}$$

The parameter $\bar{a}_1$ is chosen from a criterion similar to (6.11) guaranteeing that the impulsive boundary-value problem (6.1), (6.2), (6.30) is solvable for any $f_0(t) \in C([a,b] \setminus \{\tau_i\}_I)$, $a_i^0 \in R^n$, and $\alpha_0 \in R^m$, namely,

$$P_{Q_d^*} l \bar{K}_0(\cdot, \bar{\tau}_1)\bar{a}_1 = P_{Q_d^*}\left\{\alpha_0 - l\int_a^b K_0(\cdot,\tau) f_0(\tau)d\tau - l\sum_{i=1}^p \bar{K}_0(\cdot,\tau_i)a_i^0\right\},$$

where $K_0(t,\tau)$ is the Green matrix of the Cauchy problem (6.1), (6.30).

We use the following notation: $S = P_{Q_d^*} l \bar{K}_0(\cdot, \bar{\tau}_1)$ is a $d \times n$ matrix, S^+ is the $n \times d$ matrix pseudoinverse to the matrix S, P_{S^*} is a $d \times d$ matrix (orthoprojector) projecting the space R^d onto the space $N(S^*)$, and P_S is an $n \times n$ matrix (orthoprojector) projecting the space R^n onto the space $N(S)$. Theorem 6.2 yields the following assertion (see [135]):

Corollary 6.1. *The boundary-value problem (6.1), (6.2) which is solvable not everywhere can be made solvable for any* $f_0(t) \in C([a,b] \setminus \{\tau_i\}_I)$, $a_i^0 \in R^n$, *and* $\alpha_0 \in R^m$ *by adding another impulsive action if and only if*

$$P_{S^*} = 0 \quad or \quad \operatorname{rank} S = d.$$

The indicated additional (regularizing) pulse $\bar{a}_1$ *should be chosen as follows:*

$$\bar{a}_1 = S^+ P_{Q_d^*}\left\{\alpha_0 - l\int_a^b K_0(\cdot,\tau) f_0(\tau)d\tau - l\sum_{i=1}^p \bar{K}_0(\cdot,\tau_i)a_i^0\right\} + P_S c \quad \forall c \in R^n.$$

6.4 Conditions for the Appearance of Solutions of Weakly Perturbed Linear Boundary-Value Problems

The criterion of solvability (6.11) of impulsive boundary-value problems of the form (6.1), (6.2) enables us to propose another method of regularization of the boundary-value problem which is solvable not everywhere (without introducing additional pulses).

Consider a weakly perturbed inhomogeneous linear boundary-value problem with impulsive action of the form

$$\begin{aligned} \dot{z} &= A(t)z + \varepsilon A_1(t)z + f(t), \quad t \neq \tau_i, \\ \Delta z\Big|_{t=\tau_i} &- S_i z(\tau_i - 0) = a_i + \varepsilon A_{1i} z(\tau_i - 0), \\ lz &= \alpha + \varepsilon l_1 z, \end{aligned} \tag{6.31}$$

where $A_1(t) \in C([a,b] \setminus \{\tau_i\}_I)$ is an $n \times n$ matrix function, A_{1i} are $n \times n$ constant matrices, and l_1 is a bounded linear m-dimensional vector functional.

Assume that the generating boundary-value problem

$$\begin{aligned} \dot{z} &= A(t)z + f(t), \quad t \neq \tau_i, \\ \Delta z\Big|_{t=\tau_i} &- S_i z(\tau_i - 0) = a_i, \\ lz &= \alpha, \end{aligned} \tag{6.32}$$

obtained by setting $\varepsilon = 0$ in (6.31) is unsolvable for arbitrary inhomogeneities $f(t) \in C([a,b] \setminus \{\tau_i\}_I)$, $a_i \in R^n$, and $\alpha \in R^m$. According to Theorem 6.1, this means that the analyzed case is critical ($\operatorname{rank} Q = n_1 < m$) and, hence, the criterion of solvability (6.11) is not true for the boundary-value problem (6.32) in view of the arbitrariness of inhomogeneities.

It is of interest to analyze whether it is possible to make problem (6.32) solvable by introducing linear perturbations and (in the case of positive answer to this question) to determine the perturbations $A_1(t)$, A_{1i}, and l_1 required to make the boundary-value problem (6.31) everywhere solvable. We can answer this question with the help of the $d \times r$ matrix

$$B_0 = P_{Q_d^*}\Bigg[l_1 X_r(\cdot) - l\int_a^b K(\cdot,\tau)A_1(\tau)X_r(\tau)d\tau - l\sum_{i=1}^{p} \bar{K}(\cdot,\tau_i)A_{1i}X_r(\tau_i - 0)\Bigg] \tag{6.33}$$

constructed by using the coefficients of the original differential boundary-value problem.

Using the Vishik–Lyusternik method [19, 150], one can establish efficient conditions for the coefficients guaranteeing the appearance of solutions of the boundary-value problem (6.32) in the form of series in powers of a small parameter ε containing finitely many terms with negative powers of ε.

In what follows, we prove a theorem that enables us to solve this problem. To do this, we recall that $Q = lX(\cdot)$ is an $m \times n$ matrix, $P_{Q^*}\colon R^m \to N(Q^*)$, Q^+ is the unique matrix pseudoinverse to Q in the Moore–Penrose sense, P_{B_0} is the $r \times r$ matrix (orthoprojector) projecting the space R^r onto the null space $N(B)$ of the $d \times r$ matrix B_0, i.e., $P_{B_0}\colon R^r \to N(B_0)$, and $P_{B_0^*}$ is the $d \times d$

matrix (orthoprojector) projecting the space R^d onto the null space $N(B_0^*)$ of the $r \times d$ matrix $B_0^* = B^T$, i.e., $P_{B_0^*}\colon R^d \to N(B_0^*)$.

Theorem 6.4. *Assume that the boundary-value problem (6.31) satisfies the conditions presented above for the critical case* ($\operatorname{rank} Q = n_1 < m$) *and that the generating boundary-value problem (6.32) is unsolvable for arbitrary inhomogeneities* $f(t) \in C([a,b] \setminus \{\tau_i\}_I)$, $a_i \in R^n$, *and* $\alpha \in R^m$. *If, in addition, the condition*

$$P_{B_0^*} = 0 \tag{6.34}$$

is satisfied, then, for any $f(t) \in C([a,b] \setminus \{\tau_i\}_I)$, $a_i \in R^n$, *and* $\alpha \in R^m$, *the boundary-value problem (6.31) possesses at least one (exactly one for* $n = m$*) solution given by the series*

$$z(t,\varepsilon) = \sum_{i=k}^{\infty} \varepsilon^i z_i(t), \quad k = -1, \tag{6.35}$$

convergent for sufficiently small fixed $\varepsilon \in (0, \varepsilon_*]$.

Proof. We substitute series (6.35) in the boundary-value problem (6.31) and equate the coefficients of the same powers of ε. Thus, for ε^{-1}, we arrive at the following homogeneous boundary-value problem for the quantity $z_{-1}(t)$:

$$\begin{aligned} \dot z_{-1} &= A(t) z_{-1}, \quad t \neq \tau_i, \\ \Delta z_{-1}\Big|_{t=\tau_i} &= S_i z_{-1}(\tau_i - 0), \\ l z_{-1} &= 0. \end{aligned} \tag{6.36}$$

According to the conditions of the theorem, the homogeneous boundary-value problem (6.36) possesses the r-parameter ($r = n - n_1$) family of solutions $z_{-1}(t, c_{-1}) = X_r(t) c_{-1}$, where the r-dimensional column vector $c_{-1} \in R^r$ can be found from the condition of solvability of the problem for $z_0(t)$.

For ε^0, we get the following boundary-value problem for the unknown quantity $z_0(t)$:

$$\begin{aligned} \dot z_0 &= A(t) z_0 + A_1(t) z_{-1} + f(t), \quad t \neq \tau_i, \\ \Delta z_0\Big|_{t=\tau_i} &= S_i z_0(\tau_i - 0) + A_{1i} z_{-1}(\tau_i - 0) + a_i, \\ l z_0 &= l_1 z_{-1} + \alpha. \end{aligned} \tag{6.37}$$

As follows from Theorem 6.1, the criterion of solvability of problem (6.37) has the form

$$\begin{aligned} P_{Q_d^*}\Big\{ \alpha + l_1 X_r(\cdot) c_{-1} - l \int_a^b K(\cdot,\tau)\big(f(\tau) + A_1(\tau) X_r(\tau) c_{-1}\big) d\tau \\ - l \sum_{i=1}^{p} \bar K(\cdot,\tau_i)\big(a_i + A_{1i} X_r(\tau_i - 0) c_{-1}\big) \Big\} = 0. \end{aligned}$$

This condition yields the following system of algebraic equations for $c_{-1} \in R^n$:

$$B_0 c_{-1} = -P_{Q_d^*} \Big\{ \alpha - l \int_a^b K(\cdot,\tau) f(\tau)\, d\tau - l \sum_{i=1}^{p} \bar{K}(\cdot,\tau_i) a_i \Big\}, \tag{6.38}$$

where B_0 is the $d \times r$ matrix given by (6.33). System (6.38) is solvable for any $f(t) \in C([a,b] \setminus \{\tau_i\}_I)$, $a_i \in R^n$, and $\alpha \in R^m$ if and only if $P_{B_0^*} = 0$. In this case, system (6.38) is solvable with respect to $c_{-1} \in R^r$ to within an arbitrary vector constant $P_{B_0} c$ ($\forall c \in R^r$) and one of its solutions takes the form

$$c_{-1} = -B_0^+ P_{Q_d^*} \Big\{ \alpha - l \int_a^b K(\cdot,\tau) f(\tau)\, d\tau - l \sum_{i=1}^{p} \bar{K}(\cdot,\tau_i) a_i \Big\}.$$

If, in addition, we require that $P_{B_0} = 0$ or $n = m$, then system (6.38) becomes uniquely solvable.

Under condition (6.34), the boundary-value problem (6.37) possesses an r-parameter ($r = n - n_1$) family of solutions

$$z_0(t, c_0) = X_r(t) c_0 + \bar{z}_0(t), \tag{6.39}$$

where c_0 is an r-dimensional constant vector determined (uniquely for $n = m$) in the next stage of the process from the condition of solvability of the boundary-value problem for $z_1(t)$, $\bar{z}_0(t)$ is a special solution of problem (6.37), namely,

$$\bar{z}_0(t) = \left(G \begin{bmatrix} A_1(\tau) z_{-1}(\tau, c_{-1}) + f(\tau) \\ A_{1i} z_{-1}(\tau_i - 0, c_{-1}) + a_i \end{bmatrix} \right)(t) + X(t) Q^+ \Big\{ \alpha + l_1 z_{-1}(\cdot, c_{-1}) \Big\},$$

and $\left(G \begin{bmatrix} * \\ * \end{bmatrix} \right)(t)$ is the generalized Green operator (6.12) of the boundary-value problem (6.37).

We continue this process infinitely and prove (by induction) that, under condition (6.34), the coefficients $z_i(t)$ of series (6.35) can be found from the corresponding boundary-value problems. The fact that series (6.35) is convergent is proved by using the traditional procedures of majorization [65, 150].

Remark 6.2. In the absence of pulses, we get the inhomogeneous linear boundary-value problem with small perturbations considered in [19]:

$$\dot{z} = A(t) z + \varepsilon A_1(t) z + f(t), \tag{6.40}$$

$$lz = \alpha. \tag{6.41}$$

Theorem 6.5. *Consider the critical case* ($\operatorname{rank} Q = n_1 < n = m$) *of the boundary-value problem (6.40), (6.41) and assume that the generating boundary-value problem*

$$\dot{z} = A(t) z + f(t), \quad lz = \alpha$$

is unsolvable for arbitrary inhomogeneities $f(t) \in C[a,b]$ *and* $\alpha \in R^n$. *If, in addition,*

$$\det B_0 \neq 0,$$

where B_0 *is an* $r \times r$ *matrix given by the formula*

$$B_0 = P_{Q_d^*} l \int_a^b K(\cdot,\tau) A_1(\tau) X_r(\tau)\, d\tau,$$

then, for any $f(t) \in C[a,b]$ *and* $\alpha \in R^n$, *the boundary-value problem (6.40), (6.41) possesses a unique solution given by a series*

$$z(t,\varepsilon) = \sum_{i=-1}^{\infty} \varepsilon^i z_i(t)$$

convergent for sufficiently small fixed $\varepsilon \in (0, \varepsilon_*]$.

In what follows, we show that if condition (6.34) is not satisfied, then the conditions sufficient for the appearance of solutions of the boundary-value problem (6.31) can be established by constructing a $d \times r$ matrix B_1. For arbitrary inhomogeneities $f(t) \in C[a,b] \setminus \{\tau_i\}_I)$, $a_i \in R^n$, and $\alpha \in R^m$, the solution $z(t,\varepsilon)$ of problem (6.31) is sought in the form of series (6.35) with $k \leq -2$ convergent for $\varepsilon \in (0, \varepsilon_*]$.

Theorem 6.6. *Assume that the boundary-value problem (6.31) satisfies the conditions imposed above. Then the following equivalent statements are true:*

(i) for any $f(t) \in C[a,b] \setminus \{\tau_i\}_I)$, $a_i \in R^n$, *and* $\alpha \in R^m$, *the boundary-value problem (6.31) possesses at least one solution* $z(t,\varepsilon)$ *in the form of a series*

$$z(t,\varepsilon) = \sum_{i=k}^{\infty} \varepsilon^i z_i(t), \quad k = -2, \tag{6.42}$$

convergent for sufficiently small $\varepsilon \in (0, \varepsilon_*]$;

(ii) for any r*-dimensional constant vector* $\varphi_0 \in R^r$, *the* r*-dimensional algebraic system*

$$(B_0 + \varepsilon B_1 + \cdots) u_\varepsilon = \varphi_0 \tag{6.43}$$

has at least one solution in the form of a series

$$u_\varepsilon = \sum_{i=-1}^{\infty} \varepsilon^i u_i \tag{6.44}$$

convergent for sufficiently small fixed $\varepsilon \in (0, \varepsilon_*]$;

(iii) $P_{B_0^*} \neq 0$ *and* $P_{B_1^*} P_{B_0^*} = 0$. $\quad$ (6.45)

Proof. We substitute relation (6.42) in the boundary-value problem (6.31) and equate the coefficients of the same powers of ε.

For ε^{-2}, we get the following homogeneous boundary-value problem:

$$\dot{z}_{-2} = A(t)z_{-2}, \quad t \neq \tau_i,$$
$$\Delta z_{-2}\Big|_{t=\tau_i} = S_i z_{-2}(\tau_i - 0),$$
$$l z_{-2} = 0,$$

According to the conditions of the theorem, this system possesses the r-parameter ($r = n - n_1$, $n_1 = \operatorname{rank} Q$) family of solutions $z_{-2}(t) = X_r(t)c_{-1}$, where c_{-1} is an arbitrary r-dimensional column vector from R^r.

For ε^{-1}, we arrive at the following boundary-value problem:

$$\dot{z}_{-1} = A(t)z_{-1} + A_1(t)z_{-2}, \quad t \neq \tau_i,$$
$$\Delta z_{-1}\Big|_{t=\tau_i} = S_i z_{-1}(\tau_i - 0) + A_{1i} z_{-2}(\tau_i - 0), \tag{6.46}$$
$$l z_{-1} = l_1 z_{-2}.$$

Conditions (6.11) necessary and sufficient for the solvability of system (6.46), yield the following algebraic system for $c_{-1} \in R^r$: $B_0 c_{-1} = 0$, where B_0 is given by relation (6.33).

The boundary-value problem (6.46) possesses an r-parameter family of solutions of the form

$$z_{-1}(t) = X_r(t)c_0 + G_1(t)c_{-1},$$

where c_0 is an arbitrary r-dimensional column vector from the space R^r and

$$G_1(t)c_{-1} = \bar{z}_{-1}(t)$$
$$= \left(G\begin{bmatrix} A_1(\tau)z_{-2}(\tau, c_{-1}) \\ A_{1i} z_{-2}(\tau_i - 0, c_{-1}) \end{bmatrix}\right)(t) + X(t)Q^+ l_1 z_{-2}(\cdot, c_{-1})$$

is a special solution of problem (6.46).

Equating the coefficients of the terms with ε^0, we obtain the following boundary-value problem for $z_0(t)$:

$$\dot{z}_0 = A(t)z_0 + A_1(t)z_{-1}(t) + f(t), \quad t \neq \tau_i,$$
$$\Delta z_0\Big|_{t=\tau_i} = S_i z_0(\tau_i - 0) + A_{1i} z_{-1}(\tau_i - 0) + a_i, \tag{6.47}$$
$$l z_0 = \alpha + l_1 z_{-1}.$$

The condition of solvability (6.11) of the boundary-value problem (6.47) gives the following algebraic system for $c_0, c_{-1} \in R^r$:

$$B_0 c_0 + B_1 c_{-1} = \varphi_0,$$

where B_1 is a $d \times r$ matrix given by the formula

$$B_1 = P_{Q_d^*}\left\{ l_1 G_1(\cdot) - l \int_a^b K(\cdot,\tau) A_1(\tau) G_1(\tau)\, d\tau - l \sum_{i=1}^{p} \bar{K}(\cdot,\tau_i) A_{1i} G_1(\tau_i - 0) \right\}$$

and

$$\varphi_0 = -P_{Q_d^*}\left\{ \alpha - l \int_a^b K(\cdot,\tau) f(\tau) d\tau - l \sum_{i=1}^{p} \bar{K}(\cdot,\tau_i) a_i \right\}$$

is an arbitrary r-dimensional constant vector from R^d (in view of the arbitrariness of $f(t) \in C([a,b] \setminus \{\tau_i\}_I)$, $a_i \in R^n$, and $\alpha \in R^m$). The boundary-value problem (6.47) possesses the r-parameter family of solutions

$$z_0(t) = X_r(t) c_1 + G_1(t) c_0 + G_2(t) c_{-1} + f_1(t),$$

where c_1 is an arbitrary r-dimensional column vector from R^r,

$$G_2(t) = \left(G \begin{bmatrix} A_1(\tau) G_1(\tau) \\ A_{1i} G_1(\tau_i - 0) \end{bmatrix} \right)(t) + X(t) Q^+ l_1 G_1(\cdot),$$

and

$$f_1(t) = \left(G \begin{bmatrix} f(\tau) \\ a_i \end{bmatrix} \right)(t) + X(t) Q^+ \alpha.$$

We continue this procedure infinitely and note that series (6.42) is a solution of the boundary-value problem (6.31) for any $f(t) \in C([a,b] \setminus \{\tau_i\}_I)$, $a_i \in R^n$, and $\alpha \in R^m$ if and only if the system of algebraic equations

$$\begin{cases} B_0 c_{-1} = 0, \\ B_0 c_0 + B_1 c_{-1} = \varphi_0, \end{cases} \tag{6.48}$$

and

$$\begin{cases} B_0 c_1 + B_1 c_0 + B_2 c_{-1} = \varphi_1, \\ B_0 c_2 + B_1 c_1 + B_2 c_0 + B_3 c_{-1} = \varphi_2, \\ \ldots \end{cases} \tag{6.49}$$

is solvable with respect to $c_i \in R^r$ $(i = -1, 0, 1, \ldots)$ for any $\varphi_0 \in R^d$.

On the other hand, substituting series (6.44) in (6.43), we conclude that this series is a solution of system (6.43) if and only if the coefficients $u_i \in R^r$ satisfy a system similar to (6.48), (6.49) but with $\varphi_i = 0$, $i = 1, 2, \ldots$, i.e.,

$$\begin{cases} B_0 u_{-1} = 0, \\ B_0 u_0 + B_1 u_{-1} = \varphi_0, \end{cases} \tag{6.50}$$

and

$$\begin{cases} B_0 u_1 + B_1 u_0 + B_2 u_{-1} = 0, \\ B_0 u_2 + B_1 u_1 + B_2 u_0 + B_3 u_{-1} = 0, \\ \ldots \end{cases} \tag{6.51}$$

The system of algebraic equations (6.50) is solvable if and only if condition (6.45) is satisfied. Further, to complete the proof, it is necessary to show that the solvability of system (6.50) implies the solvability of system (6.51). Moreover, the coefficient u_{-1} can be found from the condition of solvability of the last equation in (6.50). The condition of solvability of the first equation in (6.51) (which coincides with the previous condition) gives the coefficient u_0, and so on. The proof of this part of the theorem repeats the proof of the corresponding theorem in [19] for the boundary-value problem without pulses and, hence, is omitted.

6.5 Weakly Nonlinear Boundary-Value Problems

Consider a weakly nonlinear boundary-value problem for differential equations with impulsive action

$$\begin{aligned} \dot{z} &= A(t)z + f(t) + \varepsilon Z(z,t,\varepsilon), \quad t \neq \tau_i, \quad t \in [a,b], \\ \Delta z\Big|_{t=\tau_i} &- S_i z(\tau_i - 0) = a_i + \varepsilon J_i(z(\tau_i - 0,\varepsilon),\varepsilon), \\ & i = 1,\dots,p, \quad \tau_i \in (a,b), \\ & lz = \alpha + \varepsilon J(z(\cdot,\varepsilon),\varepsilon). \end{aligned} \tag{6.52}$$

The first equation in system (6.52) is a differential equation in which $A(t)$ and $f(t) \in C([a,b]\setminus\{\tau_i\}_I)$ are $n\times n$ matrix and $n\times 1$ vector functions, respectively, and $Z(z,t,\varepsilon)$ is a nonlinear n-dimensional vector function continuously differentiable with respect to the first argument in the vicinity of solutions of the generating boundary-value problem (6.1), (6.2) (obtained from (6.52) by setting $\varepsilon = 0$), continuous (or piecewise continuous) in the second argument with discontinuities of the first kind for $t = \tau_i$ and continuous in $\varepsilon \in [0,\varepsilon_0]$.

The second equation in this system specifies the jumps of the solution, namely,

$$\Delta z\Big|_{t=\tau_i} = z(\tau_i + 0) - z(\tau_i - 0) \quad \text{for} \quad t = \tau_i,$$

S_i are $n\times n$ constant matrices such that the matrices $(E+S_i)$ are nonsingular, $a_i \in R^n$ is an n-dimensional column vector of constants, and $J_i(z(\tau_i - 0,\varepsilon),\varepsilon)$ are nonlinear vector functionals continuously differentiable with respect to z (in the Frechét sense) in the vicinity of solutions of the generating boundary-value problem (6.1), (6.2) and continuous in $\varepsilon \in [0,\varepsilon_0]$.

The third equation in system (6.52) determines the boundary conditions. Here, $l = \text{col}\,(l_1,\dots,l_m)$ is an m-dimensional bounded linear vector functional, $J(z(\cdot,\varepsilon),\varepsilon)$ is a nonlinear vector functional continuously differentiable with respect to z (in the Frechét sense) in the vicinity of solutions of the generating boundary-value problem (6.1), (6.2) and continuous in $\varepsilon \in [0,\varepsilon_0]$, and $\alpha \in R^m$ is an m-dimensional vector.

Our aim is to establish conditions required for the existence of solutions $z = z(t, \varepsilon)$ of the boundary-value problem (6.52) piecewise-continuously differentiable with respect to t with discontinuities of the first kind at the points $t = \tau_i$, i.e., $z(\cdot, \varepsilon) \in C^1([a, b] \setminus \{\tau_i\}_I)$, and continuous in $\varepsilon \in [0, \varepsilon_0]$, i.e., $z(t, \cdot) \in C[0, \varepsilon_0]$, which turn into one of the solutions $z_0(t, c_r)$ of the generating boundary-value problem (6.1), (6.2) for $\varepsilon = 0$, i.e., $z(t, 0) = z_0(t, c_r)$, $c_r \in R^r$, and to develop algorithms for the construction of solutions of this sort.

To solve the posed problem, we use the generalized Green operator of a semi-homogeneous linear impulsive boundary-value problem. The original boundary-value problem (6.52) is reduced to an equivalent operator system, which is then solved by the method of simple iterations. As in the case of systems without pulses, the form and dimensionality of the operator system are noticeably different for critical and noncritical boundary-value problems.

We first consider the critical case where $\operatorname{rank} Q = n$. In other words, we assume that the homogeneous boundary-value problem (6.3), (6.4) with impulsive action has only the trivial solution. By virtue of Theorem 6.2, the generating boundary-value problem (6.1), (6.2) possesses a unique solution of the form

$$z_0(t) = \left(G\begin{bmatrix} f \\ a_i \end{bmatrix}\right)(t) + X(t)Q^+\alpha \tag{6.53}$$

if and only if $f(t) \in C([a, b] \setminus \{\tau_i\}_I)$, $a_i \in R^n$, and $\alpha \in R^m$ satisfy the condition

$$P_{Q_d^*}\left\{\alpha - l\int_a^b K(\cdot, \tau)f(\tau)d\tau - l\sum_{i=1}^{p} \bar{K}(\cdot, \tau_i)a_i\right\} = 0, \quad d = m - n. \tag{6.54}$$

By analogy with [17, 19, 65, 99], solution (6.53) is called a generating solution of the boundary-value problem (6.52) with impulsive action.

By the (customary) change variables

$$z(t, \varepsilon) = z_0(t) + x(t, \varepsilon) \tag{6.55}$$

in problem (6.52) [65, 101], we arrive at the following boundary-value problem with impulsive action:

$$\begin{cases} \dot{x} = A(t)x + \varepsilon Z(z_0 + x, t, \varepsilon), \quad t \neq \tau_i, \\ \Delta x\Big|_{t=\tau_i} = S_i x(\tau_i - 0) + \varepsilon J_i(z_0 + x, \varepsilon), \\ lx = \varepsilon J(z_0(\cdot) + x(\cdot, \varepsilon), \varepsilon). \end{cases} \tag{6.56}$$

Since we want to construct a solution $z(t, \varepsilon)$ of problem (6.52) continuous as a function of ε for $\varepsilon \in [0, \varepsilon_0]$ and turning into the generating solution $z_0(t)$ for $\varepsilon = 0$, it is necessary to establish conditions for the existence and develop an algorithm for the construction of solutions $x(t, \varepsilon)$ of the boundary-value problem (6.56) satisfying the conditions $x(\cdot,\varepsilon) \in C^1([a,b] \setminus \{\tau_i\}_I)$ and $x(t,\cdot) \in C[0,\varepsilon_0]$ and vanishing for $\varepsilon = 0$. We apply Theorem 6.2 to problem (6.56) and formally treat the nonlinearities

$$Z(z_0 + x, t, \varepsilon), \quad J_i(z_0(\cdot) + x(\cdot,\varepsilon), \varepsilon), \quad \text{and} \quad J(z_0(\cdot) + x(\cdot,\varepsilon), \varepsilon)$$

as inhomogeneities. This enables us to conclude that problem (6.56) is solvable if and only if the nonlinearities

$$Z(z_0 + x, t, \varepsilon), \quad J_i(z_0(\cdot) + x(\cdot,\varepsilon), \varepsilon), \quad \text{and} \quad J(z_0(\cdot) + x(\cdot,\varepsilon), \varepsilon)$$

satisfy the condition

$$P_{Q_d^*}\Big\{ J(z(\cdot,\varepsilon),\varepsilon) - l\int_a^b K(\cdot,\tau)Z(z,\tau,\varepsilon)d\tau \\ - l\sum_{i=1}^{p} \bar{K}(\cdot,\tau_i)\, J_i(z(\tau_i - 0,\varepsilon),\varepsilon)\Big\} = 0, \quad d = m - n. \tag{6.57}$$

Under this condition, the boundary-value problem (6.56) possesses a unique solution $x(t,\varepsilon)$ of the form

$$x(t,\varepsilon) = \varepsilon X(t)Q^+ J(z_0(\cdot) + x(\cdot,\varepsilon),\varepsilon) \\ + \varepsilon \left(G\begin{bmatrix} Z(z_0(\tau) + x(\tau,\varepsilon),\tau,\varepsilon) \\ J_i(z_0(\tau_i - 0) + x(\tau_i - 0,\varepsilon),\varepsilon) \end{bmatrix}\right)(t). \tag{6.58}$$

System (6.58) belongs to the class of systems solvable by the method of simple iterations [65, 101]. This method yields the following algorithm for the construction of the solution $x(t,\varepsilon)$:

$$x_{k+1}(t,\varepsilon) = \varepsilon X(t)Q^+ J(z_0(\cdot) + x_k(\cdot,\varepsilon),\varepsilon) \\ + \varepsilon \left(G\begin{bmatrix} Z(z_0(\tau) + x_k(\tau,\varepsilon),\tau,\varepsilon) \\ J_i(z_0(\tau_i - 0) + x_k(\tau_i - 0,\varepsilon),\varepsilon) \end{bmatrix}\right)(t), \tag{6.59}$$

where $k = 0, 1, 2, \ldots$ and $x_0 = 0$.

In order to determine the range of convergence of the iterative process (6.59) and obtain estimates of the approximate solutions, we can use the method of finite Lyapunov majorizing equations [19]. It is well known [65, 93] that, under the indicated conditions, one can always find $\varepsilon = \varepsilon_*$ such that the iterative process (6.51) converges for $\varepsilon \in [0,\varepsilon_*]$. Hence, to establish conditions for the solvability of the boundary-value problem (6.56) with impulsive action, it suffices to determine conditions of its reducibility to the operator system (6.58). Thus, the following assertion is true:

Theorem 6.7. *Assume that the boundary-value problem (6.51) with impulsive action satisfies the conditions imposed above and the generating boundary-value problem (6.1), (6.2) with impulsive action is uniquely solvable provided that* $\operatorname{rank} Q = n$ *and condition (6.54) is satisfied. Then the boundary-value problem (6.56) with impulsive action is solvable if and only if the nonlinearities* $Z(z_0 + x, t, \varepsilon)$, $J_i(z_0 + x, \varepsilon)$, *and* $J(z_0(\cdot) + x(\cdot, \varepsilon), \varepsilon)$ *satisfy condition (6.57). Under this condition, the boundary-value problem (6.56) possesses a unique solution* $x(t, \varepsilon), x(\cdot, \varepsilon) \in C^1([a, b] \setminus \{\tau_i\}_I)$, $x(t, \cdot) \in C[\varepsilon]$, $\varepsilon \in [0, \varepsilon_*]$, *vanishing for* $\varepsilon = 0$. *This solution can be found as a result of the iterative process (6.59) convergent for* $\varepsilon \in [0, \varepsilon_*]$.

In view of the change of variables (6.55), we can state that the boundary-value problem (6.52) with impulsive action is solvable if and only if condition (6.57) is satisfied. In this case, problem (6.52) possesses a unique solution $z(t, \varepsilon), z(\cdot, \varepsilon) \in C^1([a, b] \setminus \{\tau_i\}_I)$, $z(t, \cdot) \in C[\varepsilon]$, $\varepsilon \in [0, \varepsilon_*]$, which turns into a generating solution for $\varepsilon = 0$. The solution $z(t, \varepsilon)$ is found by using the recurrence relation

$$z_k(t, \varepsilon) = z_0(t) + x_k(t, \varepsilon), \quad k = 0, 1, 2, \ldots, \quad \varepsilon \in [0, \varepsilon_*],$$

where $x_k(t, \varepsilon)$ is given by (6.59).

6.6 Critical Case. Necessary Condition for the Existence of Solutions

We now consider the critical case $\operatorname{rank} Q = n_1 < m$ in which the homogeneous boundary-value problem (6.3), (6.4) with impulsive action possesses a nontrivial solution, i.e., $\operatorname{rank} Q = n_1 < n$. Then, by virtue of Theorem 6.1, the generating boundary-value problem (6.1), (6.2) with impulsive action is solvable if and only if $f(t) \in C([a, b] \setminus \{\tau_i\}_I)$, $a_i \in R^n$, and $\alpha \in R^m$ satisfy the condition

$$P_{Q_d^*}\left\{\alpha - l\int_a^b K(\cdot, \tau) f(\tau) d\tau - l\sum_{i=1}^{p} \bar{K}(\cdot, \tau_i) a_i\right\} = 0, \quad d = m - n_1, \tag{6.60}$$

In this case, the problem possesses an r-parameter ($r = n - n_1$) family of solutions of the form

$$z_0(t, c_r) = X_r(t) c_r + \left(G\begin{bmatrix} f \\ a_i \end{bmatrix}\right)(t) + X(t) Q^+ \alpha. \tag{6.61}$$

We now consider the problem of necessary conditions for the boundary-value problem (6.52) with impulsive action to have solutions $z = z(t, \varepsilon)$ satisfying the conditions $z(\cdot, \varepsilon) \in C^1([a, b] \setminus \{\tau_i\}_I)$ and $z(t, \cdot) \in C[\varepsilon]$, $\varepsilon \in [0, \varepsilon_0]$, and turning into the generating solutions $z_0(t, c_r)$ of the generating boundary-value problem (6.1), (6.2) given by relation (6.61) for $\varepsilon = 0$. This problem is solved by the following theorem:

Theorem 6.8. *Assume that the boundary-value problem (6.52) with impulsive action satisfies the conditions imposed above and has a solution*

$$z(t,\varepsilon)\colon z(\cdot,\varepsilon)\in C^1([a,b]\setminus\{\tau_i\}_I),\ z(t,\cdot)\in C[\varepsilon],\ \varepsilon\in[0,\varepsilon_0], \tag{6.62}$$

which turns, for $\varepsilon = 0$, into the generating solution $z_0 = z_0(t,c_r^*)$ (6.61) (with a certain constant $c_r = c_r^* \in R^r$) of the generating boundary-value problem (6.1), (6.2) with impulsive action. Then the vector constant c_r^* satisfies the condition

$$P_{Q_d^*}\Big\{J(z_0(\cdot,c_r^*),0) - l\int_a^b K(\cdot,\tau)Z(z_0(\tau,c_r^*),\tau,0)d\tau - l\sum_{i=1}^{p}\bar{K}(\cdot,\tau_i)J_i(z_0(\tau_i-0,c_r^*),0)\Big\} = 0. \tag{6.63}$$

Proof. Let $\operatorname{rank} Q = n_1 < \min(n,m)$. We assume that the weakly nonlinear boundary-value problem (6.52) with impulsive action possesses a solution

$$z(t,\varepsilon)\colon z(\cdot,\varepsilon)\in C^1([a,b]\setminus\{\tau_i\}_I),\ z(t,\cdot)\in C[\varepsilon],\ \varepsilon\in[0,\varepsilon_0],$$

which turns, for $\varepsilon = 0$, into the generating solution $z_0(t,c_r^*)$ given by relation (6.61) with a constant $c_r = c_r^*$. The condition required for the existence of the generating solution (6.60) is assumed to be satisfied. Further, by applying Theorem 6.1 to the boundary-value problem (6.52), in view of condition (6.60), we conclude that, for all $\varepsilon\in[0,\varepsilon_0]$, the nonlinear vector function $Z(z(t,\varepsilon),t,\varepsilon)$ and nonlinear vector functionals $J_i(z(\tau_i-0,\varepsilon),\varepsilon)$ and $J(z(\cdot,\varepsilon),\varepsilon)$ satisfy the condition

$$P_{Q_d^*}\Big\{J(z(\cdot,\varepsilon),\varepsilon)d\tau - l\int_a^b K(\cdot,z)Z(z(\tau,\varepsilon),\tau,\varepsilon)\,d\tau - l\sum_{i=1}^{p}\bar{K}(\cdot,\tau_i)J_i(z(\tau_i-0,\varepsilon),\varepsilon)\Big\} = 0,\quad d = m-n_1. \tag{6.64}$$

We now consider the case where condition (6.63) is not satisfied. Since $z(t,\varepsilon)$ approaches $z_0(t,c_r^*)$ as $\varepsilon\to 0$ and the vector functionals $J(z(\tau_i-0,\varepsilon),\varepsilon)$ and $J(z(\cdot,\varepsilon),\varepsilon)$ are continuous in z and ε in the neighborhoods of $z_0(t,c_r^*)$ and the point $\varepsilon=0$, there exists a sufficiently small number $\varepsilon=\varepsilon_1$ such that

$$P_{Q_d^*}\Big\{J(z(\cdot,\varepsilon),\varepsilon)d\tau - l\int_a^b K(\cdot,z)Z(z(\tau,\varepsilon),\tau,\varepsilon)\,d\tau - l\sum_{i=1}^{p}\bar{K}(\cdot,\tau_i)J_i(z(\tau_i-0,\varepsilon),\varepsilon)\Big\} \neq 0$$

for $\varepsilon\in[0,\varepsilon_1]$, which contradicts condition (6.64). Thus, our assumption is not true and, hence, Theorem 6.8 is proved.

By analogy with [19, 65, 101], we introduce the following definition:

Definition 6.2. An equation

$$F(c_r) = P_{Q_d^*}\Big\{ J(z_0(\cdot, c_r), 0) - l\int_a^b K(\cdot, \tau) Z(z_0(\tau, c_r), \tau, 0) d\tau \\ - l\sum_{i=1}^{p} \bar{K}(\cdot, \tau_i) J_i(z_0(\tau_i - 0, c_r), 0)\Big\} = 0 \qquad (6.65)$$

is called the equation for generating constants for the weakly nonlinear boundary-value problem (6.52) with impulsive action.

If equation (6.65) is solvable, then the vector c_r^* specifies the generating solution $z_0(t, c_r^*)$ associated with the solution $z(t, \varepsilon)$: $z(\cdot, \varepsilon) \in C^1([a, b] \setminus \{\tau_i\}_I)$, $z(t, \cdot) \in C[\varepsilon]$, $\varepsilon \in [0, \varepsilon_0]$, $z(t, 0) = z_0(t, c_r^*)$ of the original boundary-value problem (6.52) with impulsive action. At the same time, if equation (6.65) is unsolvable, then the boundary-value problem (6.52) is also unsolvable in the class of vector functions piecewise continuously differentiable with respect to t (with discontinuities of the first kind at the points $t = \tau_i$) and continuous in $\varepsilon \in [0, \varepsilon_0]$. Since all calculations (both here and in what follows) are carried out in real numbers, the analyzed solutions of the (algebraic or transcendental) equation for generating constants (6.65) are regarded as real.

6.7 Sufficient Condition for the Existence of Solutions. Iterative Algorithm for the Construction of Solutions

In the present section, we establish sufficient conditions for the solvability of the boundary-value problem (6.52) with impulsive action in the so-called critical case of the first order. In this case, the problem of solvability of the original boundary-value problem (6.52) is solved after the analysis of the boundary-value problem used to find the first approximation to the required solution.

In problem (6.52), we perform the change of variables

$$z(t, \varepsilon) = z_0(t, c_r^*) + x(t, \varepsilon),$$

where $c_r^* \in R^r$ is a constant satisfying the equation for generating constants (6.56). As a result, we arrive at the problem of finding conditions sufficient for the solvability of the boundary-value problem with impulsive action

$$\dot{x} = A(t)x + \varepsilon Z(z_0(t, c_r^*) + x(t, \varepsilon), t, \varepsilon), \quad t \neq \tau_i,$$

$$\Delta x\Big|_{t=\tau_i} = S_i x(\tau_i - 0) + \varepsilon J_i(z_0(\tau_i - 0, c_r^*) + x(\tau_i - 0, \varepsilon), \varepsilon), \qquad (6.66)$$

$$lx = \varepsilon J(z_0(\cdot, c_r^*) + x(\cdot, \varepsilon), \varepsilon)$$

in the class of functions $x(t,\varepsilon)$ such that $x(\cdot,\varepsilon)\in C^1([a,b]\backslash\{\tau_i\}_I)$, $x(t,\cdot)\in C[\varepsilon]$, $\varepsilon \in [0,\varepsilon_0]$, and $x(t,0) = 0$.

In view of the restrictions imposed on the nonlinearities, we get the following expansions valid in a neighborhood of the point $x = 0$, $\varepsilon = 0$:

$$Z(z_0 + x, t, \varepsilon) = Z(z_0(t, c_r^*), t, 0) + A_1(t)x + R(x, t, \varepsilon),$$
$$A_1(t) = \frac{\partial Z(z,t,0)}{\partial z}\bigg|_{z=z_0(t,c_r^*)},$$
$$R(0,t,0) = 0, \quad \frac{\partial R(0,t,0)}{\partial x} = 0,$$
$$J_i(z_0+x,\varepsilon) = J_i(z_0(\tau_i, c_r^*), 0) + A_{1i}x(\tau_i-0,\varepsilon) + R_i(x(\tau_i,\varepsilon),\varepsilon), \tag{6.67}$$
$$A_{1i} = \frac{\partial J_i(z,0)}{\partial z}\bigg|_{z=z_0(\tau_i-0,c_r^*)}, \quad i = 1,\dots,p,$$
$$R_i(0,0) = 0, \quad \frac{\partial R_i(0,0)}{\partial x} = 0,$$
$$J(z_0 + x, \varepsilon) = J(z_0(\cdot, c_r^*), 0) + l_1 x(\cdot,\varepsilon) + R_0(x(\cdot,\varepsilon),\varepsilon),$$

where $l_1 x(\cdot,\varepsilon)$ is the linear part of the vector functional

$$J(z_0(\cdot, c_r^*) + x(\cdot,\varepsilon),\varepsilon),$$
$$R_0(0,0) = 0, \qquad \text{and} \qquad \frac{\partial R_0(0,0)}{\partial x} = 0.$$

We can formally treat the nonlinearities appearing in system (6.66) as inhomogeneities and apply Theorem 6.1 to this system. As a result, we get the following expression for its solution $x(t,\varepsilon)$:

$$x(t,\varepsilon) = X_r(t)c + x^{(1)}(t,\varepsilon),$$

where $c = c(\varepsilon) \in R^r$ is an unknown constant vector determined from the condition of existence of the required solution of system (6.66) similar to (6.60), namely,

$$P_{Q_d^*}\bigg\{J(z_0(\cdot,c_r^*) + x(\cdot,\varepsilon),\varepsilon) - l\int_a^b K(\cdot,\tau)Z(z_0(\tau,c_r^*) + x(\tau,\varepsilon),\tau,\varepsilon)d\tau$$
$$- l\sum_{i=1}^{p} \bar{K}(\cdot,\tau_i)J(z_0(\tau_i - 0, c_r^*) + x(\tau_i - 0,\varepsilon),\varepsilon)\bigg\} = 0,$$

and the unknown vector function $x^{(1)}(t,\varepsilon)$ is given by the formula

$$x^{(1)}(t,\varepsilon) = \varepsilon X(t)Q^+ J(z_0(\cdot,c_r^*) + x(\cdot,\varepsilon),\varepsilon)$$
$$+ \varepsilon\left(G\begin{bmatrix} Z(z_0(\tau,c_r^*) + x(\tau,\varepsilon),\tau,\varepsilon) \\ J_i(z_0(\tau_i - 0, c_r^*) + x(\tau_i - 0,\varepsilon),\varepsilon)\end{bmatrix}\right)(t).$$

By using expansions (6.67) and the fact that the vector constant $c_r^* \in R^r$ must satisfy the equation for generating constants (6.65), we get the following equivalent operator system for finding the solution $x(t,\varepsilon)$ (such that $x(\cdot,\varepsilon) \in C^1([a,b]\backslash\{\tau_i\}_I)$, $x(t,\cdot) \in C[\varepsilon]$, $\varepsilon \in [0,\varepsilon_0]$, and $x(t,0) = 0$) of the boundary-value problem (6.66) with impulsive action:

$$x(t,\varepsilon) = X_r(t)c + x^{(1)}(t,\varepsilon),$$

$$B_0 c = -P_{Q_d^*}\Big\{l_1 x^{(1)}(\cdot,\varepsilon) + R_0(x(\cdot,\varepsilon),\varepsilon) - l\int_a^b K(\cdot,\tau)[A_1(\tau)x^{(1)}(\tau,\varepsilon) + R(x(\tau,\varepsilon),\tau,\varepsilon)]d\tau - l\sum_{i=1}^{p} \bar{K}(\cdot,\tau_i)[A_{1i}x^{(1)}(\tau_i - 0,\varepsilon) + R_i(x(\tau_i - 0,\varepsilon),\varepsilon)]\Big\}, \quad (6.68)$$

$$x^{(1)}(\tau,\varepsilon) = \varepsilon\left(G\begin{bmatrix} Z(z_0(\tau,c_r^*),\tau,0) + A_1(\tau)[X_r(\tau)c + x^{(1)}(\tau,\varepsilon)] + R(x(\tau,\varepsilon),\tau,\varepsilon) \\ J_i(z_0(\tau_i-0,c_r^*),0) + A_{1i}[X_r(\tau_i-0)c + x^{(1)}(\tau_i-0,\varepsilon)] + R_i(x(\tau_i - 0,\varepsilon),\varepsilon)\end{bmatrix}\right)(t)$$
$$+ \varepsilon X(t)Q^+\big[J(z_0(\cdot,c_r^*),0) + l_1\{X_r(\cdot)c + x^{(1)}(\cdot,\varepsilon)\} + R_0(x(\cdot,\varepsilon),\varepsilon)\big],$$

where B_0 is a $d\times r$ matrix ($r = n - n_1$, $d = m - n_1$, $n_1 = \operatorname{rank} Q$) given by the formula

$$B_0 = P_{Q_d^*}\Big\{l_1 X_r(\cdot) - l\int_a^b K(\cdot,\tau)A_1(\tau)X_r(\tau)d\tau - l\sum_{i=1}^{p}\bar{K}(\cdot,\tau_i)A_{1i}X_r(\tau_i-0)\Big\}.$$

Note that this operator system is solvable if its second equation is solvable.

Let P_{B_0} be an $r\times r$ matrix (orthoprojector), i.e., $P_{B_0}\colon R^r \to N(B_0)$, and let $P_{B_0^*}$ be a $d\times d$ matrix (orthoprojector), i.e., $P_{B_0^*}\colon R^d \to N(B_0^*)$. Also let $\operatorname{rank} B_0 = d$, i.e., $P_{B_0^*} = 0$. In this case, we have $d \le r$. This means that the number m of boundary conditions is not greater than the dimension n of the differential system (6.52) because $d = m - n_1 \le r = n - n_1$. The second equation of the operator system (6.68) is solvable if and only if

$$P_{B_0^*}P_{Q_d^*}\Big\{l_1 x^{(1)}(\cdot,\varepsilon) + R_0(x(\cdot,\varepsilon),\varepsilon) - l\int_a^b K(\cdot,\tau)[A_1(\tau)x^{(1)}(\tau,\varepsilon) + R(x(\tau,\varepsilon),\tau,\varepsilon)]d\tau - l\sum_{i=1}^{p}\bar{K}(\cdot,\tau_i)[A_{1i}x^{(1)}(\tau_i - 0,\varepsilon) + R_i(x(\tau_i - 0,\varepsilon),\varepsilon)]\Big\} = 0. \quad (6.69)$$

If $P_{B_0^*} = 0$, then this condition is always satisfied. Hence, for $P_{B_0^*} = 0$, the second equation of the operator system (6.68) is solvable and its solution can be represented in the form (to within an arbitrary constant vector $P_{B_0}c \in R^r$)

$$c = -B_0^+ P_{Q_d^*}\Big\{ l_1 x^{(1)}(\cdot,\varepsilon) + R_0(x(\cdot,\varepsilon),\varepsilon) - l\int_a^b K(\cdot,\tau)[A_1(\tau)x^{(1)}(\tau,\varepsilon) + R(x(\tau,\varepsilon),\tau,\varepsilon)]d\tau - l\sum_{i=1}^p \bar{K}(\cdot,\tau_i)[A_{1i}x^{(1)}(\tau_i-0,\varepsilon) + R_i(x(\tau_i-0,\varepsilon),\varepsilon)]\Big\}, \tag{6.70}$$

where B_0^+ is the $r \times d$ matrix pseudoinverse to the matrix B_0 in the Moore–Penrose sense.

In view of relation (6.70), the operator system (6.68) equivalent to the boundary-value problem (6.66) in the set of vector functions $x(t,\varepsilon)$, $x(t,0) = 0$, piecewise continuously differentiable with respect to $t \in [a,b]$ (with discontinuities of the first kind at the points $t = \tau_i$) can be represented in the form

$$x(t,\varepsilon) = X_r(t)c + x^{(1)}(t,\varepsilon),$$

$$c = -B_0^+ P_{Q_d^*}\Big\{ l_1 x^{(1)}(\cdot,\varepsilon) + R_0(x(\cdot,\varepsilon),\varepsilon) - l\int_a^b K(\cdot,\tau)[A_1(\tau)x^{(1)}(\tau,\varepsilon) + R(x(\tau,\varepsilon),\tau,\varepsilon)]d\tau - l\sum_{i=1}^p \bar{K}(\cdot,\tau_i)[A_{1i}x^{(1)}(\tau_i-0,\varepsilon) + R_i(x(\tau_i-0,\varepsilon),\varepsilon)]\Big\}, \tag{6.71}$$

$$x^{(1)}(\tau,\varepsilon) = \varepsilon\left(G\begin{bmatrix} Z(z_0(\tau,c_r^*),\tau,0) + A_1(\tau)[X_r(\tau)c + x^{(1)}(\tau,\varepsilon)] \\ + R(x(\tau,\varepsilon),\tau,\varepsilon) \\ J_i(z_0(\tau_i-0,c_r^*),0) + A_{1i}[X_r(\tau_i-0)c + x^{(1)}(\tau_i-0,\varepsilon)] \\ + R_i(x(\tau_i-0,\varepsilon),\varepsilon) \end{bmatrix}\right)(t)$$
$$+ \varepsilon X(t)Q^+\big[J(z_0(\cdot,c_r^*),0) + l_1\{X_r(\cdot)c + x^{(1)}(\cdot,\varepsilon)\} + R_0(x(\cdot,\varepsilon),\varepsilon)\big].$$

The operator system (6.71) belongs to the class of systems solvable by the method of simple iterations [92, 148, 157]. The method of Lyapunov majorants [19] enables one to show that the iterative process is convergent and estimate the errors of approximations. By applying the method of simple iterations to the operator system (6.71), we get the following iterative algorithm for the construction of solutions $x(t,\varepsilon)$ (such that $x(\cdot,\varepsilon) \in C^1([a,b]\backslash\{\tau_i\}_I)$, $x(t,\cdot) \in C[\varepsilon]$, $\varepsilon \in [0,\varepsilon_0]$, and $x(t,0) = 0$) of the boundary-value problem (6.66) with impulsive action.

Iterative Algorithm. The first approximation $x_1^{(1)}(t,\varepsilon)$ to $x^{(1)}(t,\varepsilon)$ is set equal to

$$x_1^{(1)}(t,\varepsilon) = \varepsilon\left(G\begin{bmatrix} Z(z_0(\tau,c_r^*),\tau,0) \\ J_i(z_0(\tau_i-0,c_r^*),0)\end{bmatrix}\right)(t) + \varepsilon X(t)Q^+ J(z_0(\cdot,c_r^*),0).$$

According to the definition of the generalized Green operator $\left(G\begin{bmatrix}*\\ *\end{bmatrix}\right)(t)$, the vector function $x_1^{(1)} = x_1^{(1)}(t,\varepsilon)$ is a solution of the following boundary-value problem with impulsive action:

$$\dot{x}_1 = A(t)x_1 + \varepsilon Z(z_0(t,c_r^*),t,0), \quad t \neq \tau_i,$$

$$\Delta x_1\Big|_{t=\tau_i} = S_i x_1(\tau_i - 0) + \varepsilon J_i(z_0(\tau_i-0,c_r^*),0),$$

$$l x_1 = \varepsilon J(z_0(\cdot,c_r^*),0),$$

This solution exists due to the choice of $c_r^* \in R^r$ from the equation for generating constants (6.65). The first approximation $x_1(t,\varepsilon)$ to the required solution $x(t,\varepsilon)$ of the boundary-value problem (6.66) with impulsive action is set equal to $x_1^{(1)}(t,\varepsilon)$. The second approximation $x_2^{(1)}(t,\varepsilon)$ to the solution $x^{(1)}(t,\varepsilon)$ is sought in the form

$$x_2^{(1)}(\tau,\varepsilon) = \varepsilon\left(G\begin{bmatrix} Z(z_0(\tau,c_1^*),\tau,0) + A_1(\tau)[X_r(\tau)c_1 + x_1^{(1)}(\tau,\varepsilon)] \\ \qquad + R(x_1(\tau,\varepsilon),\tau,\varepsilon) \\ J_i(z_0(\tau_i-0,c_1^*),0) + A_{1i}[X_r(\tau_i-0)c_1 + x_1^{(1)}(\tau_i-0,\varepsilon)] \\ \qquad + R_i(x_1(\tau_i-0,\varepsilon),\varepsilon)\end{bmatrix}\right)(t)$$
$$+ \varepsilon X(t)Q^+\big[J(z_0(\cdot,c_r^*),0) + l_1\{X_r(\cdot)c_1 + x_1^{(1)}(\cdot,\varepsilon)\} + R_0(x_1(\cdot,\varepsilon),\varepsilon)\big].$$

According to the definition of the generalized Green operator $\left(G\begin{bmatrix}*\\ *\end{bmatrix}\right)(t)$, the vector function $x_2^{(1)}(t,\varepsilon)$ is a solution of the boundary-value problem with impulsive action

$$\begin{aligned} \dot{x}_2 &= A(t)x_2 + \varepsilon\Big\{Z(z_0(t,c_r^*),t,0) \\ &\quad + A_1(t)[X_r(t)c_1 + x_1^{(1)}(t,\varepsilon)] + R(x_1(t,\varepsilon),t,\varepsilon)\Big\}, \quad t \neq \tau_i, \\ \Delta x_2|_{t=\tau_i} &= S_i x_2(\tau_i-0) + \varepsilon\Big\{J_i(z_0(\tau_i-0,c_r^*),0) \\ &\quad + A_{1i}[X_r(\tau_i-0)c_1 + x_1^{(1)}(\tau_i-0,\varepsilon)] + R_i(x_1(\tau_i-0,\varepsilon),\varepsilon)\Big\}, \\ l x_2 &= \varepsilon\Big\{J(z_0(\cdot,c_r^*),0) + l_1[X_r(\cdot)c_1 + x_1^{(1)}] + R_0(x_1,\varepsilon)\Big\}. \end{aligned} \tag{6.72}$$

The necessary and sufficient condition for the solvability of the boundary-value problem (6.72) with impulsive action, i.e.,

$$B_0 c_1 + P_{Q_d^*}\Big\{ l_1 x_1^{(1)}(\cdot,\varepsilon) + R_0(x_1(\cdot,\varepsilon),\varepsilon) \\ - l\int_a^b K(\cdot,\tau)[A_1(\tau)x_1^{(1)}(\tau,\varepsilon) + R(x_1(\tau,\varepsilon),\tau,\varepsilon)]d\tau \\ - l\sum_{i=1}^{p} \bar{K}(\cdot,\tau_i)[A_{1i}x_1^{(1)}(\tau_i-0,\varepsilon) + R_i(x_1(\tau_i-0,\varepsilon),\varepsilon)]\Big\} = 0,$$

gives the first approximation c_1 to the parameter $c(\varepsilon)$. The system of algebraic equations with respect to $c_1 \in R^r$ is solvable if and only if

$$P_{B_0^*}P_{Q_d^*}\Big\{ l_1 x_1^{(1)}(\cdot,\varepsilon) + R_0(x_1(\cdot,\varepsilon),\varepsilon) \\ - l\int_a^b K(\cdot,\tau)[A_1(\tau)x_1^{(1)}(\tau,\varepsilon) + R(x_1(\tau,\varepsilon),\tau,\varepsilon)]d\tau \\ - l\sum_{i=1}^{p} \bar{K}(\cdot,\tau_i)[A_{1i}x_1^{(1)}(\tau_i-0,\varepsilon) + R_i(x_1(\tau_i-0,\varepsilon),\varepsilon)]\Big\} = 0. \quad (6.73)$$

The second approximation $x_2(t,\varepsilon)$ to the required solution $x(t,\varepsilon)$ admits the representation

$$x_2(t,\varepsilon) = X_r(t)c_1 + x_2^{(1)}(t,\varepsilon).$$

Further, the third approximation $x_3^{(1)}(t,\varepsilon)$ is sought in the form

$$x_3^{(1)}(\tau,\varepsilon) = \varepsilon\left(G\begin{bmatrix} Z(z_0(\tau,c_r^*),\tau,0) + A_1(\tau)[X_r(\tau)c_2 + x_2^{(1)}(\tau,\varepsilon)] \\ + R(x_2(\tau,\varepsilon),\tau,\varepsilon) \\ J_i(z_0(\tau_i-0,c_r^*),0) + A_{1i}[X_r(\tau_i-0)c_2 + x_2^{(1)}(\tau_i-0,\varepsilon)] \\ + R_i(x_2(\tau_i-0,\varepsilon),\varepsilon)\end{bmatrix}\right)(t) \\ + \varepsilon X(t)Q^+\big[J(z_0(\cdot,c_r^*),0) + l_1(X_r(\cdot)c_2 + x_2^{(1)}(\cdot,\varepsilon)) + R_0(x_2(\cdot,\varepsilon),\varepsilon)\big]$$

as a solution of the boundary-value problem with impulsive action

$$\dot{x}_3 = A(t)x_3 + \varepsilon\Big\{Z(z_0(t,c_r^*),t,0) \\ + A_1(t)[X_r(t)c_2 + x_2^{(1)}(t,\varepsilon)] + R(x_2(t,\varepsilon),t,\varepsilon)\Big\}, \quad t \neq \tau_i,$$

$$\Delta x_3\Big|_{t=\tau_i} = S_i x_3(\tau_i - 0) + \varepsilon J_i(z_0(\tau_i - 0, c_r^*), 0)$$
$$+ A_{1i}[X_r(\tau_i - 0)c_2 + x_2^{(1)}(\tau_i - 0, \varepsilon)] + R_i(x_2(\tau_i - 0, \varepsilon), \varepsilon),$$
$$lx_3 = \varepsilon\Big\{J(z_0(\cdot, c_r^*), 0) + l_1\{X_r(\cdot)c_2 + x_2^{(1)}(\cdot, \varepsilon)\} + R_0(x_2(\cdot, \varepsilon), \varepsilon)\Big\}.$$

The necessary and sufficient condition for the solvability of this system yields the algebraic system

$$B_0 c_2 + P_{Q_d^*}\Big\{l_1 x_2^{(1)}(\cdot, \varepsilon) + R_0(x_2(\cdot, \varepsilon), \varepsilon)$$
$$- l\int_a^b K(\cdot, \tau)[A_1(\tau)x_2^{(1)}(\tau, \varepsilon) + R(x_2(\tau, \varepsilon), \tau, \varepsilon)]d\tau$$
$$- l\sum_{i=1}^{p} \bar{K}(\cdot, \tau_i)[A_{1i}x_2^{(1)}(\tau_i - 0, \varepsilon) + R_i(x_2(\tau_i - 0, \varepsilon), \varepsilon)]\Big\} = 0.$$

The criterion of solvability of this system has the form

$$P_{B_0^*}P_{Q_d^*}\Big\{l_1 x_2^{(1)}(\cdot, \varepsilon) + R_0(x_2(\cdot, \varepsilon), \varepsilon)$$
$$- l\int_a^b K(\cdot, \tau)[A_1(\tau)x_2^{(1)}(\tau, \varepsilon) + R(x_2(\tau, \varepsilon), \tau, \varepsilon)]d\tau$$
$$- l\sum_{i=1}^{p} \bar{K}(\cdot, \tau_i)[A_{1i}x_2^{(1)}(\tau_i - 0, \varepsilon) + R_i(x_2(\tau_i - 0, \varepsilon), \varepsilon)]\Big\} = 0, \quad (6.74)$$

whence we get the second approximation c_2 to $c(\varepsilon)$:

$$c_2 = B_0^+ P_{Q_d^*}\Big\{l_1 x_2^{(1)}(\cdot, \varepsilon) + R_0(x_2(\cdot, \varepsilon), \varepsilon)$$
$$- l\int_a^b K(\cdot, \tau)[A_1(\tau)x_2^{(1)}(\tau, \varepsilon) + R(x_2(\tau, \varepsilon), \tau, \varepsilon)]d\tau$$
$$- l\sum_{i=1}^{p} \bar{K}(\cdot, \tau_i)[A_{1i}x_2^{(1)}(\tau_i - 0, \varepsilon) + R_i(x_2(\tau_i - 0, \varepsilon), \varepsilon)]\Big\}.$$

Thus, for $P_{B_0^*} = 0$, the criteria of solvability of the corresponding algebraic systems [similar to (6.69), (6.73), (6.74)] are satisfied in each stage of the iterative process. Hence, if we continue the iterative process, then we conclude that the approximation $x_{k+1}^{(1)}(t, \varepsilon)$ to the solution $x^{(1)}(t, \varepsilon)$ is given by the formula

$$x_{k+1}^{(1)}(t,\varepsilon) = \varepsilon\left(G\begin{bmatrix} Z(z_0(\tau,c_r^*),\tau,0) + A_1(\tau)[X_r(\tau)c_k + x_k^{(1)}(\tau,\varepsilon)] \\ + R(x_k(\tau,\varepsilon),\tau,\varepsilon) \\ J_i(z_0(\tau_i-0,c_r^*),0) + A_{1i}[X_r(\tau_i-0)c_k + x_k^{(1)}(\tau_i-0,\varepsilon)] \\ + R_i(x_k(\tau_i-0,\varepsilon),\varepsilon)\end{bmatrix}\right)(t)$$

$$+\,\varepsilon X(t)Q^+\big[J(z_0(\cdot,c_r^*),0) + l_1\{X_r(\cdot)c_k + x_k^{(1)}(\cdot,\varepsilon)\} + R_0(x_k(\cdot,\varepsilon),\varepsilon)\big]$$

as a solution of the boundary-value problem

$$\begin{aligned} \dot{x}_{k+1} &= A(t)x_{k+1} + \varepsilon\Big\{Z(z_0(t,c_r^*),t,0) \\ &\quad + A_1(t)[X_r(t)c_1 + x_k^{(1)}(t,\varepsilon)] + R(x_k(t,\varepsilon),t,\varepsilon)\Big\}, \quad t \neq \tau_i, \\ \Delta x_{k+1}\Big|_{t=\tau_i} &= S_i x_{k+1}(\tau_i-0) + \varepsilon\Big\{J_i(z_0(\tau_i-0,c_r),0) \\ &\quad + A_{1i}[X_r(\tau_i-0)c_k + x_k^{(1)}(\tau_i-0,\varepsilon)] + R_i(x_k(\tau_i-0,\varepsilon),\varepsilon)\Big\}, \\ l x_{k+1} &= \varepsilon\Big\{J(z_0(\cdot,c_r^*),0) + l_1[X_r(\cdot)c_k + x_k^{(1)}(\cdot,\varepsilon)] + R_0(x_k(\cdot,\varepsilon),\varepsilon)\Big\}. \end{aligned} \tag{6.75}$$

The necessary and sufficient condition for the solvability of this boundary-value problem yields the following algebraic system for the kth approximation c_k to $c(\varepsilon)$:

$$\begin{aligned} B_0 c_k + P_{Q_d^*}\Big\{ & l_1 x_k^{(1)}(\cdot,\varepsilon) + R_0(x_k(\cdot,\varepsilon),\varepsilon) \\ & - l\int_a^b K(\cdot,\tau)[A_1(\tau)x_k^{(1)}(\tau,\varepsilon) + R(x_k(\tau,\varepsilon),\tau,\varepsilon)]d\tau \\ & - l\sum_{i=1}^{p}\bar{K}(\cdot,\tau_i)[A_{1i}x_k^{(1)}(\tau_i-0,\varepsilon) + R_i(x_k(\tau_i-0,\varepsilon),\varepsilon)]\Big\}. \end{aligned} \tag{6.76}$$

The criterion of solvability of system (6.76) has the form

$$\begin{aligned} P_{B_0^*}P_{Q_d^*}\Big\{ & l_1 x_k^{(1)}(\cdot,\varepsilon) + R_0(x_k(\cdot,\varepsilon),\varepsilon) \\ & - l\int_a^b K(\cdot,\tau)[A_1(\tau)x_k^{(1)}(\tau,\varepsilon) + R(x_k(\tau,\varepsilon),\tau,\varepsilon)]d\tau \\ & - l\sum_{i=1}^{p}\bar{K}(\cdot,\tau_i)[A_{1i}x_k^{(1)}(\tau_i-0,\varepsilon) + R_i(x_k(\tau_i-0,\varepsilon),\varepsilon)]\Big\} = 0, \end{aligned} \tag{6.77}$$

and if $P_{B_0^*} = 0$, then (6.77) is satisfied in each stage of the iterative process.

Hence, the approximation $x_{k+1}^{(1)}(t,\varepsilon)$ to the required solution can be represented in the form

$$x_{k+1}(t,\varepsilon) = X_r(t)c_k + x_{k+1}^{(1)}(t,\varepsilon).$$

Thus, we arrive at the following procedure for the construction of solutions $x(t,\varepsilon)$ of the boundary-value problem (6.75) such that $x(\cdot,\varepsilon) \in C^1([a,b]\backslash\{\tau_i\}_I)$, $x(t,\cdot) \in C[\varepsilon]$, $\varepsilon \in [0,\varepsilon_0]$, and $x(t,0) = 0$:

$$c_k = -B_0^+ P_{Q_d^*}\Big\{ l_1 x_k^{(1)}(\cdot,\varepsilon) + R_0(x_k(\cdot,\varepsilon),\varepsilon) - l\int_a^b K(\cdot,\tau)[A_1(\tau)x_k^{(1)}(\tau,\varepsilon) + R(x_k(\tau,\varepsilon),\tau,\varepsilon)]d\tau - l\sum_{i=1}^{p} \bar{K}(\cdot,\tau_i)[A_{1i}x_k^{(1)}(\tau_i-0,\varepsilon) + R_i(x_k(\tau_i-0,\varepsilon),\varepsilon)]\Big\}, \quad (6.78)$$

$$x_{k+1}^{(1)}(\tau,\varepsilon) = \varepsilon\left(G\begin{bmatrix} Z(z_0(\tau,c_r^*),\tau,0) + A_1(\tau)[X_r(\tau)c_k + x_k^{(1)}(\tau,\varepsilon)] \\ + R(x_k(\tau,\varepsilon),\tau,\varepsilon) \\ J_i(z_0(\tau_i-0,c_r^*),0)+A_{1i}[X_r(\tau_i-0)c_k+x_k^{(1)}(\tau_i-0,\varepsilon)] \\ + R_i(x_k(\tau_i-0,\varepsilon),\varepsilon) \end{bmatrix}\right)(t)$$
$$+ \varepsilon X(t)Q^+\big[J(z_0(\cdot,c_r^*),0)+l_1(X_r(\cdot)c_k+x_k^{(1)}(\cdot,\varepsilon))+R_0(x_k(\cdot,\varepsilon),\varepsilon)\big],$$

$$x_{k+1}(t,\varepsilon) = X_r(t)c_k + x_{k+1}^{(1)}(t,\varepsilon),$$

$$k = 0,1,2\ldots,$$

$$x_0(t,\varepsilon) = x_0^{(1)}(t,\varepsilon) = 0.$$

To prove the convergence of the iterative process (6.78) and establish estimates characterizing this process, we can use a scheme similar to the method of Lyapunov majorants described in [19, 65]. As in the previous case, to prove the existence of solutions of the boundary-value problem (6.66) with impulsive action, it suffices to establish the conditions of its reducibility to the operator system (6.71). Therefore, the following assertion is true:

Theorem 6.9. *Assume that the boundary-value problem (6.52) with impulsive action satisfies the requirements imposed above for the critical case* (rank $Q = n_1 < m$) *and that the corresponding generating boundary-value problem (6.1), (6.2) with impulsive action possesses an r-parameter ($r = n - n_1$) family of generating solutions given by (6.61) if and only if condition (6.60) is satisfied ($d = m - n_1$). If*

$$P_{B_0^*} = 0, \quad (6.79)$$

then, for any vector $c_r = c_r^* \in R^r$ satisfying the equation for generating constants (6.65), the boundary-value problem (6.66) has at least one solution $x(t, \varepsilon)$ such that $x(\cdot, \varepsilon) \in C^1([a,b]\backslash\{\tau_i\}_I)$, $x(t, \cdot) \in C[\varepsilon]$, $\varepsilon \in [0, \varepsilon_0]$, and $x(t,0) = 0$. One of these solutions can be found by using the iterative process (6.78) convergent for $\varepsilon \in [0, \varepsilon_*]$.

Furthermore, the boundary value problem (6.52) with impulsive action has at least one solution $z(t, \varepsilon)$ that satisfies the conditions $z(\cdot, \varepsilon) \in C^1([a,b]\backslash\{\tau_i\}_I)$, $z(t, \cdot) \in C[\varepsilon]$, and $\varepsilon \in [0, \varepsilon_0]$ and turns into the generating solution $z_0(t, c_r^*)$ given by relation (6.61) for $\varepsilon = 0$. The solution $z(\cdot, \varepsilon)$ can be obtained as a result of the iterative process (6.78) by using the formula

$$z_k(t, \varepsilon) = z_0(t, c_r^*) + x_k(t, \varepsilon), \quad k = 0, 1, 2, \ldots .$$

Remark 6.3. In the case of Fredholm boundary-value problems of index zero ($m = n$), the equality $P_{B_0^*} = 0$ means that $P_{B_0} = 0$ and, hence, $\det B_0 \neq 0$. Therefore, B_0^+ in the iterative procedure (6.78) should be replaced by B_0^{-1}. The boundary-value problem (6.52) with impulsive action has exactly one solution. In particular, the periodic boundary-value problem with impulsive action for system (6.52), namely,

$$\begin{gathered} \dot{z} = A(t)z + f(t) + \varepsilon Z(z, t, \varepsilon), \quad t \neq \tau_i, \\ \Delta z\Big|_{t=\tau_i} = S_i z + a_i + \varepsilon J_i(z, \varepsilon), \\ lz = z(0) - z(T) = 0, \end{gathered} \tag{6.80}$$

is a Fredholm problem of index zero.

Here, $A(t)$ and $f(t)$ are continuous or piecewise-continuous (with discontinuities of the first kind) T-periodic $n \times n$ matrix and $n \times 1$ vector functions, respectively, $Z(z, t+T, \varepsilon) = Z(z, t, \varepsilon)$, S_i are constant $n \times n$ matrices, a_i are n-dimensional column vectors, τ_i are time points such that

$$S_{i+p} = S_i, \; a_{i+p} = a_i, \quad \tau_{i+p} = \tau_i + T \quad (\tau_0 < 0, \tau_1 < \ldots < \tau_p < T)$$

for all i, $E + S_i$ are nonsingular matrices, E is the $n \times n$ identity matrix, $J_{i+p}(z, \varepsilon) = J_i(z, \varepsilon)$, $\alpha = 0$, $J(z(\cdot, \varepsilon), \varepsilon) = 0$, and $m = n$. In this case, B_0 is an $r \times r$ matrix, $P_{B_0} = P_{B_0^*} = 0$, and $B_0^+ = B_0^{-1}$. Theorem 6.9 implies the following assertion [30]:

Theorem 6.10. *Assume that the impulsive boundary-value problem (6.80) satisfies the conditions imposed above for the critical case* ($\operatorname{rank} Q = n_1 < n$). *Then, for any simple* ($\det B_0 \neq 0$) *root $c_r = c_r^* \in R^r$ of the equation for generating constants*

$$F(c_r) = P_{Q_r^*}\Bigg\{\int_0^T K(T, \tau) Z(z_0(\tau, c_r), \tau, 0) d\tau$$

$$+\sum_{i=1}^{p} \bar{K}(T,\tau_i)J_i(z_0(\tau_i,c_r),0)\Big\} = 0, \tag{6.81}$$

the impulsive boundary-value problem (6.80) possesses a unique T-periodic solution $z(t,\varepsilon)$ in the class of functions continuously differentiable with respect to $t \in [0,T]\backslash\{\tau_i\}$ (with discontinuities of the first kind at the points $t = \tau_i$) that turns, for $\varepsilon = 0$, into the generating solution

$$z_0(t,c_r^*) = X_r(t)c_r^* + \int_0^T G(t,\tau)f(\tau)d\tau + \sum_{i=1}^{p} \bar{G}(t,\tau_i)a_i,$$

where $K(t,\tau)$ is the Green matrix of the Cauchy problem for the impulsive system (6.1) on the interval $[0,T]$,

$$K(t,\tau) = \begin{cases} X(t)X^{-1}(\tau) & \text{for } 0 \le \tau \le t \le T, \\ 0 & \text{for } 0 \le t < \tau \le T, \end{cases}$$

$$\bar{K}(t,\tau_i) = K(t,\tau_i+0),$$

and

$$G(t,\tau) = K(t,\tau) + X(t)Q^{+}K(T,\tau), \quad \bar{G}(t,\tau_i) = G(t,\tau_i+0).$$

This solution is determined by the iterative process

$$c_k = -B_0^{-1}P_{Q_d^*}\Big\{\int_0^T K(T,\tau)[A_1(\tau)x_k^{(1)}(\tau,\varepsilon) + R(x_k(\tau,\varepsilon),\tau,\varepsilon)]d\tau$$
$$+\sum_{i=1}^{p} \bar{K}(T,\tau_i)[A_{1i}x_k^{(1)}(\tau_i-0,\varepsilon) + R_i(x_k(\tau_i-0,\varepsilon),\varepsilon)]\Big\},$$

$$x_{k+1}^{(1)}(t,\varepsilon) = \varepsilon\int_0^T G(t,\tau)\Big\{Z(z_0(\tau,c_r^*),\tau,0)$$
$$+ A_1(\tau)[X_r(\tau)c_k + x_k^{(1)}(\tau,\varepsilon)] + R(x_k,\tau,\varepsilon)\Big\}d\tau$$
$$+\varepsilon\sum_{i=1}^{p} \bar{G}(t,\tau_i)\Big\{J_i(z_0(\tau_i,c_r^*),0)$$
$$+ A_{1i}[X_r(\tau_i)c_k + x_k^{(1)}(\tau_i,\varepsilon)] + R_i(x_k(\tau_i,\varepsilon),\varepsilon)\Big\},$$

$$x_{k+1}(t,\varepsilon) = X_r(t)c_k + x_{k+1}^{(1)}(t,\varepsilon), \quad z_k(t,\varepsilon) = z_0(t,c_r^*) + x_k(t,\varepsilon),$$
$$k = 0,1,2,\ldots,$$
$$x_0(t,\varepsilon) = x_0^{(1)}(t,\varepsilon) = 0,$$

convergent for sufficiently small $\varepsilon \in [0,\varepsilon_*]$.

Proof. In [30], one can find the proof of this theorem with the condition $P_{B_0^*} = 0$ replaced by the requirement that the root $c_r = c_r^*$ of the equation for generating constants (6.81) must be simple. It is necessary to show that these conditions are equivalent. Indeed, the requirement that the root of equation (6.81) is simple means that

$$\det \left[\frac{\partial F(c_r)}{\partial c_r} \right]_{c_r = c_r^*} \neq 0.$$

Consider the following expansions in a neighborhood of the point $x=0$, $\varepsilon=0$:

$$Z(z_0 + x, t, \varepsilon) = Z(z_0(t, c_r^*), t, 0) + A_1(t)x + R(x, t, \varepsilon),$$

$$A_1(t) = \left. \frac{\partial Z(z, t, 0)}{\partial z} \right|_{z = z_0(t, c_r^*)},$$

$$R(0, t, 0) = 0, \qquad \frac{\partial R(0, t, 0)}{\partial x} = 0,$$

$$J_i(z_0 + x, \varepsilon) = J_i(z_0(\tau_i, c_r^*), 0) + A_{1i} x(\tau_i, \varepsilon) + R_i(x(\tau_i, \varepsilon), \varepsilon),$$

$$A_{1i} = \left. \frac{\partial J_i(z, 0)}{\partial z} \right|_{z = z_0(\tau_i, c_r^*)},$$

$$R_i(0, 0) = 0, \qquad \frac{\partial R_i(0, 0)}{\partial x} = 0.$$

By using these expansions and the fact that the function $F(c_r)$ can be differentiated under the integral sign in a neighborhood of the point $c_r = c_r^*$, we conclude from relation (6.81) that

$$\left. \frac{\partial F(c_r)}{\partial c_r} \right|_{c_r = c_r^*} = P_{Q_r^*} \left\{ \int_0^T K(T, \tau) \left. \frac{\partial Z(z_0(\tau, c_r), \tau, 0)}{\partial c_r} \right|_{c_r = c_r^*} d\tau \right.$$
$$\left. + \sum_{i=1}^{p} \bar{K}(T, \tau_i) \left. \frac{\partial J_i(z_0(\tau_i, c_r), 0)}{\partial c_r} \right|_{c_r = c_r^*} \right\} = B_0,$$

where

$$B_0 = P_{Q_r^*} \left\{ \int_0^T K(T, \tau) A_1(\tau) X_r(\tau) d\tau + \sum_{i=1}^{p} \bar{K}(T, \tau_i) A_{1i} X_r(\tau_i) \right\}$$

is an $r \times r$ matrix because

$$\left. \frac{\partial Z(z_0(\tau, c_r), \tau, 0)}{\partial c_r} \right|_{c_r = c_r^*} = \left. \frac{\partial Z(z, \tau, 0)}{\partial z} \right|_{z = z_0(\tau, c_r^*)} \left. \frac{\partial z_0(\tau, c_r)}{\partial c_r} \right|_{c_r = c_r^*} = A_1(\tau) X_r(\tau),$$

$$\left. \frac{\partial J_i(z_0(\tau_i, c_r), 0)}{\partial c_r} \right|_{c_r = c_r^*} = \left. \frac{\partial J_i(z, 0)}{\partial z} \right|_{z = z_0(\tau_i, c_r^*)} \left. \frac{\partial z_0(\tau_i, c_r)}{\partial c_r} \right|_{c_r = c_r^*} = A_{1i} X_r(\tau_i).$$

6.8 Critical Case of the Second Order

Assume that $\operatorname{rank} B_0 < d$, i.e., $P_{B_0^*} \neq 0$. Then Theorem 6.9 is inapplicable to the analysis of the solvability of the boundary-value problem (6.52) with impulsive action.

In the present section, we consider the boundary-value problem (6.66) with impulsive action in the indicated case, which is called the critical case of the second order. In this case, to solve the problem of solvability of the original impulsive boundary-value problem, it is necessary first to analyze the boundary-value problem used to determine the second approximation. In what follows, we show that the existence of solutions of the original boundary-value problem depends on conditions deduced by using the nonlinearities and the first approximation to the exact solution.

As in the previous section, let

$$\operatorname{rank} Q = n_1 < \min(n, m)$$

and, unlike the indicated case, let

$$P_{B_0^*} \neq 0 \iff \operatorname{rank} B_0 < d.$$

To simplify calculations, we assume that the vector functions $R(\cdot, t, \varepsilon) \in C^1[x]$, $R(x, \cdot, \varepsilon) \in C[t]$, $R(x, t, \cdot) \in C[\varepsilon]$, $R_i(x(\cdot, \varepsilon), \varepsilon) \in C^1[x]$, and $R_i(x, \cdot) \in C[\varepsilon]$ in expansions (6.67) satisfy the relations

$$R(0, t, 0) = 0, \qquad \frac{\partial R(0, t, \varepsilon)}{\partial x} = 0,$$

$$R_i(0, 0) = 0, \qquad \frac{\partial R_i(0, \varepsilon)}{\partial x} = 0,$$

$$R_0(0, 0) = 0, \qquad \frac{\partial R_0(0, \varepsilon)}{\partial x} = 0.$$

In what follows, we consider the general case where

$$R(0, t, 0) = 0, \qquad \frac{\partial R(0, t, 0)}{\partial x} = 0,$$

$$R_i(0, 0) = 0, \qquad \frac{\partial R_i(0, 0)}{\partial x} = 0,$$

$$R_0(0, 0) = 0, \qquad \frac{\partial R_0(0, 0)}{\partial x} = 0.$$

The second equation in system (6.68) is solvable if and only if

$$P_{B_0^*} P_{Q_d^*} \Big\{ l_1 x^{(1)}(\cdot, \varepsilon) + R_0(x(\cdot, \varepsilon), \varepsilon)$$

$$- l \int_a^b K(\cdot,\tau)[A_1(\tau)x^{(1)}(\tau,\varepsilon) + R(x(\tau,\varepsilon),\tau,\varepsilon)]d\tau$$

$$- l \sum_{i=1}^{p} \bar{K}(\cdot,\tau_i)[A_{1i}x^{(1)}(\tau_i - 0,\varepsilon) + R_i(x(\tau_i - 0,\varepsilon),\varepsilon)]\Big\} = 0. \quad (6.82)$$

Its solution can be represented in the form of a direct sum as follows:

$$c = c^{(0)} + c^{(1)}, \quad (6.83)$$

where

$$\begin{aligned} c^{(0)} &= -B_0^+ P_{Q_d^*}\Big\{ l_1 x^{(1)}(\cdot,\varepsilon) + R_0(x(\cdot,\varepsilon),\varepsilon) \\ &\qquad - l \int_a^b K(\cdot,\tau)[A_1(\tau)x^{(1)}(\tau,\varepsilon) + R(x(\tau,\varepsilon),\tau,\varepsilon)]d\tau \\ &\qquad - l \sum_{i=1}^{p} \bar{K}(\cdot,\tau_i)[A_{1i}x^{(1)}(\tau_i - 0,\varepsilon) + R_i(x(\tau_i - 0,\varepsilon),\varepsilon)]\Big\} \\ &= (I_r - P_{B_0})c^{(0)} \in R^r \ominus N(B_0) \end{aligned}$$

and $c^{(1)}$ is an arbitrary r-dimensional constant from $N(B_0)$:

$$c^{(1)} = P_{B_0}c = P_{B_0}c^{(1)} \in N(B_0).$$

In view of representation (6.83), the third equation of the operator system (6.68) yields the following formula:

$$x^{(1)}(t,\varepsilon) = \varepsilon G_1(t) P_{B_0} c^{(1)} + x^{(2)}(t,\varepsilon), \quad (6.84)$$

where

$$G_1(t) = \left(G \begin{bmatrix} A_1(\tau)X_r(\tau) \\ A_{1i}X_r(\tau_i - 0) \end{bmatrix} \right)(t) + \varepsilon X(t) Q^+ l_1 X_r(\cdot)$$

is an $n \times r$ matrix and the vector function $x^{(2)}(t,\varepsilon)$ has the form

$$x^{(2)}(\tau,\varepsilon) = \varepsilon \left(G \begin{bmatrix} Z(z_0(\tau,c_r^*),\tau,0) + A_1(\tau)\Big[X_r(\tau)(I_r - P_{B_0})c^{(0)} \\ \qquad + \varepsilon G_1(\tau)P_{B_0}c^{(1)} + x^{(2)}(\tau,\varepsilon)\Big] \\ + R(x(\tau,\varepsilon),\tau,\varepsilon) \\ J_i(z_0(\tau_i - 0,c_r^*),0) + A_{1i}\Big[X_r(\tau_i - 0)(I_r - P_{B_0})c^{(0)} \\ \qquad + \varepsilon G_1(\tau_i - 0)P_{B_0}c^{(1)} + x^{(2)}(\tau_i - 0,\varepsilon)\Big] \\ + R_i(x(\tau - 0,\varepsilon),\varepsilon) \end{bmatrix} \right)(t)$$

$$+ \varepsilon X(t)Q^{+}\Big[J(z_0(\cdot, c_r^*), 0) + l_1[X_r(\cdot)(I_r - P_{B_0})c^{(0)} + \varepsilon G_1(\cdot)P_{B_0}c^{(1)} + x^{(2)}(\cdot, \varepsilon)] + R_0(x(\cdot, \varepsilon), \varepsilon)\Big].$$

By using representation (6.84) and the condition of solvability (6.82) of the second equation in (6.67), we arrive at the following algebraic system for $c^{(1)} \in N(B_0)$:

$$\varepsilon B_1 c^{(1)} + P_{B_0^*}P_{Q_d^*}\Big\{l_1 x^{(2)}(\cdot, \varepsilon) + R_0(x(\cdot, \varepsilon), \varepsilon) - l\int_a^b K(\cdot, \tau)[A_1(\tau)x^{(2)}(\tau, \varepsilon) + R(x(\tau, \varepsilon), \tau, \varepsilon)]d\tau - l\sum_{i=1}^{p} \bar{K}(\cdot, \tau_i)[A_{1i}x^{(2)}(\tau_i - 0, \varepsilon) + R_i(x(\tau_i - 0, \varepsilon), \varepsilon)]\Big\} = 0, \quad (6.85)$$

where B_1 is a $d \times r$ matrix of the form

$$B_1 = P_{B_0^*}P_{Q_d^*}\Big\{l_1 G_1(\cdot)P_{B_0} - l\int_a^b K(\cdot, \tau)A_1(\tau)G_1(\tau)P_{B_0}d\tau - l\sum_{i=1}^{p} \bar{K}(\cdot, \tau_i)A_{1i}G_1(\tau_i - 0)P_{B_0}\Big\}.$$

Let P_{B_1} be an $r\times r$ matrix (orthoprojector) projecting the r-dimensional Euclidean space R^r onto the null space $N(B_1)$ of the $d\times r$ matrix B_1 and let $P_{B_1^*}$ be a $d\times d$ matrix (orthoprojector) projecting the d-dimensional space R^d onto the null space $N(B_1^*)$ of the $r\times d$ matrix B_1^*. Then the necessary and sufficient condition for the solvability of system (6.85) takes the form

$$P_{B_1^*}P_{B_0}^*P_{Q_d^*}\Big\{l_1 x^{(2)}(\cdot, \varepsilon) + R_0(x(\cdot, \varepsilon), \varepsilon) - l\int_a^b K(\cdot, \tau)[A_1(\tau)x^{(2)}(\tau, \varepsilon) + R(x(\tau, \varepsilon), \tau, \varepsilon)]d\tau - l\sum_{i=1}^{p} \bar{K}(\cdot, \tau_i)[A_{1i}x^{(2)}(\tau_i - 0, \varepsilon) + R_i(x(\tau_i - 0, \varepsilon), \varepsilon)]\Big\} = 0. \quad (6.86)$$

Under this condition, system (6.85) is solvable with respect to $\varepsilon c^{(1)} \in N(B_0)$, and we get

$$\varepsilon c^{(1)} = -B_1^{+}P_{B_0^*}P_{Q_d^*}\Big\{l_1 x^{(2)}(\cdot, \varepsilon) + R_0(x(\cdot, \varepsilon), \varepsilon)$$

$$- l \int_a^b K(\cdot, \tau)[A_1(\tau)x^{(2)}(\tau, \varepsilon) + R(x(\tau, \varepsilon), \tau, \varepsilon)]d\tau$$

$$- l \sum_{i=1}^{p} \bar{K}(\cdot, \tau_i)[A_{1i}x^{(2)}(\tau_i - 0, \varepsilon) + R_i(x(\tau_i - 0, \varepsilon), \varepsilon)]\Big\} + c^{(2)}, \quad (6.87)$$

where $c^{(2)}$ is an arbitrary vector from $N(B_0) \cap N(B_1)$, namely,

$$c^{(2)} = P_{B_1}c^{(1)} = P_{B_1}P_{B_0}c^{(1)} \in N(B_0) \cap N(B_1).$$

Assume that the null spaces $N(B_0)$ and $N(B_1)$ are mutually disjoint. In this case, system (7.85) is uniquely solvable ($c^{(2)} = 0$).

In order that equality (6.86) be true, it is sufficient that the intersection of the null spaces $N(B_0^*)$ and $N(B_1^*)$ be empty, i.e.,

$$P_{B_1^*}P_{B_0^*} = 0.$$

Thus, for

$$P_{B_0^*} \neq 0 \qquad \text{and} \qquad P_{B_1^*}P_{B_0^*} = 0, \quad (6.88)$$

system (6.68) turns into the following operator system:

$$x(t, \varepsilon) = X_r(t)(I_r - P_{B_0})c^{(0)} + \varepsilon G_1(t)P_{B_0}c^{(1)} + x^{(2)}(t, \varepsilon),$$

$$c^{(0)} = -B_0^+ P_{Q_d^*}\Big\{ l_1[\varepsilon G_1(\cdot)P_{B_0}c^{(1)} + x^{(2)}(\cdot, \varepsilon)] + R_0(x(\cdot, \varepsilon), \varepsilon)$$

$$- l \int_a^b K(\cdot, \tau)\Big[A_1(\tau)[\varepsilon G_1(\tau)P_{B_0}c^{(1)} + x^{(2)}(\tau, \varepsilon)]$$

$$+ R(x(\tau, \varepsilon), \tau, \varepsilon)\Big]d\tau$$

$$- l \sum_{i=1}^{p} \bar{K}(\cdot, \tau_i)\Big[A_{1i}[\varepsilon G_i(\tau_i - 0)P_{B_0}c^{(1)} + x^{(2)}(\tau_i - 0, \varepsilon), \varepsilon)]$$

$$+ R_i(x(\tau_i - 0, \varepsilon), \varepsilon)\Big]\Big\},$$

$$\varepsilon c^{(1)} = -B_1^+ P_{B_0^*} P_{Q_d^*}\Big\{ l_1 x^{(2)}(\cdot, \varepsilon) + R_0(x(\cdot, \varepsilon), \varepsilon)$$

$$- l \int_a^b K(\cdot, \tau)[A_1(\tau)x^{(2)}(\tau, \varepsilon) + R(x(\tau, \varepsilon), \tau, \varepsilon)]d\tau$$

$$- l \sum_{i=1}^{p} \bar{K}(\cdot, \tau_i)[A_{1i}x^{(2)}(\tau_i - 0, \varepsilon) + R_i(x(\tau_i - 0, \varepsilon), \varepsilon)]\Big\}, \quad (6.89)$$

$$x^{(2)}(\tau,\varepsilon)=\varepsilon\left(G\begin{bmatrix} Z(z_0(\tau,c_r^*),\tau,0)+A_1(\tau)\Big[X_r(\tau)(I_r-P_{B_0})c^{(0)} \\ \qquad +\varepsilon G_1(\tau)P_{B_0}c^{(1)}+x^{(2)}(\tau,\varepsilon)\Big] \\ +R(x(\tau,\varepsilon),\tau,\varepsilon) \\ J_i(z_0\tau_i-0,c_r^*,0)+A_{1i}\Big[X_r(\tau_i-0)(I_r-P_{B_0})c^{(0)} \\ \qquad +\varepsilon G_1(\tau_i-0)P_{B_0}c^{(1)}+x^{(2)}(\tau_i-0,\varepsilon)\Big] \\ +R_i(x(\tau-0,\varepsilon),\varepsilon)\end{bmatrix}\right)(t)$$

$$+\,\varepsilon X(t)Q^+\Big[J(z_0(\cdot,c_r^*),0) + l_1[X_r(\cdot)(I_r-P_{B_0})c^{(0)}+\varepsilon G_1(\cdot)P_{B_0}c^{(0)}+x^{(2)}(\cdot,\varepsilon)] + R_0(x(\cdot,\varepsilon),\varepsilon)\Big].$$

For $P_{B_0^*}\neq 0$, the operator system (6.68) does not belong to the class of systems solvable by the method of simple iterations. Under conditions (6.69) and (6.88), system (6.68) can be reduced to system (6.89) (regularized) by introducing an additional variable. In this case, the r-dimensional vector constant is decomposed into the direct sum of two quantities with different definitions, and the dimension of the operator system (6.68) increases by $r=\dim N(Q)$. This enables us to reduce the $(2n+r)$-dimensional operator system to the $2(n+r)$-dimensional operator system (6.89) solvable by the method of simple iterations provided that conditions (6.88) are satisfied.

To study problem (6.66) with impulsive action in the general case where

$$R(0,t,0)=0,\qquad \frac{\partial R(0,t,0)}{\partial x}=0,$$
$$R_i(0,0)=0,\qquad \frac{\partial R_i(0,0)}{\partial x}=0,$$
$$R_0(0,0)=0,\qquad \frac{\partial R_0(0,0)}{\partial x}=0,$$

it is necessary to make expansion (6.67) more precise by using explicit representations of the linear (in x and ε) parts of the function $R(x,t,\varepsilon)$, $R_i(x,\varepsilon)$, and $R_0(x,\varepsilon)$. To do this, it is necessary to impose an additional requirement that the vector function $Z(z,t,\varepsilon)$ and vector functionals $J_i(z,\varepsilon)$ and $J(z,\varepsilon)$ must be continuously differentiable with respect to ε, i.e.,

$$Z(z,t,\cdot)\in C^1[\varepsilon],\quad J_i(z,\cdot)\in C^1[\varepsilon],\quad \text{and}\quad J(z,\cdot)\in C^1[\varepsilon].$$

We preserve the same notation for the functions $R(x,t,\varepsilon)$, $R_i(x,\varepsilon)$, and $R_0(x,\varepsilon)$ in which the terms $\varepsilon A_2(t)x$, $\varepsilon A_{2i}x$, and $\varepsilon l_2 x$, respectively, are separated. As a result, for the nonlinearities $Z(z_0+x,t,\varepsilon)$, $J_i(z_0+x,\varepsilon)$, and $J(z_0+x,\varepsilon)$, we get the following expansions similar to (6.67):

$$Z(z_0 + x, t, \varepsilon) = Z(z_0(t, c_r^*), t, 0) + A_1(t)x + \varepsilon A_2(t)x + R(x, t, \varepsilon),$$

$$J_1(z_0 + x, \varepsilon) = J_i(z_0(\tau_i - 0, c_r^*), 0) + A_{1i}x(\tau_i - 0, \varepsilon) + \varepsilon A_{2i}x(\tau_i - 0, \varepsilon) + R_i(x(\tau_i - 0, \varepsilon), \varepsilon),$$

$$J(z_0 + x, \varepsilon) = J(z_0(\cdot, c_r^*), 0) + l_1x(\cdot, \varepsilon) + \varepsilon l_2x(\cdot, \varepsilon) + R_0(x(\cdot, \varepsilon), \varepsilon),$$

where

$$A_2(t) = A_2(t, c_r^*) = \left.\frac{\partial^2 Z(z, t, \varepsilon)}{\partial\varepsilon\partial z}\right|_{\substack{z=z_0(t,c_r^*)\\ \varepsilon=0}} \in C[t],$$

$$A_{2i} = \left.\frac{\partial^2 J_i(z, \varepsilon)}{\partial\varepsilon\partial z}\right|_{\substack{z=z_0(\tau_i,c_r^*)\\ \varepsilon=0}}$$

(in the case analyzed earlier, we have $A_2(t) = 0$, $A_{2i} = 0$, and $l_2 = 0$).

By analogy with the previous case, in view of Theorem 6.1, we pass from the boundary-value problem

$$\dot{x} = A(t)x + \varepsilon\Big\{Z(z_0(t, c^*), t, 0) + A_1(t)x + \varepsilon A_2(t)x + R(x, t, \varepsilon)\Big\}, \quad t \neq \tau_i,$$

$$\Delta x\Big|_{t=\tau_i} = S_ix(\tau_i - 0) + \varepsilon\Big\{J_i(z_0(\tau_i - 0, c_r^*), 0) + A_{1i}x(\tau_i - 0, \varepsilon) + \varepsilon A_{2i}x(\tau_i - 0, \varepsilon) + R_i(x(\tau_i - 0, \varepsilon), \varepsilon)\Big\},$$

$$lx = \varepsilon\Big\{J(z_0(\cdot, c_r^*), 0) + l_1x(\cdot, \varepsilon) + \varepsilon l_2x(\cdot, \varepsilon) + R_0(x(\cdot, \varepsilon), \varepsilon)\Big\}$$

to the equivalent operator system

$$x(t, \varepsilon) = X_r(t)c + x^{(1)}(t, \varepsilon),$$

$$\begin{aligned} B_0c = -P_{Q_d^*}\Bigg\{&l_1x^{(1)}(\cdot, \varepsilon) + \varepsilon l_2[X_r(\cdot)c + x^{(1)}(\cdot, \varepsilon)] + R_0(x(\cdot, \varepsilon), \varepsilon) \\ &- l\int_a^b K(\cdot, \tau)\Big[A_1(\tau)x^{(1)}(\tau, \varepsilon) + \varepsilon A_2(\tau)[X_r(\tau)c + x^{(1)}(\tau, \varepsilon)] \\ &\qquad + R(x(\tau, \varepsilon), \tau, \varepsilon)\Big]d\tau \\ &- l\sum_{i=1}^{p} \bar{K}(\cdot, \tau_i)\Big[A_{1i}x^{(1)}(\tau_i - 0, \varepsilon) + \varepsilon A_{2i}[X_r(\tau_i - 0)c + x^{(1)}(\tau_i - 0, \varepsilon)] \\ &\qquad + R_i(x(\tau_i - 0, \varepsilon), \varepsilon)\Big]\Bigg\}, \end{aligned} \tag{6.90}$$

$$x^{(1)}(\tau, \varepsilon) = \varepsilon\left(G\begin{bmatrix} Z(z_0(\tau, c_r^*), \tau, 0) + [A_1(\tau) + \varepsilon A_2(\tau)]x(\tau, \varepsilon) \\ + R(x(\tau, \varepsilon), \tau, \varepsilon) \\ J_i(z_0(\tau_i - 0, c_r^*), 0) + [A_{1i} + \varepsilon A_{2i}]x(\tau_i - 0, \varepsilon) \\ + R_i(x(\tau_i - 0, \varepsilon), \varepsilon) \end{bmatrix}\right)(t)$$

$$+ \varepsilon X(t)Q^+[J(z_0(\cdot, c_r^*), 0) + [l_1 + \varepsilon l_2]x(\cdot, \varepsilon) + R_0(x(\cdot, \varepsilon), \varepsilon)].$$

in the space of vector functions $x(t,\varepsilon)$ piecewise continuously differentiable with respect to t (with discontinuities of the first kind at the points $t=\tau_i$) and continuous in $\varepsilon\in[0,\varepsilon_0]$.

The second equation in system (6.90) is solvable if and only if

$$P_{B_0^*}P_{Q_d^*}\Big\{l_1x^{(1)}(\cdot,\varepsilon)+\varepsilon l_2[X_r(\cdot)c+x^{(1)}(\cdot,\varepsilon)]+R_0(x(\cdot,\varepsilon),\varepsilon)$$
$$-l\int_a^b K(\cdot,\tau)\Big[A_1(\tau)x^{(1)}(\tau,\varepsilon)+\varepsilon A_2(\tau)[X_r(\tau)c+x^{(1)}(\tau,\varepsilon)]$$
$$+R(x(\tau,\varepsilon),\tau,\varepsilon)\Big]d\tau$$
$$-l\sum_{i=1}^p\bar K(\cdot,\tau_i)\Big[A_{1i}x^{(1)}(\tau_i-0,\varepsilon)+\varepsilon A_{2i}[X_r(\tau_i-0)c+x^{(1)}(\tau_i-0,\varepsilon)]$$
$$+R_i(x(\tau_i-0,\varepsilon),\varepsilon)\Big]\Big\}=0. \tag{6.91}$$

Under this condition, its solution admits a representation in the form of direct sum, namely,

$$c=c^{(0)}+c^{(1)},$$

where

$$c^{(0)}=-B_0^+P_{Q_d^*}\Big\{l_1x^{(1)}(\cdot,\varepsilon)+\varepsilon l_2x(\cdot,\varepsilon)+R_0(x(\cdot,\varepsilon),\varepsilon)$$
$$-l\int_a^b K(\cdot,\tau)\Big[A_1(\tau)x^{(1)}(\tau,\varepsilon)+\varepsilon A_2(\tau)x(\tau,\varepsilon)$$
$$+R(x(\tau,\varepsilon),\tau,\varepsilon)\Big]d\tau$$
$$-l\sum_{i=1}^p\bar K(\cdot,\tau_i)\Big[A_{1i}x(\tau_i-0,\varepsilon)+\varepsilon A_{2i}x(\tau_i-0,\varepsilon)$$
$$+R_i(x(\tau_i-0,\varepsilon),\varepsilon)\Big]\Big\}$$
$$=(I_r-P_{B_0})c^{(0)}\in R^r\ominus N(B_0),$$
$$c^{(1)}=P_{B_0}c=P_{B_0}c^{(1)}\in N(B_0).$$

From the third equation of the operator system (6.90), we obtain

$$x^{(1)}(t,\varepsilon)=\varepsilon G_1(t)P_{B_0}c^{(1)}+x^{(2)}(t,\varepsilon),$$

where $G_1(t)$ is the same $n\times n$ matrix as in relation (6.84) and the vector function $x^{(2)}(t,\varepsilon)$ has the form

$$x^{(2)}(\tau,\varepsilon)=\varepsilon\left(G\begin{bmatrix}Z(z_0(\tau,c_r^*),\tau,0)\\ \quad+[A_1(\tau)+\varepsilon A_2(\tau)]\big(X_r(\tau)(I_r-P_{B_0})c^{(0)}+x^{(1)}(\tau,\varepsilon)\big)\\ \quad+R(x(\tau,\varepsilon),\tau,\varepsilon)\\ J_i(z_0(\tau_i-0,c_r^*),0)\\ \quad+[A_{1i}+\varepsilon A_{2i}]\big(X_r(\tau_i-0)(I_r-P_{B_0})c^{(0)}+x^{(1)}(\tau_i-0,\varepsilon)\big)\\ \quad+R_i(x(\tau_i-0,\varepsilon),\varepsilon)\end{bmatrix}\right)(t)$$

$$+\varepsilon X(t)Q^+\Big[J(z_0(\cdot,c_r^*),0)$$
$$+[l_1+\varepsilon l_2]\big(X_r(\cdot)(I_r-P_{B_0})c^{(0)}+x^{(1)}(\cdot,\varepsilon)\big)+R_0(x(\cdot,\varepsilon),\varepsilon)\Big].$$

By using the representations for $x^{(1)}(t,\varepsilon)$ and c and the condition (6.91) of solvability of the second equation in (6.90), we arrive at the following algebraic system for $c^{(1)}\in N(B_0)$:

$$\begin{aligned}\varepsilon B_1c^{(1)}+P_{B_0^*}P_{Q_d^*}\Bigg\{&l_1x^{(2)}(\cdot,\varepsilon)+\varepsilon l_2[X_r(\cdot)(I_r-P_{B_0})c^{(0)}+x^{(1)}(\cdot,\varepsilon)]\\&+R_0(x(\cdot,\varepsilon),\varepsilon)\\&-l\int_a^b K(\cdot,\tau)\Big[A_1(\tau)x^{(2)}(\tau,\varepsilon)\\&\qquad+\varepsilon A_2(\tau)[X_r(\tau)(I_r-P_{B_0})c^{(0)}+x^{(1)}(\tau,\varepsilon)]\\&\qquad+R(x(\tau,\varepsilon),\tau,\varepsilon)\Big]d\tau\\&-l\sum_{i=1}^p\bar K(\cdot,\tau_i)\Big[A_{1i}x^{(2)}(\tau_i-0,\varepsilon)\\&\qquad+\varepsilon A_{2i}[X_r(\tau_i-0)(I_r-P_{B_0})c^{(0)}+x^{(1)}(\tau_i-0,\varepsilon)]\\&\qquad+R_i(x(\tau_i-0,\varepsilon),\varepsilon)\Big]\Bigg\}=0,\end{aligned}$$

where B_1 is a $d\times r$ matrix given by the formula

$$\begin{aligned}B_1=P_{B_0^*}P_{Q_d^*}\Bigg\{&[l_1G_1(\cdot)+l_2X_r(\cdot)]\\&-l\int_a^b K(\cdot,\tau)[A_1(\tau)G_1(\tau)+A_2(\tau)X_r(\tau)]d\tau\\&-l\sum_{i=1}^p\bar K(\cdot,\tau_i)[A_{1i}G_1(\tau_i-0)+A_{2i}X_r(\tau_i-0)]\Bigg\}P_{B_0}.\end{aligned}\tag{6.92}$$

As earlier, we denote $P_{B_1}\colon R^r \to N(B_1)$ and $P_{B_1^*}\colon R^d \to N(B_1^*)$. Under conditions (6.88), we can now pass from the operator system (6.90) to the system

$$x(t,\varepsilon) = X_r(t)(I_r - P_{B_0})c^{(0)} + \varepsilon G_1(t)P_{B_0}c^{(1)} + x^{(2)}(t,\varepsilon),$$

$$\begin{aligned} c^{(0)} = -B_0^+ P_{Q_d^*}\Bigg\{ & l_1[\varepsilon G_1(\cdot)P_{B_0}c^{(1)} + x^{(2)}(\cdot,\varepsilon)] + \varepsilon l_2 x(\cdot,\varepsilon) + R_0(x(\cdot,\varepsilon),\varepsilon) \\ & - l\int_a^b K(\cdot,\tau)\Big[A_1(\tau)[\varepsilon G_1(\tau)P_{B_0}c^{(1)} + x^{(2)}(\tau,\varepsilon)] \\ & \qquad + \varepsilon A_2(\tau)x(\tau,\varepsilon) + R(x(\tau,\varepsilon),\tau,\varepsilon)\Big]d\tau \\ & - l\sum_{i=1}^p \bar K(\cdot,\tau_i)\Big[A_{1i}[\varepsilon G_1(\tau_i - 0)P_{B_0}c^{(1)} + x^{(2)}(\tau_i - 0,\varepsilon)] \\ & \qquad + \varepsilon A_{2i}x(\tau_i - 0,\varepsilon) + R_i(x(\tau_i - 0,\varepsilon),\varepsilon)\Big]\Bigg\}, \end{aligned} \tag{6.93}$$

$$\begin{aligned} \varepsilon c^{(1)} = -B_1^+ P_{B_0^*}P_{Q_d^*}\Bigg\{ & l_1 x^{(2)}(\cdot,\varepsilon) + \varepsilon l_2 x(\cdot,\varepsilon) + R_0(x(\cdot,\varepsilon),\varepsilon) \\ & - l\int_a^b K(\cdot,\tau)[A_1(\tau)x^{(2)}(\tau,\varepsilon) + \varepsilon A_2 x(\tau,\varepsilon) + R(x(\tau,\varepsilon),\varepsilon)]d\tau \\ & - l\sum_{i=1}^p \bar K(\cdot,\tau_i)\Big[A_{1i}x^{(2)}(\tau_i - 0,\varepsilon) \\ & \qquad + \varepsilon A_{2i}x(\tau_i - 0,\varepsilon) + R_i(x(\tau_i - 0,\varepsilon),\varepsilon)\Big]\Bigg\}, \end{aligned}$$

$$\begin{aligned} x^{(2)}(t,\varepsilon) = \varepsilon \left(G \begin{bmatrix} Z(z_0(\tau,c_r^*),\tau,0) \\ \quad + [A_1(\tau) + \varepsilon A_2(\tau)]\Big(X_r(\tau)(I_r - P_{B_0})c^{(0)} \\ \qquad + \varepsilon G_1(\tau)P_{B_0}c^{(1)} + x^{(2)}(\tau,\varepsilon)\Big) \\ \quad + R(x(\tau,\varepsilon),\tau,\varepsilon) \\ J_i(z_0(\tau_i - 0,c_r^*),0) \\ \quad + [A_{1i} + \varepsilon A_{2i}]\Big(X_r(\tau_i - 0)(I_r - P_{B_0})c^{(0)} \\ \qquad + \varepsilon G_1(\tau_i - 0)P_{B_0}c^{(1)} + x^{(2)}(\tau_i - 0,\varepsilon)\Big) \\ \quad + R_i(x(\tau_i - 0,\varepsilon),\varepsilon) \end{bmatrix} \right)(t) \\ + \varepsilon X(t)Q^+\Big[J(z_0(\cdot,c_r^*),0) + [l_1 + \varepsilon l_2]\Big(X_r(\cdot)(I_r - P_{B_0})c^{(0)} \\ + \varepsilon G_1(\cdot)P_{B_0}c^{(1)} + x^{(2)}(\cdot,\varepsilon)\Big) + R_0(x(\cdot,\varepsilon),\varepsilon)\Big]. \end{aligned}$$

System (6.93) belongs to the class of operator systems solvable by the method of simple iterations. The procedure used to study and solve system (6.93) is similar to procedure (6.89) described earlier.

We use system (6.89) to illustrate the iterative process for system (6.93).

Iterative Algorithm. Our aim is to construct, by analogy with [19], an iterative algorithm for finding the solution $x(t,\varepsilon)$ of the boundary-value problem (6.66) with impulsive action such that $x(\cdot,\varepsilon)\in C^1([a,b]\setminus\{\tau_i\}_I)$, $x(t,\cdot)\in C[\varepsilon]$, $\varepsilon\in[0,\varepsilon_0]$, and $x(t,0)=0$ on the basis of the method of simple iterations for the operator system (6.89). The first approximation $x_1^{(2)}(t,\varepsilon)$ to $x^{(2)}(t,\varepsilon)$ is taken in the form

$$x_1^{(2)}(t,\varepsilon)=\varepsilon\left(G\begin{bmatrix} Z(z_0(\tau,c_r^*),\tau,0) \\ J_i(z_0(\tau_i-0,c_r^*),0)\end{bmatrix}\right)(t)+\varepsilon X(t)Q^+J(z_0(\cdot,c_r^*),0).$$

According to the definition of the generalized Green operator $\left(G\begin{bmatrix} * \\ * \end{bmatrix}\right)(t)$, the vector function $x_1^{(2)}=x_1^{(2)}(t,\varepsilon)$ is a solution of the boundary-value problem with impulsive action

$$\dot{x}=A(t)x+\varepsilon Z(z_0(t,c_r^*),t,0),\quad t\neq\tau_i,$$
$$\Delta x\Big|_{t=\tau_i}=S_ix(\tau_i-0)+\varepsilon J_i(z_0(\tau_i-0,c_r^*),0),$$
$$lx=\varepsilon J(z_0(\cdot,c_r^*),0).$$

This solution exists due to the choice of $c_r^*\in R^r$ from the equation for generating constants (6.65).

The first approximation $x_1(t,\varepsilon)$ to the exact solution $x(t,\varepsilon)$ of the boundary-value problem (6.66) with impulsive action is assumed to be equal to $x_1^{(2)}(t,\varepsilon)$.

The second approximation $x_2^{(2)}(t,\varepsilon)$ to $x^{(2)}(t,\varepsilon)$ is sought in the form

$$x_2^{(2)}(\tau,\varepsilon)=\varepsilon\left(G\begin{bmatrix} Z(z_0(\tau,c_r^*),\tau,0) \\ \quad+A_1(\tau)[X_r(\tau)(I_r-P_{B_0})c_1^{(0)}+x_1^{(2)}(\tau,\varepsilon)] \\ \quad+R(x_1^{(2)}(\tau,\varepsilon),\tau,\varepsilon) \\ J_i(z_0(\tau_i-0,c_r^*),0) \\ \quad+A_{1i}[X_r(\tau_i-0)(I_r-P_{B_0})c_1^{(0)}+x_1^{(2)}(\tau_i-0,\varepsilon)] \\ \quad+R_i(x_1^{(2)}(\tau_i-0,\varepsilon),\varepsilon)\end{bmatrix}\right)(t)$$
$$+\varepsilon X(t)Q^+\Big[J(z_0(\cdot,c_r^*),0)+l_1[X_r(\cdot)(I_r-P_{B_0})c_1^{(0)}+x_1^{(2)}(\cdot,\varepsilon)]$$
$$+R_0(x_1^{(2)}(\cdot,\varepsilon),\varepsilon)\Big].$$

By the definition of the generalized Green operator $\left(G\begin{bmatrix}*\\ *\end{bmatrix}\right)(t)$, the vector function $x_2^{(2)}(t,\varepsilon)$ is a solution of the boundary-value problem with impulsive action

$$\dot{x}_2 = A(t)x_2 + \varepsilon\Big\{Z(z_0(t,c_r^*),t,0) + A_1(t)[X_r(t)(I_r - P_{B_0})c_1^{(0)} + x_1^{(2)}(t,\varepsilon)] + R(x_1^{(2)}(t,\varepsilon),t,\varepsilon)\Big\}, \qquad t \neq \tau_i,$$

$$\Delta x_2|_{t=\tau_i} = S_i x_2(\tau_i - 0) + \varepsilon\Big\{J_i(z_0(\tau_i - 0, c_r^*),0) + A_{1i}[X_r(\tau_i - 0)(I_r - P_{B_0})c_1^{(0)} + x_1^{(2)}(\tau_i - 0,\varepsilon)] + R_i(x_1^{(2)}(\tau_i - 0,\varepsilon),\varepsilon)\Big\},$$

$$lx_2 = \varepsilon\Big\{J(z_0(\cdot,c_r^*),0) + l_1[X_r(\cdot)(I_r - P_{B_0})c_1^{(0)} + x_1^{(2)}(\cdot,\varepsilon)] + R_0(x_1^{(2)}(\cdot,\varepsilon),\varepsilon)\Big\}.$$

The existence of this solution is guaranteed by the choice of the vector constant $c_1^{(0)} \in R^r$ from the condition of solvability of this problem:

$$B_0 c_1^{(0)} + P_{Q_d^*}\Big\{l_1 x_1^{(2)}(\cdot,\varepsilon) + R_0(x_1^{(2)}(\cdot,\varepsilon),\varepsilon) - l\int_a^b K(\cdot,\tau)[A_1(\tau)x_1^{(2)}(\tau,\varepsilon) + R(x_1^{(2)}(\tau,\varepsilon),\tau,\varepsilon)]d\tau - l\sum_{i=1}^p \bar{K}(\cdot,\tau_i)[A_{1i}x_1^{(2)}(\tau_i - 0,\varepsilon) + R_i(x_1^{(2)}(\tau_i - 0,\varepsilon),\varepsilon)]\Big\} = 0. \quad (6.94)$$

In turn, system (6.94) is solvable if and only if

$$P_{B_0^*}P_{Q_d^*}\Big\{l_1 x_1^{(2)}(\cdot,\varepsilon) + R_0(x_1^{(2)}(\cdot,\varepsilon),\varepsilon) - l\int_a^b K(\cdot,\tau)[A_1(\tau)x_1^{(2)}(\tau,\varepsilon) + R(x_1^{(2)}(\tau,\varepsilon),\tau,\varepsilon)]d\tau - l\sum_{i=1}^p \bar{K}(\cdot,\tau_i)[A_{1i}x_1^{(2)}(\tau_i - 0,\varepsilon) + R_i(x_1^{(2)}(\tau_i - 0,\varepsilon),\varepsilon)]\Big\} = 0.$$

Under this condition, the first approximation to $c^{(0)} \in R^r$ can be found from (6.94) and has the form

$$c_1^{(0)} = -B_0^+ P_{Q_d^*}\Big\{l_1 x_1^{(2)}(\cdot,\varepsilon) + R_0(x_1^{(2)}(\cdot,\varepsilon),\varepsilon)$$

$$- l \int_a^b K(\cdot, \tau)[A_1(\tau)x_1^{(2)}(\tau, \varepsilon) + R(x_1^{(2)}(\tau, \varepsilon), \tau, \varepsilon)]d\tau$$
$$- l \sum_{i=1}^{p} \bar{K}(\cdot, \tau_i)[A_{1i}x_1^{(2)}(\tau_i - 0, \varepsilon) + R_i(x_1^{(2)}(\tau_i - 0, \varepsilon), \varepsilon)]\Big\}.$$

The second approximation $x_2(t, \varepsilon)$ to the required solution admits the representation

$$x_2(t, \varepsilon) = X_r(t)(I_r - P_{B_0})c_1^{(0)} + x_1^{(2)}(t, \varepsilon).$$

The third approximation is sought as a solution of the boundary-value problem with impulsive action

$$\begin{aligned}
\dot{x}_3 = A(t)x_3 + \varepsilon\Big\{&Z(z_0(t, c^*), t, 0) \\
&+ A_1(t)[X_r(t)(I_r - P_{B_0})c_2^{(0)} + \varepsilon G_1(t)P_{B_0}c_1^{(1)} + x_2^{(2)}(t, \varepsilon)] \\
&+ R(x_2(t, \varepsilon), t, \varepsilon)\Big\}, \qquad t \neq \tau_i,
\end{aligned}$$

$$\begin{aligned}
\Delta x_3|_{t=\tau_i} = S_i x_3(\tau_i - 0) + \varepsilon\Big\{&J_i(z_0(\tau_i - 0, c_r^*), 0) \\
&+ A_{1i}\Big[X_r(\tau_i - 0)(I_r - P_{B_0})c_2^{(0)} + \varepsilon G_1(\tau_i - 0)P_{B_0}c_1^{(1)} \\
&+ x_2^{(2)}(\tau_i - 0, \varepsilon)\Big] + R_i(x_2(\tau_i - 0, \varepsilon), \varepsilon)\Big\},
\end{aligned}$$

$$\begin{aligned}
lx_3 = \varepsilon\Big\{&J(z_0(\cdot, c_r^*), 0) + l_1[X_r(\cdot)(I_r - P_{B_0})c_2^{(0)} + \varepsilon G_1(\cdot)P_{B_0}c_1^{(1)} + x_2^{(2)}(\tau_i - 0, \varepsilon)] \\
&+ R_0(x_2(\tau_i - 0, \varepsilon), \varepsilon)\Big\},
\end{aligned}$$

where the constants $c_2^{(0)}$ and $c_1^{(1)}$ are found from the condition of solvability of this system:

$$\begin{aligned}
B_0 c_2^{(0)} + P_{Q_d^*}\Big\{&l_1\Big[\varepsilon G_1(\cdot)P_{B_0}c_1^{(1)} + x_2^{(2)}(\cdot, \varepsilon)\Big] + R_0(x_2(\cdot, \varepsilon), \varepsilon) \\
&- l \int_a^b K(\cdot, \tau)\Big[A_1(\tau)[\varepsilon G_1(\tau)P_{B_0}c_1^{(1)} + x_2^{(2)}(\tau, \varepsilon)] + R(x_2(\tau, \varepsilon), \tau, \varepsilon)\Big]d\tau \\
&- l \sum_{i=1}^{p} \bar{K}(\cdot, \tau_i)\Big[A_{1i}[\varepsilon G_1(\tau_i - 0)P_{B_0}c_1^{(1)} + x_2^{(2)}(\tau_i - 0, \varepsilon)] \\
&+ R_i(x_2(\tau_i - 0, \varepsilon), \varepsilon)\Big]\Big\} = 0.
\end{aligned}$$

The necessary and sufficient condition of solvability of this system with respect to $c_2^{(0)} = (I_r - P_{B_0})c_2^{(0)} \in R^r \ominus N(B_0)$ yields the following algebraic system for $\varepsilon c^{(1)} \in N(B_0)$:

$$\varepsilon B_1 c_1^{(1)} = -P_{B_0^*} P_{Q_d^*} \Big\{ l_1 x_2^{(2)}(\cdot,\varepsilon) + R_0(x_2(\cdot,\varepsilon),\varepsilon) - l \int_a^b K(\cdot,\tau)[A_1(\tau)x_2^{(2)}(\tau,\varepsilon) + R(x_2(\tau,\varepsilon),\tau,\varepsilon)]d\tau - l \sum_{i=1}^p \bar{K}(\cdot,\tau_i)[A_{1i}x_2^{(2)}(\tau_i-0,\varepsilon) + R_i(x_2,\varepsilon)] \Big\}.$$

Since the condition $P_{B_1^*} P_{B_0^*} = 0$ is assumed to be satisfied, the last two systems enable us to find the first approximation $c_1^{(1)}$ to $c^{(1)}(\varepsilon)$ and the second approximation $c_2^{(0)}$ to $c^{(0)}(\varepsilon)$ by the formulas

$$\varepsilon c_1^{(1)} = -B_1^+ P_{B_0^*} P_{Q_d^*} \Big\{ l_1 x_2^{(2)}(\cdot,\varepsilon) + R_0(x_2(\cdot,\varepsilon),\varepsilon) - l \int_a^b K(\cdot,\tau)[A_1(\tau)x_2^{(2)}(\tau,\varepsilon) + R(x_2(\tau,\varepsilon),\tau,\varepsilon)]d\tau - l \sum_{i=1}^p \bar{K}(\cdot,\tau_i)[A_{1i}x_2^{(2)}(\tau_i-0,\varepsilon) + R_i(x_2(\tau_i-0,\varepsilon),\varepsilon)] \Big\},$$

$$c_2^{(0)} = -B_0^+ P_{Q_d^*} \Big\{ l_1[\varepsilon G_1(\cdot)P_{B_0}c_1^{(1)} + x_2^{(2)}(\cdot,\varepsilon)] + R_0(x_2(\cdot,\varepsilon),\varepsilon) - l \int_a^b K(\cdot,\tau)\Big[A_1(\tau)[\varepsilon G_1(\tau)P_{B_0}c_1^{(1)} + x_2^{(2)}(\tau,\varepsilon)] + R(x_2(\tau,\varepsilon),\tau,\varepsilon)\Big]d\tau - l \sum_{i=1}^p \bar{K}(\cdot,\tau_i)\Big[A_{1i}[\varepsilon G_1(\tau_i-0)P_{B_0}c_1^{(1)} + x_2^{(2)}(\tau_i-0,\varepsilon)] + R_i(x_2(\tau_i-0,\varepsilon),\varepsilon)\Big] \Big\}.$$

For the third approximations $x_3^{(2)}(t,\varepsilon)$ to $x^{(2)}(t,\varepsilon)$ and $x_3(t,\varepsilon)$ to the required solution $x(t,\varepsilon)$, we find

$$x_3^{(2)}(\tau,\varepsilon) = \varepsilon \left(G \begin{bmatrix} Z(z_0(\tau,c_r^*),\tau,0) + A_1(\tau)\Big[X_r(\tau)(I_r - P_{B_0})c_2^{(0)} + \varepsilon G_1(\tau)P_{B_0}c_1^{(1)} + x_2^{(2)}(\tau,\varepsilon)\Big] + R(x_2(\tau,\varepsilon),\tau,\varepsilon) \\ J_i(z_0(\tau_i-0,c_r^*),0) + A_{1i}\Big[X_r(\tau_i-0)(I_r - P_{B_0})c_2^{(0)} + \varepsilon G_1(\tau_i-0)P_{B_0}c_1^{(1)} + x_2^{(2)}(\tau_i-0,\varepsilon)\Big] + R_i(x_2(\tau_i-0,\varepsilon),\varepsilon) \end{bmatrix} \right)(t)$$

$$+ \varepsilon X(t)Q^+\Big[J(z_0(\cdot, c_r^*), 0)$$
$$+ l_1\big[X_r(\cdot)(I_r - P_{B_0})c_2^{(0)} + \varepsilon G_1(\cdot)P_{B_0}c_1^{(1)} + x_2^{(2)}(\cdot, \varepsilon)\big]$$
$$+ R_0(x_2(\cdot, \varepsilon), \varepsilon)\Big],$$

$$x_3(t, \varepsilon) = X_r(t)(I_r - P_{B_0})c_2^{(0)} + \varepsilon G_1(t)P_{B_0}c_1^{(1)} + x_3^{(2)}(t, \varepsilon).$$

If we continue this process infinitely, then we get the following iterative procedure for finding the solution $x(t, \varepsilon)$ of the impulsive boundary-value problem (6.66) such that $x(\cdot, \varepsilon) \in C^1([a, b]\backslash\{\tau_i\}_I)$, $x(t, \cdot) \in C[\varepsilon]$, $\varepsilon \in [0, \varepsilon_0]$, and $x(t, 0) = 0$:

$$\varepsilon c_{k-1}^{(1)} = -B_1^+ P_{B_0^*} P_{Q_d^*}\Big\{l_1 x_k^{(2)}(\cdot, \varepsilon) + R_0(x_k(\cdot, \varepsilon), \varepsilon)$$
$$- l\int_a^b K(\cdot, \tau)[A_1(\tau)x_k^{(2)}(\tau, \varepsilon) + R(x_k(\tau, \varepsilon), \varepsilon)]d\tau$$
$$- l\sum_{i=1}^{p} \bar{K}(\cdot, \tau_i)[A_{1i}x_k^{(2)}(\tau_i - 0, \varepsilon) + R_i(x_k(\tau_i - 0, \varepsilon), \varepsilon)]\Big\},$$

$$c_k^{(0)} = -B_0^+ P_{Q_d^*}\Big\{l_1[\varepsilon G_1(\cdot)P_{B_0}c_{k-1}^{(1)} + x_k^{(2)}(\cdot, \varepsilon)] + R_0(x_k(\cdot, \varepsilon), \varepsilon)$$
$$- l\int_a^b K(\cdot, \tau)\Big[A_1(\tau)[\varepsilon G_1(\tau)P_{B_0}c_{k-1}^{(1)} + x_k^{(2)}(\tau, \varepsilon)] + R(x_k(\tau, \varepsilon), \tau, \varepsilon)\Big]d\tau$$
$$- l\sum_{i=1}^{p} \bar{K}(\cdot, \tau_i)\Big[A_{1i}[\varepsilon G_1(\tau_1 - 0)P_{B_0}c_{k-1}^{(1)} + x_k^{(2)}(\tau_i - 0, \varepsilon)]$$
$$+ R_i(x_k(\tau_i - 0, \varepsilon), \varepsilon)\Big]\Big\}, \tag{6.95}$$

$$x_{k+1}^{(2)}(\tau, \varepsilon) = \varepsilon\left(G\begin{bmatrix} Z(z_0(\tau, c_r^*), \tau, 0) \\ +A_1(\tau)\Big[X_r(\tau)(I_r - P_{B_0})c_k^{(0)} + \varepsilon G_1 P_{B_0}c_{k-1}^{(1)} \\ +x^{(2)}(\tau, \varepsilon)\Big] + R(x_k(\tau, \varepsilon), \tau, \varepsilon) \\ J_i(z_0(\tau_i - 0, c_r^*), 0) \\ +A_{1i}\Big[X_r(\tau_i - 0)(I_r - P_{B_0})c_k^{(0)} + \varepsilon G_1(\tau_i - 0)P_{B_0}c_{k-1}^{(1)} \\ +x_k^{(2)}(\tau_i - 0, \varepsilon)\Big] + R_i(x_k(\tau_i - 0, \varepsilon), \varepsilon) \end{bmatrix}\right)(t)$$

$$+ \varepsilon X(t)Q^+\Big[J(z_0(\cdot, c_r^*), 0)$$

$$+ l_1\big[X_r(\cdot)(I_r - P_{B_0})c_k^{(0)} + \varepsilon G_1(\cdot)P_{B_0}c_{k-1}^{(1)} + x_k^{(2)}(\cdot,\varepsilon)\big]$$
$$+ R_0(x_k(\cdot,\varepsilon),\varepsilon)\Big],$$

$$x_{k+1}(t,\varepsilon) = X_r(t)(I_r - P_{B_0})c_k^{(0)} + \varepsilon G_1(t)P_{B_0}c_{k-1}^{(1)} + x_{k+1}^{(2)}(t,\varepsilon),$$
$$k = 0, 1, 2, \ldots,$$
$$x_0(t,\varepsilon) = x_0^{(2)}(t,\varepsilon) = 0.$$

To prove that the iterative process (6.95) converges, one can use the method of Lyapunov majorizing equations described in [19]. To prove that the boundary-value problem (6.66) with impulsive action is solvable (both in the noncritical case and in the critical case of the first order), it suffices to establish conditions required for the reducibility of the problem to the operator system (6.93). Thus, the following theorem is true:

Theorem 6.11. *Assume that the boundary-value problem (6.52) with impulsive action satisfies the conditions imposed above for the critical case* ($\operatorname{rank} Q = n_1 < m$) *and that the corresponding generating boundary-value problem (6.1), (6.2) with impulsive action possesses an r-parameter ($r = n - n_1$) family of generating solutions given by (6.61) if and only if condition (6.60) is satisfied with $d = m - n_1$. If, in addition,*

$$P_{B_0^*} \neq 0, \quad P_{B_1^*}P_{B_0^*} = 0, \tag{6.96}$$

and

$$P_{B_0^*}P_{Q_d^*}\Big\{l_1 x_1^{(2)}(\cdot,\varepsilon) + R_0(x_1(\cdot,\varepsilon),\varepsilon)$$
$$- l\int_a^b K(\cdot,\tau)[A_1(\tau)x_1^{(2)}(\tau,\varepsilon) + R(x_1(\tau,\varepsilon),\tau,\varepsilon)]d\tau$$
$$- l\sum_{i=1}^{p} \bar{K}(\cdot,\tau_i)[A_{1i}x_1^{(2)}(\tau_i - 0,\varepsilon) + R_i(x_1(\tau_i - 0,\varepsilon),\varepsilon)]\Big\} = 0, \tag{6.97}$$

then, for any $c_r = c_r^* \in R^r$ satisfying the equation for generating constants (6.65), the boundary-value problem (6.66) with impulsive action has at least one solution $x(t,\varepsilon)$ such that $x(\cdot,\varepsilon) \in C^1([a,b]\backslash\{\tau_i\}_I)$, $x(t,\cdot) \in C[\varepsilon]$, $\varepsilon \in [0,\varepsilon_0]$, and $x(t,0) = 0$. One of these solutions can be obtained as a result of the iterative process (6.95) convergent for $\varepsilon \in [0,\varepsilon_*] \subseteq [0,\varepsilon_0]$. In this case, the boundary-value problem (6.52) with impulsive action possesses at least one solution $z(t,\varepsilon)$ that satisfies the conditions $z(\cdot,\varepsilon) \in C^1([a,b]\backslash\{\tau_i\}_I)$ and $z(t,\cdot) \in C[\varepsilon]$, $\varepsilon \in [0,\varepsilon_0]$, *and turns into the generating solution $z_0(t, c_r^*)$ given by (6.61). This solution can be found as a result of the iterative process (6.95) by using the formula*

$$z_k(t, \varepsilon) = z_0(t, c_r^*) + x_k(t, \varepsilon), \quad k = 0, 1, 2, \dots .$$

Remark 6.4. In the analyzed critical case (6.96), condition (6.97) is necessary and sufficient for the existence of the second approximation to the required solution. If condition (6.97) is not satisfied, then the boundary-value problem (6.66) with impulsive action is unsolvable by the method of simple iterations in the class of functions such that $x(\cdot, \varepsilon) \in C^1([a, b] \backslash \{\tau_i\}_I)$, $x(t, \cdot) \in C[\varepsilon]$, $\varepsilon \in [0, \varepsilon_0]$, and $x(t, 0) = 0$. In the analyzed case, the sufficient condition (6.96) is the sole restriction imposed on the problem and guaranteeing the existence of solution and the possibility of its construction with the help of the iterative analog of the Lyapunov–Poincaré method of small parameter.

If we now assume that $P_{B_0} P_{B_1} = 0$, then we get the following corollary of Theorem 6.11:

Corollary 6.2. *Assume that the boundary-value problem (6.52) with impulsive action satisfies the conditions imposed above for the critical case* ($\operatorname{rank} Q = n_1 < m$) *and that the corresponding generating boundary-value problem (6.1), (6.2) with impulsive action possesses the r-parameter ($r = n - n_1$) family of generating solutions given by (6.61) if and only if condition (6.60) is satisfied with $d = m - n_1$. If, in addition,*

$$P_{B_0^*} \neq 0, \quad P_{B_1^*} P_{B_0^*} = 0, \quad P_{B_0} P_{B_1} = 0, \tag{6.98}$$

and condition (6.97) is satisfied, then, for any $c_r = c_r^* \in R^r$ satisfying the equation for generating constants (6.65), the boundary-value problem (6.66) with impulsive action possesses a unique solution $x(t, \varepsilon)$ such that $x(\cdot, \varepsilon) \in C^1([a, b] \backslash \{\tau_i\}_I)$, $x(t, \cdot) \in C[\varepsilon]$, $\varepsilon \in [0, \varepsilon_0]$, and $x(t, 0) = 0$. This solution can be obtained as a result of the iterative process (6.95) convergent for $\varepsilon \in [0, \varepsilon_*] \subseteq [0, \varepsilon_0]$. The boundary-value problem (6.52) with impulsive action possesses a unique solution $z(t, \varepsilon)$ that satisfies the conditions $z(\cdot, \varepsilon) \in C^1([a, b] \backslash \{\tau_i\}_I)$ and $z(t, \cdot) \in C[\varepsilon]$, $\varepsilon \in [0, \varepsilon_0]$, turns into the generating solution $z_0(t, c_r^*)$ given by (6.61), and can be found as a result of the iterative process (6.95) by the formula

$$z_k(t, \varepsilon) = z_0(t, c_r^*) + x_k(t, \varepsilon), \quad k = 0, 1, 2, \dots .$$

6.9 Degenerate Systems of Differential Equations with Impulsive Action

In the present section, we study singular systems of ordinary differential equations with impulsive action by using the relationship between the analyzed systems and so-called interface boundary-value problems. An approach that combines the theory of impulsive differential equations with the well-known results from the theory of singular Fredholm boundary-value problems is applied. The main aim of the present section is to establish necessary and sufficient conditions for the existence of solutions of singular systems of ordinary differential equations with impulsive action in a properly chosen space. To this end, we use the well-known results from the theory of singular Fredholm boundary-value problems (section 5.8) [177].

Consider the problem of existence and construction of the solutions of a singular linear system of ordinary differential equations with impulsive action at fixed points of time

$$B(t)\dot{x} = A(t)x + f(t), \qquad t \in [a, b], \tag{6.99}$$

$$\Delta E_i x\Big|_{t=\tau_i} = S_i x(\tau_i - 0) + \gamma_i, \quad \tau_i \in (a, b), \quad i = 1, \ldots, p, \tag{6.100}$$

under the assumption that the unperturbed singular differential systems can be reduced to the central canonical form (see section 5.7); here, $A(t)$ and $B(t)$ are $n \times n$ matrices, $\det B(t) = 0, \forall t \in [a, b]$, and $f(t)$ is an n-dimensional column vector function. Equation (6.100) specifies the jumps of the solution, namely,

$$\Delta E_i x\Big|_{t=\tau_i} := E_i\big(x(\tau_i+) - x(\tau_i-)\big),$$

where E_i, S_i are $m_i \times n$ real constant matrices and γ_i $(i = 1, \ldots, p)$ are m_i-dimensional real constant column vectors, i.e., $\gamma_i \in \mathbb{R}^{m_i}$.

As in the previous chapter (Section 5.7), we assume that the components of the matrices $A(t)$ and $B(t)$ and of the vector $f(t)$ are real functions continuously differentiable sufficiently many times on the interval $[a, b]$: $A(t), B(t) \in C^{3q-2}[a; b]$ and $f(t) \in C^{q-1}[a; b]$, where $q = \max\limits_i s_i$ and s_i is the dimension of a nilpotent Jordan cell I_i in the central canonical form.

The solution $x(t)$ is sought in the space of n-dimensional piecewise continuously differentiable vector functions $x(t) \in C^1\big([a, b] \setminus \{\tau_1, \ldots, \tau_p\}_I\big)$ with discontinuities of the first kind for $t = \tau_i$. The norms in the spaces

$$C\big([a, b] \setminus \{\tau_1, \ldots, \tau_p\}_I\big), \quad C^1\big([a, b] \setminus \{\tau_1, \ldots, \tau_p\}_I\big)$$

are introduced in a standard way, by analogy with [69, 139].

Note that if, e.g.,

(a) for all $i = 1, \ldots, p$, $E_i := E$ are the $n \times n$ identity matrices, S_i are $n \times n$ matrices and, hence, γ_i are n-dimensional column vectors, then the conditions (6.100) take the form of standard impulsive conditions [139]:

$$\Delta x\Big|_{t=\tau_i} := x(\tau_i+) - x(\tau_i-) = S_i x(\tau_i-) + \gamma_i, \quad i = 1, \ldots, p;$$

(b) $E_1 := (1,0,\ldots,0), \ldots, E_i := (0,\ldots,0,1,0,\ldots,0), \ldots \quad (i=1,\ldots,p,\ p \leq n)$ and $S_i \ (i=1,\ldots,p)$ are $1 \times n$ vectors, then conditions (6.100) take the form of impulsive conditions only for the corresponding components $x_i(t)$ of the vector $x(t) = \operatorname{col}\big(x_1(t),\ldots,x_i(t),\ldots,x_n(t)\big) \ (i=1,\ldots,p)$:

$$\Delta E_i x\Big|_{t=\tau_i} := x_i(\tau_i+) - x_i(\tau_i-), \quad i=1,\ldots,p.$$

Main Results. In this section, we establish the relationship between the considered problem (6.99), (6.100) and the so-called interface boundary-value problem [195] and obtain the conditions of solvability for these problems.

Relationship with the Interface Boundary-Value Problem. By using the following notation:

$$\begin{cases} \ell_1 x := E_1 x(\tau_1+) - (E_1+S_1)x(\tau_1-), \\ \ell_2 x := E_2 x(\tau_2+) - (E_2+S_2)x(\tau_2-), \\ \quad \vdots \\ \ell_p x := E_p x(\tau_p+) - (E_p+S_p)x(\tau_p-), \end{cases} \tag{6.101}$$

the systems of singular differential equations with impulsive action (6.99), (6.100) can be transformed into the following equivalent interface boundary-value problem:

$$B(t)\dot{x} = A(t)x + f(t), \quad t \in [a,b], \tag{6.102}$$

$$\ell x(\cdot) = \gamma, \tag{6.103}$$

where $\gamma := \operatorname{col}(\gamma_1,\ldots,\gamma_{m_p}) \in \mathbb{R}^m \ (m := m_1 + \ldots + m_p)$, and ℓ is an m-dimensional linear vector functional

$$\ell := \operatorname{col}(\ell_1,\ell_2,\ldots,\ell_p) \colon C^1\big([a,b] \setminus \{\tau_1,\ldots,\tau_p\}_I\big) \to \mathbb{R}^m;$$

$$\ell_i \colon C^1\big([a,b] \setminus \{\tau_i\}_I\big) \to \mathbb{R}^{m_i}, \quad i=1,2,\ldots,p.$$

The solution $x(t)$ of problem (6.102), (6.103) and, hence, the solution of the initial problem (6.99), (6.100) for a singular linear systems of ordinary differential equations with impulsive action at fixed points of time is sought in the space $C^1\big([a,b] \setminus \{\tau_1,\ldots,\tau_p\}_I\big)$ of n-dimensional piecewise continuously differentiable vector functions with discontinuities of the first kind at $t = \tau_i$. We use the general solution (see section 5.7)

$$x\,(t,\ c) = X_{n-s}(t)c + \widetilde{x}\,(t)\,, \quad \forall c \in \mathbb{R}^{n-s} \tag{6.104}$$

of the singular differential system (6.99) to find the condition of solvability and the form of the general solution of the linear inhomogeneous interface boundary-value problem (6.102), (6.103) and, hence, the solution of problem (6.99), (6.100) for the singular linear system of ordinary differential equations with impulsive action at fixed points of time. The solution (6.104) of the singular differential system (6.99) is a solution of the boundary-value problem (6.102), (6.103) if and only if it satisfies the boundary conditions (6.103).

This means that the algebraic system

$$Qc = \gamma - \ell\widetilde{x}\,(\cdot) \tag{6.105}$$

must be solvable with respect to $c \in \mathbb{R}^{n-s}$, where Q is an $m \times (n-s)$ constant matrix

$$Q := \operatorname{col}\left(-S_1 X_{n-s}(\tau_1), \dots, -S_p X_{n-s}(\tau_p)\right). \tag{6.106}$$

The algebraic system (6.105) is solvable if and only if its right-hand side belongs to the orthogonal complement $N(Q^*) = R(Q)$ of the kernel $N(Q^*) = \ker Q^*$ of the adjoint matrix Q^*, i.e., if the following condition is satisfied:

$$P_{Q^*}\{\gamma - \ell\widetilde{x}\,(\cdot)\} = 0, \tag{6.107}$$

where $P_{Q^*} := E_m - QQ^+$ is the $m \times m$ matrix (orthogonal projection) projecting the space $\mathbb{R}^m$ onto $\ker Q^*$. Let $\operatorname{rank} Q := n_1 < n-s$. Since $\operatorname{rank} P_{Q^*} = d = m - n_1$, by $P_{Q_d^*}$ we denote a $d \times m$ matrix formed by d linearly independent rows of the $m \times m$ matrix P_{Q^*} and Q^+ is the unique $(n-s) \times m$ Moore–Penrose pseudoinverse matrix for the matrix Q. As a result, the criterion (6.107) consists of d linearly independent conditions,

$$P_{Q_d^*}\{\gamma - \ell\widetilde{x}\,(\cdot)\} = o. \tag{6.108}$$

It is well known (see [174]) that system (6.105) has an r-parameter family of linearly independent solutions

$$c = Q^+\{\gamma - \ell\widetilde{x}\,(\cdot)\} + P_{Q_r} c_r, \quad \forall\, c_r \in \mathbb{R}^r, \tag{6.109}$$

where $P_Q := E_{n-s} - Q^+Q$ is an $(n-s) \times (n-s)$ matrix (orthogonal projection) projecting the space $\mathbb{R}^{n-s}$ onto the $\ker Q$. Since $\operatorname{rank} P_Q = r = (n-s) - n_1$, by P_{Q_r} we denote an $(n-s) \times r$ matrix formed by r linearly independent columns of the matrix P_Q. Substituting solutions (6.109) in the general solution, we conclude that the singular linear inhomogeneous boundary-value problem (6.102), (6.103) has an r-parameter family of linearly independent solutions

$$\begin{aligned} x\,(t,\, c_r) &= X_{n-s}(t) P_{Q_r} c_r + X_{n-s}(t) Q^+\{\gamma - \ell\widetilde{x}\,(\cdot)\,\} \\ &\quad + \int_a^t X_{n-s}\,(t)\, Y_{n-s}^*\,(\tau)\, f(\tau) d\tau \\ &\quad - \Phi(t) \sum_{k=0}^{q-1} \mathrm{I}^k \frac{d^k}{dt^k} \left([\Psi^*(t) L(t) \Phi\,(t)]^{-1}\, \Psi^*(t) f(t) \right) \end{aligned} \tag{6.110}$$

if and only if condition (6.108) is satisfied. Thus, we have proved the following statement:

Theorem 6.12. *The interface singular boundary-value problem (6.102),(6.103) and, hence, the impulsive differential system (6.99), (6.100) is solvable if and only if the inhomogeneities* $f(t) \in C^{q-1}([a,b])$ *and* $\gamma_i \in \mathbb{R}^{m_i}$ *satisfy d linearly independent conditions:*

$$P_{Q_d^*}\{\gamma - \ell\widetilde{x}(\cdot)\} = 0.$$

Then the corresponding homogeneous system has exactly r *linearly independent solutions* $x(t,\, c_r) = X_r(t)c_r$ *and the inhomogeneous system (6.99), (6.100) possesses an* r*-parameter family of linearly independent solutions of the form*

$$x(t,\, c_r) = X_r(t)c_r + (G[f,\gamma])(t). \tag{6.111}$$

In (6.111), $X_r(t) = X_{n-s}(t)P_{Q_r}$ is an $n \times r$ matrix and $(G[f,\gamma])(t)$ is the generalized Green operator of the singular boundary-value problem (6.102), (6.103) acting upon a vector function $f(t) \in C^{q-1}([a,b])$ and $\gamma \in \mathbb{R}^m$ as follows:

$$\begin{aligned}(G[f,\gamma])(t) := &\int_a^t X_{n-s}(t)\, Y_{n-s}^*(\tau)\, f(\tau)d\tau \\ &- \Phi(t)\sum_{k=0}^{q-1} \mathrm{I}^k \frac{d^k}{dt^k}\Big([\Psi^*(t)L(t)\Phi(t)]^{-1}\,\Psi^*(t)f(t)\Big) \\ &+ X_{n-s}(t)Q^+\gamma - X_{n-s}(t)Q^+\ell\widetilde{x}(\cdot).\end{aligned} \tag{6.112}$$

Bifurcation Conditions. By using the results of the previous chapter (Section 5.7), we now establish sufficient conditions for the bifurcation of solutions of a linear singular Fredholm differential system with impulsive action and a small parameter. It is also assumed that the unperturbed singular differential system can be reduced to the central canonical form.

The crucial assumption in Theorem 6.12 is the so-called solvability criterion (6.108). According to this criterion, if the inhomogeneities $f \in C^{q-1}[a,b]$, $\gamma_i \in \mathbb{R}^{m_i}$, $i = 1,\ldots,p$, in problem (6.99), (6.100) are such that (6.108) is not satisfied, then the solution of the problem does not exist. In this case, we can modify system (6.99) by a linear perturbation in order to guarantee that the perturbed singular boundary-value problem

$$B(t)\dot{x} = A(t)x + f(t) + \varepsilon A_1(t)x, \quad t \in [a,b], \tag{6.113}$$

$$\Delta E_i x\Big|_{t=\tau_i} = S_i x(\tau_i - 0) + \gamma_i, \quad \tau_i \in (a,b), \quad i = 1,\ldots,p, \tag{6.114}$$

where $A(t), A_1(t), B(t) \in C^{3q-2}[a,b]$, $\det B(t) = 0$, $\gamma_i \in \mathbb{R}^{m_i}$, $i = 1,\ldots,p$, and $\varepsilon > 0$ is a small parameter, becomes solvable for any inhomogeneities.

Therefore, it is of interest to analyze whether the problem (6.99), (6.100) can be made solvable by introducing linear perturbations, and if this is possible, then for what kind of perturbations $A_1(t) \in C^{3q-2}[a,b]$, the boundary-value problem (6.113), (6.114) becomes solvable for all inhomogeneities $f \in C^{q-1}[a,b]$, $\gamma := \mathrm{col}\,(\gamma_1, \ldots, \gamma_p) \in \mathbb{R}^m$?

By using the Vishik–Lyusternik method and the technique proposed in the previous sections, we can suggest an algorithm for finding a family of linearly independent solutions of these problems in the general case.

To simplify the formulation of our results, it is convenient to use the following notation for the $d \times r$ matrix B_0:

$$B_0 := -P_{Q_d^*} \ell \Big\{ \int_a^{\cdot} X_r(\cdot) Y_{n-s}^*(\tau) A_1(\tau) X_r(\tau) d\tau - \Phi(\cdot) \sum_{k=0}^{q-1} \mathrm{I}^k \frac{d^k}{dt^k} \Big(\big[\Psi^*(t) L(t) \Phi(t)\big]^{-1} \Psi^*(t) A_1(t) X_r(t) \Big)(\cdot) \Big\}, \tag{6.115}$$

where $\ell \colon C^1\big([a,b] \setminus \{\tau_1, \ldots, \tau_p\}_I\big) \to R^m$. Thus, we can formulate the following statement:

Theorem 6.13. *Suppose that the singular impulsive differential system (6.99), (6.100) has no solutions for some inhomogeneities $f(t) \in C^{q-1}[a,b]$, $\gamma \in R^m$. If, in addition,*

$$\mathrm{rank}\, B_0 = d, \tag{6.116}$$

then there exists a ρ-parameter ($\rho := r - d = n - s - m$) family of linearly independent solutions of the perturbed singular impulsive differential system (6.113), (6.114) in the form of a part of the series in powers of the parameter ε:

$$x(t,\varepsilon) = \sum_{i=-1}^{\infty} \varepsilon^i x_i(t, c_\rho), \quad \forall c_\rho \in \mathbb{R}^\rho, \tag{6.117}$$

convergent for fixed $\varepsilon \in (0, \varepsilon_*]$, where ε_* is an appropriate constant characterizing the domain of convergence of series (6.117) and the coefficients $x_i(t, c_\rho)$ are determined from the corresponding problems.

The proof of these results is similar to the reasoning used in Section 5.7 or in [167, 168, 177]. Therefore, it is omitted. As in [175], we can study the nonlinear perturbed singular impulsive differential systems.

Chapter 7

SOLUTIONS OF DIFFERENTIAL AND DIFFERENCE SYSTEMS BOUNDED ON THE ENTIRE REAL AXIS

The well-known Palmer lemma [115] on the Fredholm properties of the solutions of linear systems of ordinary differential equations bounded on the entire real axis enables us to apply the theory developed in the previous chapters to the analysis of conditions required for the existence of solutions of linear and nonlinear differential and difference systems bounded on the entire real axis. This direction was extensively developed as a qualitative theory of differential equations by Holmes [76], Chow, Hale, and Mallet-Paret [47], Coppel [49], Henry [75], Guckenheimer and Holmes [68], Gruendler [66], Melnikov [105], Mitropol'skii, Samoilenko, and Kulik [108], and Sacker and Sell [133]. The application of the theory of pseudoinverse matrices to the investigation of the indicated problems enables one to formulate the well-known results more exactly and establish new facts some of which were known for the Fredholm operators of index zero of the original linear problems.

7.1 Solutions of Linear Weakly Perturbed Systems Bounded on the Entire Real Axis

Unperturbed Problem. Let $BC(J)$ be the Banach space of real continuous vector functions $x\colon J \to R^n$ bounded on the interval J with the norm $\|x\| = \sup_{t\in J} |x(t)|$ and let $BC^1(J)$ be the Banach space of real continuous vector functions $x\colon J \to R^n$ differentiable on J and bounded together with their derivative. The norm in this space is defined as follows: $\|x\| = \sup_{t\in J} |x(t)| + \sup_{t\in J} |\dot{x}(t)|$, where $|x(t)| := \|x\|_{R^n}$.

It is known [49] that the system

$$\dot{x} = A(t)x, \quad A(\cdot) \in BC(J), \tag{7.1}$$

is exponentially dichotomous (e-dichotomous) on the interval J if there exist a projector P $(P^2 = P)$ and constants $K \geq 1$ and $\alpha > 0$ such that, for any $t, s \in J$, the following inequalities are satisfied:

$$\begin{aligned} \|X(t)PX^{-1}(s)\| &\leq Ke^{-\alpha(t-s)}, \qquad t \geq s, \\ \|X(t)(I-P)X^{-1}(s)\| &\leq Ke^{-\alpha(s-t)}, \qquad s \geq t, \end{aligned} \tag{7.2}$$

where $X(t)$ is the normal ($X(0) = I$) fundamental matrix of system (7.1) and $A(t)$ is an $n \times n$ matrix whose components belong to the Banach space $BC(J)$. In what follows, as J, we use one of the following intervals: $J = R = (-\infty, +\infty)$, $J = R_+ = [0, +\infty)$, or $J = R_- = (-\infty, 0]$.

We consider the problem of existence and structure of the solutions

$$x \colon R \to R^n, \quad x(\cdot) \in BC^1(R),$$

of an inhomogeneous system

$$\dot{x} = A(t)x + f(t), \quad f \colon R \to R^n, \quad f(\cdot) \in BC(R), \quad A(\cdot) \in BC(R), \tag{7.3}$$

in the case where system (7.1) has nontrivial solutions bounded on R. In the noncritical [115] or regular [108] case where the homogeneous system (7.1) is e-dichotomous on R and, hence, does not have nontrivial solutions bounded on R, the inhomogeneous system (7.3) has a unique solution bounded on R for any $f(\cdot) \in BC(R)$. The critical or resonance case where the homogeneous system (7.1) has nontrivial solutions bounded on R and, hence, is not an e-dichotomous system on R was studied by Palmer who established necessary [116] and sufficient conditions [115] for the analyzed problem to possess the Fredholm property. By using these results and the theory of pseudoinverse matrices, we can formulate the following assertion [21]:

Lemma 7.1. *Assume that the linear operator*

$$(Lx)(t) = \dot{x}(t) - A(t)x(t) \colon BC^1(R) \to BC(R) \tag{7.4}$$

is e-dichotomous on the semiaxes $J = R_+$ and $J = R_-$ with projectors P and Q, respectively. Then the following assertions are true:

(a) system (7.1) possesses an r-parameter $\big(r = \operatorname{rank}[PP_{N(D)}] = \operatorname{rank}[(I - Q)P_{N(D)}]\big)$ family of solutions $X_r(t)c_r$ bounded on R for any $c_r \in R^r$, where

$$X_r(t) = X(t)[PP_{N(D)}]_r = X(t)[(I - Q)P_{N(D)}]_r$$

is an $n \times r$ matrix whose columns form a complete system of r linearly independent solutions of system (7.1) bounded on R;

(b) the system

$$\dot{x} = -A^*(t)x, \quad A(\cdot) \in BC(J), \tag{7.5}$$

conjugate to system (7.1) possesses a d-parameter $\big(d = \operatorname{rank}[P_{N(D^*)}(I - P)] = \operatorname{rank}[P_{N(D^*)}Q]\big)$ family of solutions $H_d(t)c_d$ bounded on R for any $c_d \in R^d$, where

$$H_d(t) = X^{*-1}(t)[Q^* P_{N(D^*)}]_d = X^{*-1}(t)[(I - P^*)P_{N(D^*)}]_d$$

is an $n \times d$ matrix whose columns form a complete system of d linearly independent solutions of system (7.5) conjugate to system (7.1) bounded on R;

(c) *the operator L is a Fredholm operator whose index is given by the formula*

$$\begin{aligned}\operatorname{ind} L &= \operatorname{rank}[PP_{N(D)}] - \operatorname{rank}[P_{N(D^*)}(I-P)] \\ &= \operatorname{rank}[(I-Q)P_{N(D)}] - \operatorname{rank}[P_{N(D^*)}Q] = r-d;\end{aligned}$$

furthermore, $f \in \operatorname{Im} L$ if and only if

$$\int_{-\infty}^{\infty} H_d^*(s)f(s)ds = 0, \tag{7.6}$$

where $H_d^*(t) = [X^{*-1}(t)[Q^* P_{N(D^*)}]_d]^*$ is a $d \times n$ matrix whose rows form a complete system of d linearly independent solutions of system (7.5) bounded on R;

(d) *if condition (7.6) is satisfied, then the inhomogeneous system (7.3) has the following r-parameter family of solutions bounded on R:*

$$x(t, c_r) = X_r(t)c_r + (G[f])(t) \quad \textit{for any} \quad c_r \in R^r, \tag{7.7}$$

where

$$(G[f])(t) = X(t)\begin{cases} \displaystyle\int_0^t PX^{-1}(s)f(s)ds - \int_t^{\infty}(I-P)X^{-1}(s)f(s)ds \\ \displaystyle\qquad + PD^+\Big\{\int_{-\infty}^0 QX^{-1}(s)f(s)ds \\ \displaystyle\qquad\quad + \int_0^{\infty}(I-P)X^{-1}(s)f(s)ds\Big\}, \quad t \geq 0, \\ \displaystyle\int_{-\infty}^t QX^{-1}(s)f(s)ds - \int_t^0 (I-Q)X^{-1}(s)f(s)ds \\ \displaystyle\qquad + (I-Q)D^+\Big\{\int_{-\infty}^0 QX^{-1}(s)f(s)ds \\ \displaystyle\qquad\quad + \int_0^{\infty}(I-P)X^{-1}(s)f(s)ds\Big\}, \quad t \leq 0, \end{cases} \tag{7.8}$$

is the generalized Green operator of the problem of finding solutions of system (7.3) bounded on R with the following properties:

$$(LG[f])(t) = f(t), \quad t \in R,$$

$$(G[f])(0+0) - (G[f])(0-0) = -\int_{-\infty}^{\infty} H^*(s)f(s)ds,$$

$$H^*(t) = [X^{*-1}(t)Q^* P_{N(D^*)}]^* = [X^{*-1}(t)(I-P^*)P_{N(D^*)}]^*$$
$$= P_{N(D^*)}QX^{-1}(t) = P_{N(D^*)}(I-P)X^{-1}(t),$$

where $D = P - (I - Q)$ *is an* $n \times n$ *matrix,* D^+ *is its Moore–Penrose pseudoinverse matrix [38, 124],* $P_{N(D)}$ *and* $P_{N(D^*)}$ *are* $n \times n$ *matrices (orthoprojectors), i.e.,* $P^2_{N(D)} = P_{N(D)} = P^*_{N(D)}$ *and* $P^2_{N(D^*)} = P_{N(D^*)} = P^*_{N(D^*)}$, *projecting* R^n *onto the kernel* $N(D) = \ker D$ *and the cokernel* $N(D^*) = \operatorname{coker} D = \ker D^*$ *of the matrix* D, *respectively, and* $[*]_d$ *is an* $n \times d$ *matrix formed of* d *linearly independent columns of the matrix* $*$ *in the brackets.*

Proof [21, 23]. The general solution of problem (7.3) bounded on the semiaxes has the form

$$x(t,\xi) = \begin{cases} X(t)P\xi + \displaystyle\int_0^t X(t)PX^{-1}(s)f(s)ds \\ \qquad - \displaystyle\int_t^{\infty} X(t)(I-P)X^{-1}(s)f(s)ds, \quad t \geq 0, \\ X(t)(I-Q)\xi + \displaystyle\int_{-\infty}^t X(t)QX^{-1}(s)f(s)ds \\ \qquad - \displaystyle\int_t^0 X(t)(I-Q)X^{-1}(s)f(s)ds, \quad t \leq 0. \end{cases} \tag{7.9}$$

Solution (7.9) is bounded on R if and only if the vector constant $\xi \in R^n$ satisfies the condition

$$P\xi - \int_0^{\infty} (I-P)X^{-1}(s)f(s)ds = (I-Q)\xi + \int_{-\infty}^0 QX^{-1}(s)f(s)ds.$$

Thus, the constant $\xi \in R^n$ is determined from the algebraic system

$$[P-(I-Q)]\xi = \int_{-\infty}^0 QX^{-1}(s)f(s)ds + \int_0^{\infty} (I-P)X^{-1}(s)f(s)ds. \tag{7.10}$$

System (7.3) has solutions bounded on R if and only if the algebraic system (7.10) is solvable with respect to $\xi \in R^n$. To this end, it is necessary and sufficient that the free term of system (7.10) belong to the orthogonal complement $N^{\perp}(D^*) = R(D)$ of the subspace $N(D^*)$. In view of the notation introduced above, this is equivalent to the condition

$$P_{N(D^*)}\Big\{ \int_{-\infty}^{0} QX^{-1}(s)f(s)ds + \int_{0}^{\infty}(I-P)X^{-1}(s)f(s)ds \Big\} = 0. \qquad (7.11)$$

In this case, the general solution of system (7.3) bounded on R takes the form (7.9), where the constant $\xi \in R^n$ is given by system (7.10) as follows:

$$\xi = D^+\Big\{ \int_{-\infty}^{0} QX^{-1}(s)f(s)ds + \int_{0}^{\infty}(I-P)X^{-1}(s)f(s)ds \Big\} + P_{N(D)}c \quad \text{for any } c \in R^n.$$

In other words, $f \in \operatorname{Im}(L)$ if and only if condition (7.11) is satisfied and the general solution of system (7.3) bounded on the entire real axis R has the form

$$x(t,c) = X(t)\begin{cases} PP_{N(D)}c + \int_{0}^{t} PX^{-1}(s)f(s)ds - \int_{t}^{\infty}(I-P)X^{-1}(s)f(s)ds \\ \quad + PD^+\Big\{ \int_{-\infty}^{0} QX^{-1}(s)f(s)ds + \int_{0}^{\infty}(I-P)X^{-1}(s)f(s)ds \Big\}, & t \geq 0, \\ (I-Q)P_{N(D)}c + \int_{-\infty}^{t} QX^{-1}(s)f(s)ds - \int_{t}^{0}(I-Q)X^{-1}(s)f(s)ds \\ \quad + (I-Q)D^+\Big\{ \int_{-\infty}^{0} QX^{-1}(s)f(s)ds + \int_{0}^{\infty}(I-P)X^{-1}(s)f(s)ds \Big\}, & t \leq 0. \end{cases}$$

Since $DP_{N(D)} = 0$ [38], we have $PP_{N(D)} = (I - Q)P_{N(D)}$. Let

$$r = \text{rank}\,[PP_{N(D)}] = \text{rank}\,[(I - Q)P_{N(D)}].$$

Then we have $\dim N(L) = r$, and vice versa. Let $[PP_{N(D)}]_r = [(I - Q)P_{N(D)}]_r$ be an $n \times r$ matrix whose columns form a complete system of r linearly independent columns of the matrix $PP_{N(D)}$ or of the matrix $(I - Q)P_{N(D)}$. Then

$$X_r(t) = X(t)[PP_{N(D)}]_r = X(t)[(I - Q)P_{N(D)}]_r$$

is an $n \times r$ matrix whose columns form a complete system of r linearly independent solutions of system (7.3) bounded on R. Therefore, the general solution of system (7.3) bounded on R can be represented in the form (7.7), where the generalized Green operator $(G(f))(t)$ of the problem of finding solutions of system (7.3) bounded on the entire real axis has the form (7.8) and satisfies the conditions imposed above, which can easily be verified by calculations.

Since $P_{N(D^*)}D = 0$ [38], we have $P_{N(D^*)}Q = P_{N(D^*)}(I - P)$. Therefore, condition (7.11) is equivalent to one of the following conditions:

$$P_{N(D^*)} \int\limits_{-\infty}^{\infty} QX^{-1}(s)f(s)ds = 0,$$
$$P_{N(D^*)} \int\limits_{-\infty}^{\infty} (I - P)X^{-1}(s)f(s)ds = 0. \tag{7.12}$$

Let $d = \text{rank}\,[P_{N(D^*)}(I - P)] = \text{rank}\,[P_{N(D^*)}Q]$. Then each condition in (7.12) consists of d linearly independent conditions. Indeed, let $[Q^*P_{N(D^*)}]_d$ be an $n \times d$ matrix whose columns coincide with d linearly independent columns of the matrix $[Q^*P_{N(D^*)}]$. Since $X^{*-1}(t)$ is the fundamental matrix of system (7.5), which is e-dichotomous on R_+ with projector $I - P^*$ and on R_- with projector $I - Q^*$ [115, p. 246], we get the $n \times d$ matrix

$$H_d(t) = X^{*-1}(t)[Q^*P_{N(D^*)}]_d = X^{*-1}(t)[(I - P^*)P_{N(D^*)}]_d$$

whose columns form a complete system of d linearly independent solutions of system (7.5) conjugate to system (7.1) bounded on R and the $d \times n$ matrix $H_d^*(t) = [X^{*-1}(t)[Q^*P_{N(D^*)}]_d]^*$ whose rows form a complete system of $d = \text{rank}\,[P_{N(D^*)}Q]$ linearly independent solutions of system (7.5) bounded on R. Thus, we arrive at condition (7.6), which completes the proof of the lemma.

Corollary 7.1. *Assume that the homogeneous system (7.1) is e-dichotomous on R_+ and R_- with projectors P and Q, respectively, such that $PQ = QP = Q$. In this case, system (7.1) is e-trichotomous [53, p. 363] on R and the inhomogeneous system (7.3) has at least one solution bounded on R for all $f \in BC(R)$ [53, p. 371].*

In other words, this is the so-called weakly regular case [108, p. 37], where the intersection of stable and unstable hyperplanes defined by the orthoprojectors P and Q, respectively, is transverse [119, p. 170]. In this case, Lemma 7.1 can be formulated as follows:

Lemma 7.1.1. *Suppose that system (7.1) is e-dichotomous on R_+ and R_- with projectors P and Q, respectively, and such that $PQ = QP = Q$. Then the following assertions are true:*

(a) the homogeneous system (7.1) possesses an r-parameter family of solutions bounded on R:

$$X_r(t)c_r = X(t)[PP_{N(D)}]_r c_r = X(t)[(I-Q)P_{N(D)}]_r c_r \quad \forall c_r \in R^r$$

$(r = \operatorname{rank}[PP_{N(D)}] = \operatorname{rank}[(I-Q)P_{N(D)}])$;

(b) the only solution of system (7.5) conjugate to system (7.1) bounded on R is trivial;

(c) the operator L is a Fredholm operator with index

$$\operatorname{ind} L = \operatorname{rank}[PP_{N(D)}] = \operatorname{rank}[(I-Q)P_{N(D)}] = r,$$

and, in addition, $f \in \operatorname{Im}(L)$ for any $f \in BC(R)$;

(d) the inhomogeneous system (7.3) possesses an r-parameter family of solutions bounded on R:

$$x(t, c_r) = X_r(t)c_r + (G[f])(t) \quad \forall c_r \in R^r,$$

where $(G[f])(t)$ is the generalized Green operator (7.8) of the problem of finding solutions of system (7.3) bounded on R and such that

$$(LG[f])(t) = f(t), \ t \in R, \quad \text{and} \quad (G[f])(0+0) - (G[f])(0-0) = 0.$$

Proof. Indeed, since $P_{N(D^*)}D = 0$ and

$$DP = (P - (I-Q))P = QP = Q,$$

we find

$$P_{N(D^*)}Q = P_{N(D^*)}DP = 0.$$

Thus, the necessary and sufficient condition (7.6) for the existence of solutions of system (7.3) bounded on the entire real axis is satisfied for all $f \in BC(R)$.

Note that the case $PQ = QP = Q$ is equivalent to the case of transverse intersection of the stable and unstable subspaces of system (7.1) [119, p. 170].

Corollary 7.2. *Assume that the homogeneous system (7.1) is e-dichotomous on R_+ and R_- with projectors P and Q, respectively, such that $PQ=QP=P$. In this case, system (7.5) conjugate to (7.1) is e-trichotomous on R and the inhomogeneous system (7.3) possesses a unique solution bounded on R but not for all $f \in BC(R)$.*

In this case, Lemma 7.1 can be reformulated as follows:

Lemma 7.1.2. *Assume that the linear operator (7.4) is e-dichotomous on the semiaxes $J = R_+$ and $J = R_-$ with projectors P and Q, respectively, and, in addition, $PQ = QP = P$. Then the following assertions are true:*

(a) the only solution of system (7.1) bounded on R is trivial ($r = \operatorname{rank}[PP_{N(D)}] = \operatorname{rank}[(I-Q)P_{N(D)}] = 0$);

(b) system (7.5) conjugate to (7.1) is trichotomous (weakly regular) and possesses a d-parameter ($d = \operatorname{rank}[P_{N(D^*)}(I-P)] = \operatorname{rank}[P_{N(D^*)}Q]$) family of solutions $H_d(t)c_d$ bounded on R for any $c_d \in R^d$, where

$$H_d(t) = X^{*-1}(t)[Q^*P_{N(D^*)}]_d = X^{*-1}(t)[(I-P^*)P_{N(D^*)}]_d$$

is an $n \times d$ matrix whose columns form a complete system of d linearly independent solutions of system (7.5) bounded on R;

(c) the operator L is a Fredholm operator with index

$$\operatorname{ind} L = -\operatorname{rank}[P_{N(D^*)}(I-P)] = -\operatorname{rank}[P_{N(D^*)}Q] = -d$$

and, in addition, $f \in \operatorname{Im} L$ if and only if condition (7.6) is satisfied, i.e.,

$$\int_{-\infty}^{\infty} H_d^*(s)f(s)ds = 0;$$

(d) if condition (7.6) is satisfied, then the inhomogeneous system (7.3) possesses a unique solution $x(t) = (G[f])(t)$ bounded on R, where $(G[f])(t)$ is the generalized Green operator (7.8) of the problem of finding solutions of system (7.1) bounded on R and such that

$$(LG[f])(t) = f(t), t \in R$$

and

$$(G[f])(0+0) - (G[f])(0-0) = -\int_{-\infty}^{\infty} H^*(s)f(s)ds.$$

Proof. To prove the e-trichotomy of system (7.5), it suffices to show that the projectors $(I-P^*)$ and $(I-Q^*)$ specifying the e-dichotomy of system (7.5) on the semiaxes satisfy the condition

$$(I - P^*)(I - Q^*) = (I - Q^*)(I - P^*) = I - Q^*.$$

This is true because, in view of the fact that $PQ = QP = P$, we have

$$Q^*P^* = P^*Q^* = P^*$$

and, hence,

$$(Q^* - I)P^* = 0 \quad \text{and} \quad P^*(Q^* - I) = 0.$$

By adding the term $I - Q^*$ to both sides of these equalities, we arrive at the required condition, which proves the e-trichotomy of system (7.5) [53].

Further, since $DP_{N(D)} = 0$ and

$$PD = P(P - (I - Q)) = PQ = P,$$

we get $PP_{N(D)} = PDP_{N(D)} = 0$.

By virtue of Lemma 7.1, we have $r = 0$, and the only solution of the homogeneous system (7.1) bounded on the entire real axis is trivial. Moreover, the inhomogeneous system (7.3) possesses a unique solution bounded on R for all $f(\cdot) \in BC(R)$ satisfying condition (7.6).

Corollary 7.3. *If the homogeneous system (7.1) is e-dichotomous on R_+ and R_- with projectors P and Q, respectively, such that $PQ = QP = P = Q$, then it is e-dichotomous on R and the inhomogeneous system (7.3) possesses a unique solution bounded on R for any $f \in BC(R)$.*

In other words, this is the so-called regular case [108, p. 37]. Combining Lemmas 7.1.1 and 7.1.2, we arrive at the following well-known statement:

Lemma 7.1.3. *Assume that system (7.1) is e-dichotomous on R_+ and R_- with projectors P and Q, respectively, such that $PQ = QP = P = Q$. Then the following assertions are true:*

(a) *the only solution of the homogeneous system (7.1) bounded on R is trivial* ($r = \operatorname{rank}[PP_{N(D)}] = \operatorname{rank}[(I - Q)P_{N(D)}] = 0$);

(b) *the only solution of system (7.5) conjugate to (7.1) bounded on R is trivial* ($d = \operatorname{rank}[P_{N(D^*)}(I - P)] = \operatorname{rank}[P_{N(D^*)}Q] = 0$);

(c) *the operator L defined by relation (7.4) is a Fredholm operator of index zero; furthermore, $f \in \operatorname{Im}(L)$ for any $f \in BC(R)$;*

(d) *the homogeneous system (7.1) is e-dichotomous on R and the inhomogeneous system (7.3) possesses a unique solution bounded on R that can be represented in the form $x(t) = (G[f])(t)$, where $(G[f])(t)$ is the Green operator (7.8) ($P = Q, D^+ = D^{-1}$) [24, 108] of the problem of finding solutions of system (7.3) bounded on R.*

The results presented above specify the Palmer lemma [115, p. 245] more precisely, include a formula for the index of the operator L that differs from the formula proposed in [132], and are necessary to establish new conditions for the existence of solutions of weakly perturbed linear [136] and nonlinear systems [21, 137] bounded on the entire real axis.

Perturbed Problem. Consider a weakly perturbed linear inhomogeneous system

$$\dot{x} = A(t)x + \varepsilon A_1(t)x + f(t), \quad A(\cdot), A_1(\cdot) \in BC(R). \tag{7.13}$$

Assume that the generating system (7.3) (obtained from system (7.13) for $\varepsilon = 0$ and satisfying the conditions of the formulated lemma) does not have solutions bounded on the entire real axis for arbitrary inhomogeneities $f \in BC(R)$. By the lemma, this means that we have the critical case and the criterion of solvability (7.6) of problem (7.3) is not satisfied (due to the arbitrariness of $f \in BC(R)$).

It is of interest to analyze whether it is possible to reduce problem (7.3) to a solvable problem with the help of linear perturbations. If this is possible, then it is important to know what kind of perturbations $A_1(t)$ must be used to make system (7.13) always solvable in the class of functions bounded on the entire real axis. In other words, it is necessary to specify perturbations for which the analyzed homogeneous system

$$\dot{x} = A(t)x + \varepsilon A_1(t)x, \quad A(\cdot), A_1(\cdot) \in BC(R), \tag{7.14}$$

turns into a system e-trichotomous or e-dichotomous on the entire real axis.

Let us show that this problem can be solved by using the $d \times r$ matrix

$$B_0 = \int_{-\infty}^{\infty} H_d^*(\tau) A_1(\tau) X_r(\tau) d\tau \tag{7.15}$$

constructed from the coefficients of system (7.13). Thus, by using the Vishik–Lyusternik method [150], it is possible to establish sufficient conditions for the appearance of a solution of problem (7.13) in the class of functions bounded on the entire real axis in the form of a Laurent series in powers of the small parameter ε containing finitely many terms with negative powers of ε.

Prior to formulating a theorem that solves the indicated problem, we recall that $P_{N(B_0)}$ is an $r \times r$ matrix (orthoprojector) projecting R^r onto the null space $N(B_0)$ of the matrix B_0, i.e., $P_{B_0} \colon R^r \to N(B_0)$ and $P_{N(B_0^*)}$ is a $d \times d$ matrix (orthoprojector) projecting R^d onto the null space $N(B_0^*)$ of the $r \times d$ matrix $B_0^* = B_0^t$, i.e., $P_{N(B_0^*)} \colon R^d \to N(B_0^*)$.

Theorem 7.1. *Suppose that system (7.13) satisfies the conditions imposed above for the critical case, the generating system (7.1) is e-dichotomous on R_+ and R_- with projectors P and Q, respectively, and, for arbitrary inhomo-*

geneities $f \in BC(R)$, *system (7.2) does not have solutions bounded on the entire real axis. If, in addition,*

$$P_{N(B_0^*)} = 0, \tag{7.16}$$

then system (7.14) is e-trichotomous on R *and, for any* $f \in BC(R)$, *system (7.13) possesses at least one solution bounded on* R *in the form of a series*

$$x(t, \varepsilon) = \sum_{i=-1}^{\infty} \varepsilon^i x_i(t) \tag{7.17}$$

uniformly convergent for sufficiently small fixed $\varepsilon \in (0, \varepsilon_*]$.

Proof. We substitute series (7.17) in system (7.13) and equate the coefficients of the same powers of ε. The problem of determination of the coefficient $x_{-1}(t)$ of the term with ε^{-1} in series (7.17) is reduced to the problem of finding solutions of the homogeneous system

$$\dot{x}_{-1} = A(t)x_{-1}. \tag{7.18}$$

bounded on the entire real axis. By virtue of the lemma, the homogeneous problem (7.18) possesses an r-parameter family of solutions $x_{-1}(t, c_{-1}) = X_r(t)c_{-1}$ bounded on R, where the r-dimensional vector column $c_{-1} \in R^r$ is determined from the condition of solvability of the problem used to determine the coefficient $x_0(t)$ of series (7.17).

The problem used to find the coefficient $x_0(t)$ of the term with ε^0 in series (7.17) reduces to the problem of finding solutions of the system

$$\dot{x}_0 = A(t)x_0 + A_1(t)x_{-1} + f(t). \tag{7.19}$$

bounded on the entire real axis.

By virtue of the lemma, the criterion of solvability of problem (7.19) takes the form

$$\int_{-\infty}^{\infty} H_d^*(\tau)(A_1(\tau)x_{-1}(\tau, c_{-1}) + f(\tau))d\tau = 0.$$

This criterion and formula (7.15) yield the following algebraic system for $c_{-1} \in R^r$:

$$B_0 c_{-1} = -\int_{-\infty}^{\infty} H_d^*(\tau)f(\tau)d\tau.$$

In order that this system be solvable for any $f \in BC(R)$, it is necessary and sufficient that condition (7.16) be satisfied. This condition is equivalent to the condition

$$\operatorname{rank} B_0 = d \leq r.$$

The algebraic system is solvable with respect to $c_{-1} \in R^r$ to within an arbitrary vector constant $P_{N(B_0)}c$, $c \in R^r$, from the null space of the matrix B_0 [38]. One of its solutions has the form

$$c_{-1} = -B_0^+ \int_{-\infty}^{\infty} H_d^*(\tau) f(\tau) d\tau,$$

where B_0^+ is the unique $r \times d$ matrix pseudoinverse to B_0.

If condition (7.16) is satisfied, then system (7.19) possesses an r-parameter family of solutions bounded on R:

$$x_0(t, c_0) = X_r(t)c_0 + (G[A_1(\cdot)x_{-1}(\cdot, c_{-1}) + f(\cdot)])(t),$$

where c_0 is an r-dimensional constant vector determined in the next stage of the process from the condition of solvability of the problem of determination of the coefficient $x_1(t)$ of series (7.17) and $(G[*])(t)$ is the generalized Green operator (7.8) of the problem of finding bounded solutions of system (7.3).

The problem of determination of the coefficient $x_1(t)$ of the term with ε^1 in series (7.17) reduces to the problem of determination of solutions of the system

$$\dot{x}_1 = A(t)x_1 + A_1(t)x_0 \tag{7.20}$$

bounded on the entire real axis. In the case where condition (7.16) is satisfied, by using the criterion of solvability of this system, i.e.,

$$\int_{-\infty}^{\infty} H_d^*(\tau)A_1(\tau)X_r(\tau)d\tau c_0 + \int_{-\infty}^{\infty} H_d^*(\tau)A_1(\tau)(G[A_1(\cdot)x_{-1}(\cdot, c_{-1}) + f(\cdot)])(\tau)d\tau = 0,$$

we determine the vector $c_0 \in R^r$ (to within an arbitrary vector constant $P_{N(B_0)}c$, $c \in R^r$, from the null space of the matrix B_0) as

$$c_0 = -B_0^+ \int_{-\infty}^{\infty} H_d^*(\tau)A_1(\tau)(G[A_1(\cdot)x_{-1}(\cdot, c_{-1}) + f(\cdot)])(\tau)d\tau.$$

Thus, under condition (7.16), system (7.20) possesses an r-parameter family of solutions bounded on R:

$$x_1(t, c_1) = X_r(t)c_1 + (G[A_1(\cdot)x_0(\cdot, c_0)])(t).$$

As above, we can easily show (by induction) that, under condition (7.16), the problem of determination of the coefficients $x_i(t)$ of the terms with ε^i in series (7.17) is reduced to the problem of finding solutions of the system

$$\dot{x}_i = A(t)x_i + A_1(t)x_{i-1} \tag{7.21}$$

bounded on the entire real axis. If condition (7.16) is satisfied, then the solution of system (7.21) bounded on the entire real axis takes the form

$$x_i(t, c_i) = X_r(t)c_i + (G[A_1(\cdot)x_{i-1}(\cdot, c_{i-1})])(t), \tag{7.22}$$

where the vector constant $c_i \in R^r$ is given by the formula

$$c_i = -B_0^+ \int\limits_{-\infty}^{\infty} H_d^*(\tau)A_1(\tau)(G[A_1(\cdot)x_{i-1}(\cdot, c_{i-1})])(\tau)d\tau \quad (i = 1, 2, \ldots) \tag{7.23}$$

(to within an arbitrary constant $P_{N(B_0)}c$, $c \in R^r$, from the null space of the matrix B_0).

We now show that, for sufficiently small $\varepsilon \in (0, \varepsilon_*]$, series (7.17) with coefficients specified by (7.22) converges uniformly in t and ε. First, we recall that the following estimates are true on the semiaxes:

$$\begin{aligned} \|X(t)PX^{-1}(s)\| &\le K_1 e^{-\alpha_1(t-s)}, \quad t \ge s, \\ \|X(t)(I-P)X^{-1}(s)\| &\le K_1 e^{-\alpha_1(s-t)}, \quad t \le s \quad (\forall t, s \in R_+), \\ \|X(t)QX^{-1}(s)\| &\le K_2 e^{-\alpha_2(t-s)}, \quad t \ge s, \\ \|X(t)(I-Q)X^{-1}(s)\| &\le K_2 e^{-\alpha_2(s-t)}, \quad t \le s \quad (\forall t, s \in R_-). \end{aligned} \tag{7.24}$$

In view of relations (7.24), we find

$$\|X(t)P\| \le K_1 e^{-\alpha_1 t}, \ t \ge 0, \quad \text{and} \quad \|X(t)(I-Q)\| \le K_2 e^{\alpha_2 t}, \ t \le 0,$$

and, therefore,

$$\begin{aligned} \|X_r(t)\| &\le \|X(t)PP_{N(D)}\| = \|X(t)(I-Q)P_{N(D)}\| \\ &\le \begin{cases} K_1 p e^{-\alpha_1 t}, & t \ge 0, \\ K_2 p e^{\alpha_2 t}, & t \le 0, \end{cases} \end{aligned} \tag{7.25}$$

where $p = \max(\|P_{N(D)}\|, \|P_{N(D^*)}\|)$.

It is known that the fundamental matrix $Y(t) = X^{*-1}(t)$ of system (7.5) conjugate to system (7.1) is e-dichotomous on the semiaxes R_+ and R_- with the projectors $I-P^*$ and $I-Q^*$, respectively. Inequalities similar to (7.24) and characterizing the e-dichotomy of the conjugate system (7.5) on the semiaxes have the form

$$
\begin{aligned}
\|X^{*-1}(t)P^*X^*(s)\| &= \|[X(s)PX^{-1}(t)]^*\| = \|X(s)PX^{-1}(t)\| \\
&\le K_1 e^{-\alpha_1(s-t)}, \quad s \ge t, \\
\|X^{*-1}(t)(I-P^*)X^*(s)\| &= \|[X(s)(I-P)X^{-1}(t)]^*\| \\
&\le K_1 e^{-\alpha_1(t-s)}, \quad s \le t \quad (\forall t, s \in R_+), \\
\|X^{*-1}(t)Q^*X^*(s)\| &\le K_2 e^{-\alpha_2(s-t)}, \quad s \ge t, \\
\|X^{*-1}(t)(I-Q^*)X^*(s)\| &\le K_2 e^{-\alpha_2(t-s)}, \quad s \le t \quad (\forall t, s \in R_-),
\end{aligned}
\tag{7.26}
$$

and, therefore,

$$
\begin{aligned}
\|H_d^*(t)\| &= \|H_d(t)\| \\
&\le \|X^{*-1}(t)Q^*P_{N(D^*)}\| = \|X^{*-1}(t)(I-P^*)P_{N(D^*)}\| \\
&\le \begin{cases} K_2 p e^{\alpha_2 t}, & t \le 0, \\ K_1 p e^{-\alpha_1 t}, & t \ge 0. \end{cases}
\end{aligned}
\tag{7.27}
$$

By using inequalities (7.24), we obtain the following estimate for the generalized Green operator (7.7):

$$
\begin{aligned}
\|(G[f])(t)\| &\le \begin{cases}
\Big\{ \int\limits_0^t K_1 e^{-\alpha_1(t-s)} ds + \int\limits_t^\infty K_1 e^{\alpha_1(t-s)} ds \\
\quad + K_1 N \Big[\int\limits_{-\infty}^0 K_2 e^{\alpha_2 s} ds + \int\limits_0^\infty K_1 e^{-\alpha_1 s} ds \Big] \Big\} \|f(t)\|, \quad t \ge 0, \\
\Big\{ \int\limits_{-\infty}^t K_2 e^{-\alpha_2(t-s)} ds + \int\limits_t^0 K_2 e^{\alpha_2(t-s)} ds \\
\quad + K_2 N \Big[\int\limits_{-\infty}^0 K_2 e^{\alpha_2 s} ds + \int\limits_0^\infty K_1 e^{-\alpha_1 s} ds \Big] \Big\} \|f(t)\|, \quad t \le 0,
\end{cases} \\
&\le K\|f(t)\|,
\end{aligned}
$$

where

$$
N = \|D^+\|, \quad K = \max_{i=1,2} K_i \left(\frac{1}{\alpha_i} + N\kappa \right), \quad \text{and} \quad \kappa = \frac{K_2}{\alpha_2} + \frac{K_1}{\alpha_1}.
$$

We use inequalities (7.24)–(7.27) to estimate the coefficients $x_i(t, c_i)$ (7.22) and c_i (7.23) of series (7.17). Indeed, for any $t \in R$, we can write

$$\|c_1\| \leq baK\left\{ \int_{-\infty}^{0} K_2 p e^{\alpha_2 s} ds + \int_{0}^{\infty} K_1 p e^{-\alpha_1 s} ds \right\} \|A_1(t) x_0(t, c_0)\|$$
$$= bpa^2 \kappa K \|x_0(t, c_0)\|,$$

$$\|x_1(t, c_1)\| \leq k\|c_1\| + Ka\|x_0(t, c_0)\| \leq aK(abpk\kappa + 1)\|x_0(t, c_0)\|,$$

where $b = \|B_0^+\|$, $a = \|A_1(t)\|$, and $k = \max(K_1 p, K_2 p)$.

Further, we have

$$\|c_2\| \leq bpa^2 \kappa K \|x_1(t, c_1)\| \leq (aK)^2 abp\kappa(abpk\kappa + 1)\|x_0(t, c_0)\|,$$
$$\|x_2(t, c_2)\| \leq k\|c_2\| + aK\|x_1(t, c_1)\| \leq [aK(abpk\kappa + 1)]^2 \|x_0(t, c_0)\|.$$

We continue this process infinitely and conclude that the following estimates are true for the coefficients $c_i \in R^r, x_i(t, c_i)$ of series (7.17):

$$\|c_i\| \leq (aK)^i abp\kappa(abpk\kappa + 1)^{i-1}\|x_0(t, c_0)\|,$$
$$\|x_i(t, c_i)\| \leq [aK(abpk\kappa + 1)]^i \|x_0(t, c_0)\| \quad (i = 1, 2, \ldots).$$

Thus, for all $t \in R$, series (7.17) is majorized by the series

$$\varepsilon^{-1}\|x_{-1}(t, c_{-1})\| + \sum_{i=0}^{+\infty} [\varepsilon aK(abpk\kappa + 1)]^i \|x_0(t, c_0)\|.$$

In this case, $\|x_{-1}(t, c_{-1})\|$ and $\|x_0(t, c_0)\|$ are bounded. Therefore, for $t \in R$ and all fixed $\varepsilon \in (0, \varepsilon_*]$, where $\varepsilon_* < [aK(abpk\kappa+1)]^{-1}$, series (7.17) is uniformly convergent. This completes the proof of Theorem 7.1.

In the case where the number $r = \operatorname{rank}[PP_{N(D)} = (I - Q)P_{N(D)}]$ of linearly independent solutions of system (7.1) bounded on R is equal to the number $d = \operatorname{rank}[P_{N(D^*)}(I - P) = P_{N(D^*)}Q]$ of linearly independent solutions of system (7.5) conjugate to system (7.1) bounded on R, Theorem 7.1 yields the following assertion:

Theorem 7.2. *Suppose that system (7.13) satisfies the conditions imposed above for the critical case, the generating system (7.1) is e-dichotomous on R_+ and R_- with projectors P and Q, respectively, and, for arbitrary inhomogeneities $f \in BC(R)$, system (7.3) does not have solutions bounded on the entire real axis. If, in addition,*

$$\det B_0 \neq 0 \quad (r = d), \tag{7.28}$$

then system (7.14) is e-dichotomous on R and, for any $f \in BC(R)$, system (7.13) possesses a unique solution bounded on R in the form of series (7.17) uniformly convergent for sufficiently small fixed $\varepsilon \in (0, \varepsilon_*]$.

Proof. If $\operatorname{ind} L = r - d = 0$, then B_0 is a square matrix. Hence, it follows from condition (7.16) that $P_{N(B_0)} = P_{N(B_0^*)} = 0$, which is equivalent to condition (7.28), i.e., $\det B_0 \neq 0$. In this case, the constant vectors $c_i \in R^r$ are uniquely determined from relations (7.23). Therefore, the coefficients of series (7.17) are also uniquely determined and, for any $f \in BC(R)$, system (7.13) possesses a unique solution bounded on R. This, in turn, means that system (7.14) is e-dichotomous on R.

Examples. We now illustrate the assertions proved above.

1. Consider system (7.13) with

$$A(t) = \operatorname{diag}\{-\tanh t, -\tanh t, \tanh t\}$$

and

$$A_1(t) = \{a_{ij}(t)\}_{i,j=1}^3 \in BC(R).$$

It is easy to see that

$$X(t) = \operatorname{diag}\left\{\frac{2}{(e^t + e^{-t})}, \frac{2}{(e^t + e^{-t})}, \frac{(e^t + e^{-t})}{2}\right\}$$

and the homogeneous system (7.1) is e-dichotomous on the semiaxes R_+ and R_- with the projectors $P = \operatorname{diag}\{1,1,0\}$ and $Q = \operatorname{diag}\{0,0,1\}$, respectively. Thus, we have

$$D = 0, \quad D^+ = 0, \quad P_{N(D)} = P_{N(D^*)} = I_3,$$

$$r = \operatorname{rank} P P_{N(D)} = 2, \quad d = \operatorname{rank} P_{N(D^*)} Q = 1,$$

$$X_r(t) = \begin{pmatrix} 2/(e^t + e^{-t}) & 0 \\ 0 & 2/(e^t + e^{-t}) \\ 0 & 0 \end{pmatrix},$$

and

$$H_d(t) = \begin{pmatrix} 0 \\ 0 \\ 2/(e^t + e^{-t}) \end{pmatrix}.$$

The inhomogeneous system (7.3) with the matrix $A(t)$ specified above has solutions bounded on the entire real axis not for all possible inhomogeneities. In order that this system be solvable, the inhomogeneities

$$f(t) = \operatorname{col}\{f_1(t), f_2(t), f_3(t)\} \in BC(R)$$

must satisfy condition (7.6) in Lemma 7.1. In the analyzed case, this condition takes the form

$$\int_{-\infty}^{+\infty} f_3(s)/(e^s + e^{-s})ds = 0 \quad \forall f_1(t) \in BC(R) \quad \text{and} \quad \forall f_2(t) \in BC(R).$$

System (7.14) with the coefficients specified above is e-trichotomous on R if the coefficients $a_{31}(t), a_{32}(t) \in BC(R)$ of the perturbing matrix $A_1(t)$ satisfy condition (7.16), i.e., $P_{N(B_0^*)} = 0$, where

$$B_0 = \int_{-\infty}^{+\infty} H_d^*(t) A_1(t) X_r(t) dt = 4 \int_{-\infty}^{+\infty} \left[\frac{a_{31}(t)}{(e^t + e^{-t})^2}, \frac{a_{32}(t)}{(e^t + e^{-t})^2} \right] dt.$$

If $a_{31}(t), a_{32}(t) \in BC(R)$ are such that at least one of the inequalities

$$\int_{-\infty}^{+\infty} a_{31}(t)/(e^t + e^{-t})^2 dt \neq 0 \quad \text{and} \quad \int_{-\infty}^{+\infty} a_{32}(t)/(e^t + e^{-t})^2 dt \neq 0$$

is satisfied, then either condition (7.16) or the equivalent condition $\operatorname{rank} B_0 = d = 1$ from Theorem 7.1 is satisfied and system (7.14) is e-trichotomous on R. Thus, for $a_{31}(t) = \text{const} \neq 0$ or $a_{32}(t) = \text{const} \neq 0$, one of these inequalities and, hence, condition (7.16) are always satisfied. In this case, the coefficients $a_{11}(t)$, $a_{12}(t)$, $a_{13}(t)$, $a_{21}(t)$, $a_{22}(t)$, $a_{23}(t)$, and $a_{33}(t)$ are arbitrary functions from the space $BC(R)$. Moreover, for any

$$f(t) = \operatorname{col}\{f_1(t), f_2(t), f_3(t)\} \in BC(R),$$

a solution of system (7.13) bounded on R is given by series (7.17) (to within a constant from the null space $N(B_0)$, $\dim N(B_0) = r - \operatorname{rank} B_0 = 1$).

2. Consider system (7.13) with

$$A(t) = \operatorname{diag}\{-\tanh t, \tanh t\}, \quad A_1(t) = \{a_{ij}(t)\}_{i,j=1}^2 \in BC(R).$$

It is easy to see that

$$X(t) = \operatorname{diag}\left\{\frac{2}{(e^t + e^{-t})}, \frac{(e^t + e^{-t})}{2}\right\}$$

and the homogeneous system (7.1) is e-dichotomous on the semiaxes R_+ and R_- with the projectors $P = \operatorname{diag}\{1, 0\}$ and $Q = \operatorname{diag}\{0, 1\}$, respectively. Thus, we have

$$D = 0, \quad D^+ = 0, \quad P_{N(D)} = P_{N(D^*)} = I,$$
$$r = \operatorname{rank} P P_{N(D)} = 1, \quad d = \operatorname{rank} P_{N(D^*)} Q = 1,$$
$$X_r(t) = \operatorname{col}\{2/(e^t + e^{-t}), 0\}, \quad H_d^*(t) = \{0, 2/(e^t + e^{-t})\}.$$

After elementary transformations, we conclude that the inhomogeneous system (7.3) with the matrix $A(t)$ specified above has solutions bounded on the entire real axis not for all possible inhomogeneities. In the analyzed case, this system possesses the required solutions only for the inhomogeneities $f(t) = \operatorname{col}\{f_1(t), f_2(t)\} \in BC(R)$ satisfying condition (7.6), which takes the form

$$\int_{-\infty}^{+\infty} \frac{f_2(s)}{(e^s + e^{-s})} ds = 0 \quad \forall f_1(t) \in BC(R).$$

System (7.14) with the coefficients specified above is e-dichotomous on R if the coefficient $a_{21}(t)$ of the perturbing matrix $A_1(t)$ satisfies condition (7.28) in Theorem 7.2:

$$B_0 = \int_{-\infty}^{+\infty} H_d^*(t) A_1(t) X_r(t) dt = 4 \int_{-\infty}^{+\infty} a_{21}(t)/(e^t + e^{-t})^2 dt \neq 0.$$

Thus, for $a_{21}(t) = \text{const} = 1 \neq 0$ [108, p. 48], this condition is always satisfied and the coefficients $a_{11}(t)$, $a_{12}(t)$, and $a_{22}(t)$ are arbitrary functions from the space $BC(R)$. In this case, for any $f(t) = \operatorname{col}\{f_1(t), f_2(t)\} \in BC(R)$, the solution of system (7.13) bounded on R is uniquely defined by series (7.17).

Remark 7.1. In the case where the operator L is a Fredholm operator of index zero ($\operatorname{ind} L = 0$, $r = d$), the condition of e-dichotomy (7.28) of system (7.14) on the entire real axis can be found in [156]. In this case, the property of e-dichotomy is proved by using another approach which does not enable one construct a solution of system (7.13) bounded on the entire real axis. If condition (7.16) (or condition (7.28)) is not satisfied, then it is necessary to seek a solution of system (7.13) bounded on the entire real axis in the form of series (7.17), namely, $\sum_{i=-k}^{\infty} \varepsilon^i x_i(t)$ with $k > 1$.

7.2 Nonlinear Systems

For a weakly nonlinear system

$$\dot{x} = A(t)x + f(t) + \varepsilon Z(x, t, \varepsilon), \tag{7.29}$$

we establish conditions required for the existence of its solutions $x = x(t, \varepsilon)$ bounded on R and such that

$$x(\cdot, \varepsilon)\colon R \to R^n, \quad x(\cdot, \varepsilon) \in BC^1(R), \quad \text{and} \quad x(t, \cdot) \in C[0, \varepsilon_0],$$

where $\varepsilon \in R$ is a sufficiently small parameter, which turns into one of generating solutions $x_0(t, c_r)$ of system (7.3) given by relation (7.7) for $\varepsilon = 0$. The nonlinear vector function $Z(x, t, \varepsilon)$ is such that

$$Z(\cdot, t, \varepsilon) \in C^1[\|x - x_0\| \leq q], \quad Z(x, \cdot, \varepsilon) \in BC(R), \quad \text{and} \quad Z(x, t, \cdot) \in C[0, \varepsilon_0],$$

where q is a sufficiently small constant.

Theorem 7.3 (necessary condition). *Assume that system (7.1) is e-dichotomous on R_+ and R_- with projectors P and Q, respectively, and system (7.29) has a solution $x(t,\varepsilon)$ bounded on R and such that $x(\cdot,\varepsilon)\colon R\to R^n$, $x(\cdot,\varepsilon)\in BC^1(R)$, and $x(t,\cdot)\in C[0,\varepsilon_0]$, which turns into one of generating solutions $x_0(t,c_r)$ (7.7) of system (7.3) with a vector constant $c_r=c_r^0\in R^r$ for $\varepsilon=0$. In this case, the vector c_r^0 satisfies the equation*

$$F(c_r^0)=\int\limits_{-\infty}^{\infty}H_d^*(s)Z(x_0(s,c_r^0),s,0)ds=0. \tag{7.30}$$

Proof. Condition (7.6) for the existence of generating solutions $x_0(t,c_r)$ (7.7) bounded on R is assumed to be satisfied. We consider the nonlinearity in system (7.29) as an inhomogeneity and apply Lemma 7.1. This gives

$$\int\limits_{-\infty}^{\infty}H_d^*(s)Z(x(s,\varepsilon),s,\varepsilon)ds=0.$$

In this integral, we pass to the limit as $\varepsilon\to 0$ and arrive at the required condition (7.30).

By analogy with the case of periodic problems [38, p. 184], it is natural to say that equation (7.30) is the equation for generating constants of the problem of finding solutions of system (7.29) bounded on the entire real axis R. If equation (7.30) is solvable, then the vector constant $c_r^0\in R^r$ specifies the generating solution $x_0(t,c_r^0)$ corresponding to the solution $x=x(t,\varepsilon)$ of the original problem (7.29) bounded on R and such that

$$x(\cdot,\varepsilon)\colon R\to R^n,\quad x(\cdot,\varepsilon)\in BC^1(R),$$

$$x(t,\cdot)\in C[0,\varepsilon_0],\quad\text{and}\quad x(t,0)=x_0(t,c_r^0).$$

At the same time, if equation (7.30) is unsolvable, then problem (7.29) does not have solutions bounded on R in the analyzed space. Note that, here and in what follows, all expressions are obtained in the real form and, hence, we are interested in real solutions of equation (7.30), which can be algebraic or transcendental.

By the change of variables in system (7.29)

$$x(t,\varepsilon)=x_0(t,c_r^0)+y(t,\varepsilon),$$

we arrive at the problem of finding sufficient conditions for the existence of solutions $y=y(t,\varepsilon)$ of the problem

$$\dot y=A(t)y+\varepsilon Z(x_0(t,c_r^0)+y,t,\varepsilon) \tag{7.31}$$

bounded on R and such that

$$y(\cdot,\varepsilon)\colon R\to R^n,\quad y(\cdot,\varepsilon)\in BC^1(R),\quad y(t,\cdot)\in C[0,\varepsilon_0],\quad y(t,0)=0$$

Since the vector function $Z(x,t,\varepsilon)$ is continuously differentiable with respect to x and continuous in ε in the neighborhood of the point $x_0(t,c_r^0)$, $\varepsilon=0$, we can separate the term linear as a function of y and terms of order zero with respect to ε:

$$Z(x_0(t,c_r^0)+y,t,\varepsilon)=f_0(t,c_r^0)+A_1(t)y+R(y(t,\varepsilon),t,\varepsilon), \tag{7.32}$$

where

$$f_0(t,c_r^0)=Z(x_0(t,c_r^0),t,0),\quad f_0(\cdot,c_r^0)\in BC(R),$$

$$A_1(t)=A_1(t,c_r^0)=\frac{\partial Z(x,t,0)}{\partial x}\Big|_{x=x_0(t,c_r^0)},\quad A_1(\cdot)\in BC(R),$$

$$R(0,t,0)=0,\frac{\partial R(0,t,0)}{\partial y}=0,\quad R(y,\cdot,\varepsilon)\in BC(R).$$

We now formally consider the vector function $Z(x_0+y,t,\varepsilon)$ in system (7.31) as an inhomogeneity and apply Lemma 7.1 to this system. As a result, we obtain the following representation for the solution of system (7.31) bounded on R:

$$y(t,\varepsilon)=X_r(t)c+y^{(1)}(t,\varepsilon).$$

In this expression, the unknown vector of constants $c=c(\varepsilon)\in R^r$ is determined from a condition similar to (7.30) for the existence of a bounded solution of system (7.31):

$$B_0c=-\int_{-\infty}^{\infty}H_d^*(\tau)[A_1(\tau)y^{(1)}(\tau,\varepsilon)+R(y(\tau,\varepsilon),\tau,\varepsilon)]d\tau, \tag{7.33}$$

where

$$B_0=\int_{-\infty}^{\infty}H_d^*(\tau)A_1(\tau)X_r(\tau)d\tau$$

is a $d\times r$ matrix,

$$r=\operatorname{rank}[PP_{N(D)}]=\operatorname{rank}[(I-Q)P_{N(D)}],$$

and

$$d=\operatorname{rank}[P_{N(D^*)}(I-P)]=\operatorname{rank}[P_{N(D^*)}Q].$$

The unknown vector function $y^{(1)}(t,\varepsilon)$ is determined by using the generalized Green operator (7.8) as follows:

$$y^{(1)}(t,\varepsilon)=\varepsilon\left(G\left[Z(x_0(\tau,c_r^0)+y,\tau,\varepsilon)\right]\right)(t).$$

Let $P_{N(B_0)}$ be an $r \times r$ matrix orthoprojector $R^r \to N(B_0)$ and let $P_{N(B_0^*)}$ be a $d \times d$ matrix-orthoprojector $R^d \to N(B_0^*)$. Equation (7.31) is solvable with respect to $c \in R^r$ if and only if

$$P_{N(B_0^*)} \int_{-\infty}^{\infty} H_d^*(\tau)[A_1(\tau)y^{(1)}(\tau,\varepsilon) + R(y(\tau,\varepsilon),\tau,\varepsilon)]d\tau = 0. \tag{7.34}$$

For

$$P_{N(B_0^*)} = 0,$$

condition (7.34) is always satisfied and equation (7.33) is solvable with respect to $c \in R^r$ to within an arbitrary vector constant $P_{N(B_0)}c$, $c \in R^r$, from the null space of the matrix B_0:

$$c = -B_0^+ \int_{-\infty}^{\infty} H_d^*(\tau)[A_1(\tau)y^{(1)}(\tau,\varepsilon) + R(y(\tau,\varepsilon),\tau,\varepsilon)]d\tau + P_{N(B_0)}c.$$

To find a solution $y = y(t,\varepsilon)$ of problem (7.31) bounded on R and such that

$$y(\cdot,\varepsilon)\colon R \to R^n, \quad y(\cdot,\varepsilon) \in BC^1(R), \quad y(t,\cdot) \in C[0,\varepsilon_0], \quad \text{and} \quad y(t,0) = 0,$$

it is necessary to solve the following operator system:

$$y(t,\varepsilon) = X_r(t)c + y^{(1)}(t,\varepsilon), \tag{7.35}$$

$$c = -B_0^+ \int_{-\infty}^{\infty} H_d^*(\tau)[A_1(\tau)y^{(1)}(\tau,\varepsilon) + R(y(\tau,\varepsilon),\tau,\varepsilon)]d\tau,$$

$$y^{(1)}(t,\varepsilon) = \varepsilon\left(G\left[Z(x_0(\tau,c_r^0) + y,\tau,\varepsilon)\right]\right)(t).$$

The operator system (7.35) belongs to the class of systems [38, p. 188] solvable by the method of simple iterations convergent for sufficiently small $\varepsilon \in [0,\varepsilon_*] \subseteq [0,\varepsilon_0]$. Indeed, system (7.35) can be rewritten in the form

$$z = L^{(1)}z + Fz, \tag{7.36}$$

where $z = \operatorname{col}(y^{(1)}(t,\varepsilon), c(\varepsilon), y(t,\varepsilon))$ is a $(2n+r)$-dimensional column vector, $L^{(1)}$ and F are, respectively, linear and nonlinear operators bounded on R,

$$L^{(1)} = \begin{pmatrix} 0 & 0 & 0 \\ L_1 & 0 & 0 \\ I_n & X_r & 0 \end{pmatrix}, \qquad L_1 * = -B_0^+ \int_{-\infty}^{\infty} H_d^*(\tau)A_1(\tau) * d\tau,$$

and

$$Fz = \operatorname{col}\left[\varepsilon G\left[Z(x_0(\tau, c_r^0) + y(\tau,\varepsilon), \tau, \varepsilon)\right], \int_{-\infty}^{\infty} H_d^*(\tau) R(y(\tau,\varepsilon),\tau,\varepsilon)d\tau, 0\right].$$

In view of the structure of the operator $L^{(1)}$ containing zero blocks on the principal diagonal and above, the operator $(I_s - L^{(1)})^{-1}$ exists. System (7.36) admits the following transformation:

$$z = Sz \quad (S := (I_s - L^{(1)})^{-1}F, \;\; s = 2n + r), \tag{7.37}$$

where S is a contraction operator in a sufficiently small neighborhood of the point $x_0(t, c_r^0)$, $\varepsilon = 0$. Thus, the solvability of the operator system (7.37) can be established by using one of the existing versions of the fixed-point principle [87] applicable to the system for sufficiently small $\varepsilon \in [0, \varepsilon_*]$. By using the method of simple iterations to find the solution of the operator system (7.37) and, hence, the solutions of the original system (7.29) bounded on R, we arrive at the following assertion [137]:

Theorem 7.4 (sufficient condition). *Assume that the weakly nonlinear system (7.29) satisfies the conditions imposed above (and, thus, the corresponding homogeneous system (7.1) is dichotomous on the semiaxes R_+ and R_-) and the corresponding generating linear system (7.3) possesses the r-parameter set (7.7) of generating solutions $x_0(t, c_r)$ bounded on R. Also let*

$$P_{N(B_0^*)} = 0. \tag{7.38}$$

Then, for any vector $c_r = c_r^0 \in R^r$ satisfying the equation for generating constants (7.30), there exists at least one solution of system (7.29) bounded on R. The indicated solution $x(t,\varepsilon)$ is such that $x(\cdot,\varepsilon) \in BC^1(R)$ and $x(t,\cdot) \in C[0,\varepsilon_0]$, turns into the generating solution $x(t,0) = x_0(t, c_r^0)$ given by relation (7.7) for $\varepsilon = 0$, and can be found by the method of simple iterations convergent for sufficiently small $\varepsilon \in [0,\varepsilon_*] \subseteq [0,\varepsilon_0]$, namely,

$$y_{k+1}^{(1)}(t,\varepsilon) = \varepsilon\left(G\left[Z(x_0(\tau, c_r^0) + y_k, \tau, \varepsilon)\right]\right)(t), \tag{7.39}$$

$$c_k = -B_0^+ \int_{-\infty}^{\infty} H_d^*(\tau)[A_1(\tau) y_k^{(1)}(\tau,\varepsilon) + R(y_k(\tau,\varepsilon),\tau,\varepsilon)]d\tau,$$

$$y_{k+1}(t,\varepsilon) = X_r(t)c_k + y_{k+1}^{(1)}(t,\varepsilon),$$

$$x_k(t,\varepsilon) = x_0(t, c_r^0) + y_k(t,\varepsilon), \quad k = 0,1,2,\ldots, \quad y_0(t,\varepsilon) = 0,$$

$$x(t,\varepsilon) = \lim_{k\to\infty} x_k(t,\varepsilon).$$

In the case where the number $r = \operatorname{rank}[PP_{N(D)} = (I-Q)P_{N(D)}]$ of linear independent solutions of system (7.1) bounded on R is equal to the number $d = \operatorname{rank}[P_{N(D^*)}(I-P) = P_{N(D^*)}Q]$ of linear independent bounded (on R) solutions of system (7.8) conjugate to (7.1), the condition $P_{N(B_0^*)} = 0$ implies that $P_{N(B_0)} = 0$ and, hence, $\det B_0 \neq 0$. In this case, Theorem 7.4 yields the following assertion [23]:

Theorem 7.5 (sufficient condition). *Assume that the weakly nonlinear system (7.29) satisfies the conditions imposed above (and, hence, the corresponding homogeneous system (7.1) is dichotomous on the semiaxes R_+ and R_-) and the corresponding generating linear system (7.3) possesses the r-parameter set (7.7) of generating solutions $x_0(t, c_r)$ bounded on R. Also let*

$$\det B_0 \neq 0 \quad (r = d). \tag{7.40}$$

Then, for any vector $c_r = c_r^0 \in R^r$ satisfying the equation for generating constants (7.30), there exists a unique solution of system (7.29) bounded on R. This solution $x(t, \varepsilon)$ is such that $x(t, \cdot) \in C[0, \varepsilon_0]$, turns into the generating solution $x(t, 0) = x_0(t, c_r^0)$ (7.7) for $\varepsilon = 0$, and can be found by the method of simple iterations (7.39) convergent for sufficiently small $\varepsilon \in [0, \varepsilon_*] \subseteq [0, \varepsilon_0]$.

Conclusions. The required estimates for ε_* and the error of approximation of the iterative process can be obtained in a standard way [87].

Condition (7.40) means [38] that the constant $c_r^0 \in R^r$ is a simple root of the equation (7.30) for generating constants of the problem of finding solutions of system (7.29) bounded on the entire real axis R. By using the procedure proposed in [38, p. 193] with some simplifying assumptions, we can generalize the proposed method to the case of multiple roots of equation (7.30).

If L is a Fredholm operator of index zero and $r = 1$, then Theorem 7.5 yields the well-known Palmer's result [115, p. 248]. If L is a Fredholm operator and, in addition, it possesses the exponential trichotomy on R, then Theorem 7.4 turns into the result obtained earlier by Elaidy and Hajek [53].

Examples. **1.** Consider a system

$$\dot{x} = A(t)x + f(t) + \varepsilon A_1(t)x, \tag{7.41}$$

where

$$A(t) = \operatorname{diag}\{-\tanh t, -\tanh t, \tanh t\}, \quad A_1(t) = \{a_{ij}(t)\}_{i,j=1}^{3} \in BC(R).$$

One can easily see that

$$X(t) = \operatorname{diag}\left\{\frac{2}{(e^t + e^{-t})}, \frac{2}{(e^t + e^{-t})}, \frac{(e^t + e^{-t})}{2}\right\}.$$

The homogeneous system $\dot{x} = A(t)x$ is e-dichotomous on the semiaxes R_+ and R_- with the projectors $P = \text{diag}\{1, 1, 0\}$ and $Q = \text{diag}\{0, 0, 1\}$. Then

$$D = 0, \quad D^+ = 0, \quad P_{N(D)} = P_{N(D^*)} = I_3,$$

$$r = \text{rank}\, PP_{N(D)} = 2, \quad d = \text{rank}\, P_{N(D^*)}Q = 1,$$

$$X_r(t) = \begin{pmatrix} 2/(e^t + e^{-t}) & 0 \\ 0 & 2/(e^t + e^{-t}) \\ 0 & 0 \end{pmatrix}, \quad H_d(t) = \begin{pmatrix} 0 \\ 0 \\ 2/(e^t + e^{-t}) \end{pmatrix}.$$

In order that the inhomogeneous system $\dot{x} = A(t)x + f(t)$ possess a two-parameter family $x_0(t, c_r) = X_r(t)c_r + (G[f])(t)$, $c_r \in R^2$ of solutions bounded on R, it is necessary that the inhomogeneity $f(t) = \text{col}\{f_1(t), f_2(t), f_3(t)\} \in BC(R)$ satisfy the condition

$$\int_{-\infty}^{+\infty} f_3(s)/(e^s + e^{-s})ds = 0 \quad \forall f_1(t) \in BC(R) \quad \text{and} \quad \forall f_2(t) \in BC(R).$$

By using Theorems 7.3 and 7.4, we get the following result valid for system (7.41): Let $P_{N(B_0^*)} = 0$. Then, for any vector $c_r = c_r^0 \in R^2$ satisfying the equation for generating constants (7.30), i.e.,

$$B_0 c_r^0 = -\int_{-\infty}^{\infty} H_d^*(s)A_1(s)(Gf)(s)ds,$$

where

$$B_0 = \int_{-\infty}^{+\infty} H_d^*(t)A_1(t)X_r(t)dt = 4\int_{-\infty}^{+\infty} [\frac{a_{31}(t)}{(e^t + e^{-t})^2}, \frac{a_{32}(t)}{(e^t + e^{-t})^2}]dt,$$

there exists a one-parameter family of solutions of system (7.41) bounded on R ($\rho = \text{rank}\, P_{N(B_0)} = r - \text{rank}\, B_0 = r - d = 1$). These solutions $x(t, \varepsilon)$ are such that $x(t, \cdot) \in C[0, \varepsilon_0]$ and turn into the generation solution $x(t, 0) = x_0(t, c_r^0)$ for $\varepsilon = 0$.

If $a_{31}(t)$ or $a_{32}(t) \in BC(R)$ satisfies one of the conditions

$$\int_{-\infty}^{+\infty} a_{31}(t)/(e^t + e^{-t})^2 dt \neq 0 \quad \text{and} \quad \int_{-\infty}^{+\infty} a_{32}(t)/(e^t + e^{-t})^2 dt \neq 0,$$

then condition (7.38) is satisfied. Thus, if $a_{31}(t) = \text{const} \neq 0$ or $a_{32}(t) = \text{const} \neq 0$, then one of these inequalities is definitely true and condition (7.38) is satisfied. In this case, the coefficients $a_{11}(t)$, $a_{12}(t)$, $a_{13}(t)$, $a_{21}(t)$, $a_{22}(t)$, $a_{23}(t)$, and $a_{33}(t)$ are arbitrary functions from the space $BC(R)$.

2. Consider system (7.41) in which

$$A(t) = \operatorname{diag}\{-\tanh t, \tanh t\} \quad \text{and} \quad A_1(t) = \{a_{ij}(t)\}_{i,j=1}^{2} \in BC(R).$$

One can easily show that

$$X(t) = \operatorname{diag}\left\{\frac{2}{(e^t + e^{-t})}, \frac{(e^t + e^{-t})}{2}\right\},$$

and the homogeneous system $\dot{x} = A(t)x$ is e-dichotomous on the semiaxes R_+ and R_- with the projectors $P = \operatorname{diag}\{1, 0\}$ and $Q = \operatorname{diag}\{0, 1\}$, respectively. Then

$$D = 0, \quad D^+ = 0, \quad P_{N(D)} = P_{N(D^*)} = I,$$
$$r = \operatorname{rank} PP_{N(D)} = 1, \quad d = \operatorname{rank} P_{N(D^*)}Q = 1,$$
$$X_r(t) = \operatorname{col}\{2/(e^t + e^{-t}), 0\}, \quad H_d^*(t) = \{0, 2/(e^t + e^{-t})\}.$$

In order that the inhomogeneous system $\dot{x} = A(t)x + f(t)$ possess a one-parameter family of solutions bounded on R, $x_0(t, c_r) = X_r(t)c_r + (G[f])(t)$, $c_r \in R$, it is necessary that $f(t) = \operatorname{col}\{f_1(t), f_2(t)\} \in BC(R)$ satisfy the condition

$$\int_{-\infty}^{+\infty} f_2(s)/(e^s + e^{-s})ds = 0 \quad \forall f_1(t) \in BC(R).$$

In this case, according to Theorem 7.5, we get the following result valid for system (7.41): Assume that condition (7.40) is satisfied for

$$B_0 = \int_{-\infty}^{+\infty} H_d^*(t)A_1(t)X_r(t)dt = 4\int_{-\infty}^{+\infty} a_{21}(t)/(e^t + e^{-t})^2 dt \neq 0 \quad (r = d = 1).$$

Then, for any constant $c_r = c_r^0 \in R$ satisfying the equation for generating constants (7.30), i.e.,

$$B_0 c_r^0 = -\int_{-\infty}^{\infty} H_d^*(s)A_1(s)(Gf)(s)ds,$$

there exists a unique solution of system (7.41) bounded on R. This solution $x(t, \varepsilon)$ is such that $x(t, \cdot) \in C[0, \varepsilon_0]$ and turns into the generating solution $x(t, 0) = x_0(t, c_r^0)$ for $\varepsilon = 0$.

Thus, if $a_{21}(t) = \text{const} = 1$ [108, p. 48], then the last inequality is true and condition (7.40) is satisfied. In this case, the coefficients $a_{11}(t)$, $a_{12}(t)$, and $a_{22}(t)$ are arbitrary functions from the space $BC(R)$.

7.3 Solutions of Linear and Nonlinear Difference Equations Bounded on the Entire Real Axis[1]

By $B(J)$ we denote the Banach space of vector-valued functions $x\colon J \to R^N$ bounded on J with the norm $\|x\| = \sup_{n\in J} |x(n)|$, where $|x(n)| := \|x\|_{R^N}$ and J, as a rule, stands either for the set of integers Z, or for the set of nonnegative integers Z_+, or for the set of nonpositive integers Z_-.

Consider a system

$$x(n+1) = A(n)x(n), \tag{7.42}$$

where $A(n)$, $n \in Z$, is an invertible $N \times N$ matrix whose elements are real-valued functions bounded on the entire real axis Z and $A(\cdot) \in B(Z)$. As in the differential case [49, 115], it is known that system (7.42) is dichotomous on Z if there exists a projector P $(P^2 = P)$ and constants $K \geq 1$ and $0 < \lambda < 1$ such that

$$\|X(n)PX^{-1}(m)\| \leq K\lambda^{n-m}, \quad n \geq m,$$

$$\|X(n)(I-P)X^{-1}(m)\| \leq K\lambda^{m-n}, \quad m \geq n,$$

for all $m, n \in J = Z$, where $X(n)$ is the normal $(X(0) = I)$ fundamental $N \times N$ matrix of system (7.42).

Consider the problem of finding solutions

$$x\colon Z \to R^N, \quad x(\cdot) \in B(Z),$$

of the inhomogeneous system

$$x(n+1) = A(n)x + f(n), \quad A(\cdot), f(\cdot) \in B(Z), \tag{7.43}$$

bounded on Z. In the nonresonance case, the homogeneous system (7.42) is dichotomous on Z and, hence, the only solution of system (7.42) bounded on Z is trivial. At the same time, the inhomogeneous system (7.43) possesses a unique solution bounded on Z for every $f(\cdot) \in B(Z)$. The resonance case in which system (7.42) has nontrivial solutions bounded on Z is studied much less completely [75]. We now consider some facts concerning the Fredholm property of the analyzed problem. This property is used in what follows for the investigation of weakly nonlinear systems [22, 31].

Linear Systems. Assume that system (7.42) is dichotomous on Z_+ and Z_- with projectors P and Q, respectively. It is easy to see that the general solution of (7.43) bounded on the semiaxes Z_+ and Z_- is given by the formula

[1] The research presented in this section was partially supported by the Slovak Grant Agency (Grant VEGA 1/0026/03).

$$x(n,\xi) = X(n)\begin{cases} P\xi + \sum_{k=0}^{n-1} PX^{-1}(k+1)f(k) \\ \qquad - \sum_{k=n}^{+\infty}(I-P)X^{-1}(k+1)f(k), \quad n \geq 0, \\ (I-Q)\xi + \sum_{k=-\infty}^{n-1} QX^{-1}(k+1)f(k) \\ \qquad - \sum_{k=n}^{-1}(I-Q)X^{-1}(k+1)f(k), \quad n \leq 0. \end{cases} \tag{7.44}$$

In order that solution (7.44) be bounded on Z, it is necessary that the vector constant $\xi \in R^N$ satisfy the condition

$$P\xi - \sum_{k=0}^{+\infty}(I-P)X^{-1}(k+1)f(k) = (I-Q)\xi + \sum_{k=-\infty}^{-1} QX^{-1}(k+1)f(k).$$

In this case, for a constant $\xi \in R^N$, we get the following algebraic system:

$$[P-(I-Q)]\xi = \sum_{k=-\infty}^{-1} QX^{-1}(k+1)f(k) + \sum_{k=0}^{+\infty}(I-P)X^{-1}(k+1)f(k). \tag{7.45}$$

Let $D = P - (I-Q)$ be an $N \times N$ matrix and let D^+ be the $N \times N$ matrix pseudoinverse to D in the Moore–Penrose sense [38, 124]. By $P_{N(D)}$ and $P_{N(D^*)}$ we denote the $N \times N$ matrix orthoprojectors ($P_{N(D)}^2 = P_{N(D)} = P_{N(D)}^*$ and $P_{N(D^*)}^2 = P_{N(D^*)} = P_{N(D^*)}^*$) projecting R^N onto the kernel $\ker D = N(D)$ and cokernel $\ker D^* = N(D^*)$ of the matrix D (the symbol $*$ denotes the operation of transposition).

In order that system (7.43) have solutions bounded on Z, it is necessary that the algebraic system (7.45) be solvable for $\xi \in R^N$. At the same time, for the solvability of system (7.45), it is necessary and sufficient that its right-hand side belong to the orthogonal complement $N^{\perp}(D^*) = R(D)$ of the subspace $N(D^*)$. Thus, we can write

$$P_{N(D^*)}\left\{ \sum_{k=-\infty}^{-1} QX^{-1}(k+1)f(k) + \sum_{k=0}^{+\infty}(I-P)X^{-1}(k+1)f(k) \right\} = 0. \tag{7.46}$$

Therefore, the general solution of system (7.43) bounded on Z has the form (7.44), where the constant $\xi \in R^N$ is determined from system (7.45) as follows:

$$\xi = D^+\left\{ \sum_{k=-\infty}^{-1} QX^{-1}(k+1)f(k) + \sum_{k=0}^{+\infty}(I-P)X^{-1}(k+1)f(k) \right\} + P_{N(D)}c \qquad \forall c \in R^N. \tag{7.47}$$

In other words, the general solution of system (7.43) bounded on the entire real axis Z has the form

$$x(n,c) = X(n) \begin{cases} PP_{N(D)}c + \sum_{k=0}^{n-1} PX^{-1}(k+1)f(k) \\ \quad - \sum_{k=n}^{+\infty}(I-P)X^{-1}(k+1)f(k) \\ \quad + PD^{+}\Big\{ \sum_{k=-\infty}^{-1} QX^{-1}(k+1)f(k) \\ \qquad + \sum_{k=0}^{+\infty}(I-P)X^{-1}(k+1)f(k)\Big\}, \quad n \geq 0, \\ (I-Q)P_{N(D)}c + \sum_{k=-\infty}^{n-1} QX^{-1}(k+1)f(k) \\ \quad - \sum_{k=n}^{-1}(I-Q)X^{-1}(k+1)f(k) \\ \quad + (I-Q)D^{+}\Big\{ \sum_{k=-\infty}^{-1} QX^{-1}(k+1)f(k) \\ \qquad + \sum_{k=0}^{+\infty}(I-P)X^{-1}(k+1)f(k)\Big\}, \quad n \leq 0, \end{cases} \tag{7.48}$$

only in the case where condition (7.46) is satisfied.

By using the properties of the matrix D and the orthoprojectors onto the kernel and cokernel of this matrix, we get the following assertions:

Since $DP_{N(D)} = 0$ [38, p. 90], we have

$$PP_{N(D)} = (I-Q)P_{N(D)}.$$

Let

$$[PP_{N(D)}]_r = [(I-Q)P_{N(D)}]_r$$

be an $N \times r$ matrix whose columns form a complete set of r linearly independent columns of the matrix

$$PP_{N(D)} = (I-Q)P_{N(D)},$$

$r = \operatorname{rank}[PP_{N(D)}] = \operatorname{rank}[(I-Q)P_{N(D)}]$. Then

$$X_r(n) = X(n)[PP_{N(D)}]_r = X(n)[(I-Q)P_{N(D)}]_r$$

is an $N \times r$ matrix whose columns form a complete set of r linearly independent solutions of system (7.43) bounded on Z.

Since $P_{N(D^*)}D = 0$ [38, p. 90], we have $P_{N(D^*)}Q = P_{N(D^*)}(I-P)$. Therefore, condition (7.5) is equivalent to one of the following conditions:

$$\begin{aligned} P_{N(D^*)} \sum_{k=-\infty}^{+\infty} QX^{-1}(k+1)f(k) = 0, \\ P_{N(D^*)} \sum_{k=-\infty}^{+\infty} (I-P)X^{-1}(k+1)f(k) = 0. \end{aligned} \tag{7.49}$$

Let

$$d = \operatorname{rank}[P_{N(D^*)}(I-P)] = \operatorname{rank}[P_{N(D^*)}Q].$$

Then each condition in (7.49) contains only d linearly independent conditions. Indeed, let $[Q^*P_{N(D^*)}]_d$ be an $N \times d$ matrix formed by d linearly independent columns of the matrix $[Q^*P_{N(D^*)}]$. Note that $X^{*-1}(t)$ is the fundamental matrix of the system

$$x(n+1) = A^{*-1}(n)x(n), \tag{7.50}$$

conjugate to (7.42). It is easy to see that system (7.50) is dichotomous on Z_+ and Z_- with the projectors $I - P^*$ and $I - Q^*$, respectively. Then, as above,

$$H(n) = X^{*-1}(n)[Q^*P_{N(D^*)}] = X^{*-1}(n)[(I-P^*)P_{N(D^*)}]$$

is an $N \times N$ matrix formed by d solutions (columns) of system (7.50) bounded on Z. Hence,

$$H_d(n) = X^{*-1}(n)[Q^*P_{N(D^*)}]_d = X^{*-1}(n)[(I-P^*)P_{N(D^*)}]_d$$

is an $N \times d$ matrix whose columns form a complete set of d linearly independent solutions of system (7.50) conjugate to (7.42) bounded on Z and, therefore, $H_d^*(n)$ is a $d \times N$ matrix whose rows form a complete set of d linearly independent solutions of system (7.50) bounded on Z. This result can be formulated as follows:

Lemma 7.2. *Assume that system (7.42) is dichotomous on Z_+ and Z_- with projectors P and Q, respectively. Then the following assertions are true:*

(a) the operator

$$(Lx)(n) \overset{\text{def}}{=} x(n+1) - A(n)x(n)\colon B(Z) \to B(Z) \tag{7.51}$$

is a Fredholm operator and

$$\begin{aligned} \operatorname{ind} L &= \operatorname{rank}[PP_{N(D)}] - \operatorname{rank}[P_{N(D^*)}(I-P)] \\ &= \operatorname{rank}[(I-Q)P_{N(D)}] - \operatorname{rank}[P_{N(D^*)}Q] = r - d; \end{aligned}$$

(b) *the homogeneous system (7.42) possesses an r-parameter family of solutions bounded on Z:*

$$X_r(n)c_r = X(n)[PP_{N(D)}]_r c_r = X(n)[(I-Q)P_{N(D)}]_r c_r \quad \forall c_r \in R^r,$$
$$(r = \dim N(L) = \operatorname{rank}[PP_{N(D)}] = \operatorname{rank}[(I-Q)P_{N(D)}]);$$

(c) *system (7.50) conjugate to (7.42) possesses a d-parameter set of solutions bounded on Z:*

$$\begin{aligned} H_d(n)c_d &= X^{*-1}(n)[Q^* P_{N(D^*)}]_d c_d \\ &= X^{*-1}(n)[(I-P^*)P_{N(D^*)}]_d c_d \quad \forall c_d \in R^d, \end{aligned}$$
$$(d = \dim N(L^*) = \operatorname{rank}[P_{N(D^*)}(I-P)] = \operatorname{rank}[P_{N(D^*)}Q]);$$

(d) $f \in \operatorname{Im}(L)$ *only in the case where*

$$\sum_{k=-\infty}^{+\infty} H_d^*(k+1)f(k) = 0; \tag{7.52}$$

(e) *the inhomogeneous system (7.43) possesses an r-parameter family of solutions bounded on Z; the general solution of system (7.43) bounded on Z admits the representation*

$$x_0(n, c_r) = X_r(n)c_r + (G[f])(n) \quad \forall\, c_r \in R^r, \tag{7.53}$$

where

$$(G[f])(n) = X(n) \begin{cases} \displaystyle\sum_{k=0}^{n-1} PX^{-1}(k+1)f(k) \\ \displaystyle\quad - \sum_{k=n}^{+\infty}(I-P)X^{-1}(k+1)f(k) \\ \displaystyle\quad + PD^+\Big\{ \sum_{k=-\infty}^{-1} QX^{-1}(k+1)f(k) \\ \displaystyle\qquad + \sum_{k=0}^{+\infty}(I-P)X^{-1}(k+1)f(k)\Big\}, \quad n \ge 0, \\ \displaystyle\sum_{k=-\infty}^{n-1} QX^{-1}(k+1)f(k) \\ \displaystyle\quad - \sum_{k=n}^{-1}(I-Q)X^{-1}(k+1)f(k) \\ \displaystyle\quad + (I-Q)D^+\Big\{ \sum_{k=-\infty}^{-1} QX^{-1}(k+1)f(k) \\ \displaystyle\qquad + \sum_{k=0}^{+\infty}(I-P)X^{-1}(k+1)f(k)\Big\}, \quad n \le 0, \end{cases} \tag{7.54}$$

is the generalized Green operator of the problem of finding solutions of system (7.43) bounded on the entire real axis Z and such that

$$(LG[f])(n) = f(n), \quad n \in Z,$$

and

$$(G[f])(0+0) - (G[f])(0-0) = -\sum_{k=-\infty}^{+\infty} H^*(k+1)f(k).$$

Corollary 7.4. *Assume that the homogeneous system (7.42) is dichotomous on Z_+ and Z_- with projectors P and Q, respectively, such that $PQ = QP = Q$. In this case, system (7.42) is trichotomous [53, p. 363] on Z and the inhomogeneous system (7.43) possesses at least one solution bounded on Z for any $f \in BC(R)$ [53, p. 371].*

In other words, this is the so-called weakly regular case [108, p. 37]. In this case, Lemma 7.2 can be formulated as follows:

Assume that system (7.42) is dichotomous on Z_+ and Z_- with projectors P and Q, respectively, such that $PQ = QP = Q$. Then the following assertions are true:

(a) the operator L defined by relation (7.51) is a Fredholm operator and

$$\operatorname{ind} L = \dim N(L) = \operatorname{rank}[PP_{N(D)}] = \operatorname{rank}[(I-Q)P_{N(D)}] = r;$$

(b) the homogeneous system (7.42) possesses an r-parameter family of solutions bounded on Z:

$$X_r(n)c_r = X(n)[PP_{N(D)}]_r c_r = X(n)[(I-Q)P_{N(D)}]_r c_r \quad \forall c_r \in R^r$$

$$(r = \dim N(L) = \operatorname{rank}[PP_{N(D)}] = \operatorname{rank}[(I-Q)P_{N(D)}]);$$

(c) the only solution of system (7.50) (conjugate to (7.42)) bounded on Z is trivial;

(d) $f \in \operatorname{Im}(L)$ for all $f \in BC(R)$;

(e) the inhomogeneous system (7.43) possesses an r-parameter family of solutions bounded on Z and the general solution of system (7.43) bounded on Z can be represented in the form (7.53), where $(G[f])(n)$ is the generalized Green operator (7.54) of the problem of finding solutions of system (7.43) bounded on the entire real axis Z and such that

$$(LG[f])(n) = f(n), \quad n \in Z, \quad \text{and} \quad (G[f])(0+0) - (G[f])(0-0) = 0.$$

Proof. Indeed, since

$$P_{N(D^*)}D = 0 \qquad \text{and} \qquad DP = (P-(I-Q))P = QP = Q,$$

we have

$$P_{N(D^*)}Q = P_{N(D^*)}DP = 0.$$

Thus, the necessary and sufficient condition (7.52) for the existence of solutions of equation (7.43) bounded on Z is satisfied for all $f \in B(Z)$.

Corollary 7.5. *Assume that the homogeneous system (7.42) is dichotomous on Z_+ and Z_- with projectors P and Q, respectively, such that $PQ = QP = P$. In this case, system (7.50) is trichotomous on Z, and the inhomogeneous system (7.43) has only one solution bounded on Z but not for all $f \in B(Z)$.*

In this case, Lemma 7.2 can be formulated as follows:

Let system (7.42) be dichotomous on Z_+ and Z_- with projectors P and Q, respectively, such that $PQ = QP = P$. Then the following assertions are true:

(a) the operator L given by relation (7.51) is a Fredholm operator and

$$\operatorname{ind} L = -\dim N(L^*) = -\operatorname{rank}[P_{N(D^*)}(I-P)]$$

$$= -\operatorname{rank}[P_{N(D^*)}Q] = -d;$$

(b) the only solution of the homogeneous system (7.42) bounded on Z is trivial $(r = \dim N(L) = \operatorname{rank}[PP_{N(D)}] = \operatorname{rank}[(I-Q)P_{N(D)}] = 0)$;

(c) system (7.50) conjugate to (7.42) possesses a d-parameter set of solutions bounded on Z:

$$H_d(n)c_d = X^{*-1}(n)[Q^* P_{N(D^*)}]_d c_d$$

$$= X^{*-1}(n)[(I-P^*)P_{N(D^*)}]_d c_d, \quad \forall c_d \in R^d$$

$$(d = \dim N(L^*) = \operatorname{rank}[P_{N(D^*)}(I-P)] = \operatorname{rank}[P_{N(D^*)}Q]);$$

(d) $f \in \operatorname{Im}(L)$ only in the case where condition (7.52) is satisfied for $f \in B(Z)$;

(e) the inhomogeneous system (7.43) possesses a unique solution bounded on Z; this solution admits the representation $x_0(n) = (G[f])(n)$, where $(G[f])(n)$ is the generalized Green operator (7.54) of the problem of finding solutions of system (7.43) bounded on the entire real axis Z and such that

$$(LG[f])(n) = f(n), \quad n \in Z,$$

and

$$(G[f])(0+0) - (G[f])(0-0) = -\sum_{k=-\infty}^{\infty} H^*(k+1)f(k).$$

Proof. Indeed, since

$$DP_{N(D)} = 0 \quad \text{and} \quad PD = P(P-(I-Q)) = PQ = P,$$

we have

$$PP_{N(D)} = PDP_{N(D)} = 0.$$

Hence, $r = 0$, and the only solution of the homogeneous system (7.42) bounded on Z is trivial. The inhomogeneous system (7.43) possesses a unique solution bounded on Z.

Corollary 7.6. *Assume that the homogeneous system (7.42) is dichotomous on Z_+ and Z_- with projectors P and Q, respectively, such that $PQ = QP = P = Q$. In this case, system (7.42) is dichotomous on Z and the inhomogeneous system (7.43) has only one solution bounded on Z for any $f \in B(Z)$.*

In this case, Lemma 7.2 can be formulated as follows:

Let system (7.42) be dichotomous on Z_+ and Z_- with projectors P and Q, respectively, such that $PQ = QP = P = Q$. Then the following assertions are true:

(a) *the operator L specified by relation (7.51) is a Fredholm operator and* $\operatorname{ind} L = 0$;

(b) *the only solution of the homogeneous system (7.42) bounded on Z is trivial* $(r = \operatorname{rank}[PP_{N(D)}] = \operatorname{rank}[(I - Q)P_{N(D)}] = 0)$;

(c) *the only solution of system (7.50) (conjugate to (7.42)) bounded on Z is trivial* $(d = \operatorname{rank}[P_{N(D^*)}(I - P)] = \operatorname{rank}[P_{N(D^*)}Q] = 0)$;

(d) $f \in \operatorname{Im}(L)$ *for all* $f \in B(Z)$;

(e) *the inhomogeneous system (7.43) possesses a unique solution bounded on Z; this solution can be represented in the form $x_0(n) = (G[f])(n)$, where $(G[f])(n)$ is the generalized Green operator (7.54) $(D^+ = D^{-1})$ of the problem of finding solutions of system (7.43) bounded on the entire real axis Z and such that*

$$(LG[f])(n) = f(n), \quad n \in Z, \quad \text{and} \quad (G[f])(0+0) - (G[f])(0-0) = 0.$$

Nonlinear Systems. Consider a weakly nonlinear system

$$x(n+1) = A(n)x(n) + f(n) + \varepsilon Z(x, n, \varepsilon), \tag{7.55}$$

where ε is a small parameter.

Our aim is to establish conditions required for the existence of solutions $x = x(n, \varepsilon)$ of this system bounded on Z and such that

$$x(\cdot, \varepsilon)\colon Z \to R^N, \quad x(\cdot, \varepsilon) \in B(Z), \quad \text{and} \quad x(n, \cdot) \in C[0, \varepsilon_0].$$

These solutions must turn into one of the generating solutions $x_0(n, c_r)$ (7.53) of system (7.43) for $\varepsilon = 0$. The vector-valued function $Z(x, n, \varepsilon)$ is such that

$Z(\cdot, n, \varepsilon) \in C^1[\|x - x_0\| \le q]$, $Z(x, \cdot, \varepsilon) \in B(Z)$, and $Z(x, n, \cdot) \in C[0, \varepsilon_0]$, where q is a sufficiently small constant.

Theorem 7.6 (necessary condition). *Assume that system (7.42) is dichotomous on Z_+ and Z_- with projectors P and Q, respectively, and system (7.55) possesses a solution $x(n, \varepsilon)$ bounded on Z and such that $x(\cdot, \varepsilon)\colon Z \to R^N$, $x(\cdot, \varepsilon) \in B(Z)$, and $x(n, \cdot) \in C[0, \varepsilon_0]$, which turns into one of generating solutions $x_0(n, c_r)$ (7.53) of system (7.43) with a vector constant $c_r = c_r^0 \in R^r$ for $\varepsilon = 0$. Then the vector c_r^0 is a solution of the equation*

$$F(c_r^0) = \sum_{k=-\infty}^{\infty} H_d^*(k+1) Z(x_0(k, c_r^0), k, 0) = 0. \tag{7.56}$$

Proof. Condition (7.52) for the existence of generating solutions $x_0(n, c_r)$ (7.53) bounded on Z is assumed to be satisfied. We now treat the vector function $Z(x, n, \varepsilon)$ in system (7.55) as an inhomogeneity and apply Lemma 7.2 to this system. As a result, we obtain

$$\sum_{k=-\infty}^{\infty} H_d^*(k+1) Z(x(k, \varepsilon), k, \varepsilon) = 0.$$

Passing to the limit as $\varepsilon \to 0$ in this equality, we arrive at the required condition (7.56).

By analogy with the case of periodic problems [38, p. 184], it is natural to say that (7.56) is the equation for generating constants of the problem of finding solutions of system (7.55) bounded on the entire real axis Z. If equation (7.55) is solvable, then the vector constant $c_r^0 \in R^r$ specifies the generating solution $x_0(n, c_r^0)$ corresponding to the solution $x = x(n, \varepsilon)$ of the original problems (7.55) bounded on Z and such that $x(\cdot, \varepsilon)\colon Z \to R^N$, $x(\cdot, \varepsilon) \in B(Z)$, $x(n, \cdot) \in C[0, \varepsilon_0]$, and $x(n, 0) = x_0(n, c_r^0)$. At the same time, if equation (7.56) is unsolvable, then problem (7.55) does not have solutions bounded on Z from the analyzed space. Recall that we are interested in real solutions of equation (7.56), which can be algebraic or transcendental.

By the change of variables in system (7.55)

$$x(n, \varepsilon) = x_0(n, c_r^0) + y(n, \varepsilon),$$

we arrive at the problem of finding sufficient conditions for the existence of solutions $y = y(n, \varepsilon)$ of the problem

$$y(n+1) = A(n) y(n) + \varepsilon Z(x_0(n, c_r^0) + y, n, \varepsilon). \tag{7.57}$$

bounded on Z and such that $y(\cdot, \varepsilon)\colon Z \to R^N$, $y(\cdot, \varepsilon) \in B(Z)$, $y(n, \cdot) \in C[0, \varepsilon_0]$, and $y(n, 0) = 0$. In view of the fact that the vector-valued function $Z(x, n, \varepsilon)$ is continuously differentiable with respect to x and continuous in ε in the neighborhood of the point $x_0(n, c_r^0)$, $\varepsilon = 0$, we can separate the term linear as a function of y and terms of order zero with respect to ε:

$$Z(x_0(n, c_r^0) + y, n, \varepsilon) = f_0(n, c_r^0) + A_1(n)y + R(y(n, \varepsilon), n, \varepsilon), \tag{7.58}$$

where

$$f_0(n, c_r^0) = Z(x_0(n, c_r^0), n, 0), \quad f_0(\cdot, c_r^0) \in B(Z),$$
$$A_1(n) = A_1(n, c_r^0) = \frac{\partial Z(x, n, 0)}{\partial x}\Big|_{x=x_0(n,c_r^0)}, \quad A_1(\cdot) \in B(Z),$$
$$R(0, n, 0) = 0, \frac{\partial R(0, n, 0)}{\partial y} = 0, \quad R(y, \cdot, \varepsilon) \in B(Z).$$

We now formally treat the vector function $Z(x_0 + y, n, \varepsilon)$ in system (7.57) as an inhomogeneity and apply Lemma 7.2 to this system. As a result, we obtain the following representation of the solution of system (7.57) bounded on Z:

$$y(n, \varepsilon) = X_r(n)c + y^{(1)}(n, \varepsilon).$$

In this expression, the unknown constant vector $c = c(\varepsilon) \in R^r$ is determined from a condition similar to condition (7.52) for the existence of the required solution of system (7.57):

$$B_0 c = - \sum_{k=-\infty}^{+\infty} H_d^*(k+1)[A_1(k)y^{(1)}(k, \varepsilon) + R(y(k, \varepsilon), k, \varepsilon)], \tag{7.59}$$

where

$$B_0 = \sum_{k=-\infty}^{+\infty} H_d^*(k+1)A_1(k)X_r(k)$$

is a $d \times r$ matrix,

$$r = \text{rank}\,[PP_{N(D)}] = \text{rank}\,[(I - Q)P_{N(D)}],$$

and

$$d = \text{rank}\,[P_{N(D^*)}(I - P)] = \text{rank}\,[P_{N(D^*)}Q].$$

The unknown vector function $y^{(1)}(n, \varepsilon)$ is determined with the help of the generalized Green operator (7.54) by the following formula:

$$y^{(1)}(n, \varepsilon) = \varepsilon \left(G\left[Z(x_0(\cdot, c_r^0) + y(\cdot, \varepsilon), \cdot, \varepsilon)\right]\right)(n).$$

Let $P_{N(B_0)}$ be an $r \times r$ matrix orthoprojector $R^r \to N(B_0)$ and let $P_{N(B_0^*)}$ be a $d \times d$ matrix orthoprojector $R^d \to N(B_0^*)$. Equation (7.59) is solvable with respect to $c \in R^r$ if and only if

$$P_{N(B_0^*)} \sum_{k=-\infty}^{+\infty} H_d^*(k+1)[A_1(k)y^{(1)}(k, \varepsilon) + R(y(k, \varepsilon), k, \varepsilon)] = 0. \tag{7.60}$$

If

$$P_{N(B_0^*)} = 0,$$

then condition (7.60) is always satisfied and equation (7.59) is solvable with respect to $c \in R^r$ to within an arbitrary vector constant $P_{N(B_0)}c$, $c \in R^r$, from the null space of the matrix B_0:

$$c = -B_0^+ \sum_{k=-\infty}^{+\infty} H_d^*(k+1)[A_1(k)y^{(1)}(k,\varepsilon) + R(y(k,\varepsilon),k,\varepsilon)] + P_{N(B_0)}c.$$

To find one of solutions $y = y(n,\varepsilon)$ of problem (7.57) bounded on Z and such that

$$y(\cdot,\varepsilon)\colon Z \to R^N, \quad y(\cdot,\varepsilon) \in B(Z), \quad y(n,\cdot) \in C[0,\varepsilon_0], \quad \text{and} \quad y(n,0) = 0,$$

we get the following operator system:

$$y(n,\varepsilon) = X_r(n)c + y^{(1)}(n,\varepsilon), \tag{7.61}$$

$$c = -B_0^+ \sum_{k=-\infty}^{+\infty} H_d^*(k+1)[A_1(k)y^{(1)}(k,\varepsilon) + R(y(k,\varepsilon),k,\varepsilon)],$$

$$y^{(1)}(n,\varepsilon) = \varepsilon\left(G\left[Z(x_0(\cdot,c_r^0) + y,\cdot,\varepsilon)\right]\right)(n).$$

The operator system (7.61) belongs to the class of systems [38, p. 188] solvable by the method of simple iterations. Indeed, system (7.61) can be rewritten as

$$z = L^{(1)}z + Fz, \tag{7.62}$$

where $z = \text{col}\,(y^{(1)}(n,\varepsilon), c(\varepsilon), y(n,\varepsilon))$ is a $(2N+r)$-dimensional column vector, $L^{(1)}$ and F are, respectively, linear and nonlinear operators bounded on Z,

$$L^{(1)} = \begin{pmatrix} 0 & 0 & 0 \\ L_1 & 0 & 0 \\ I_N & X_r & 0 \end{pmatrix}, \quad L_1* = -B_0^+ \sum_{k=-\infty}^{+\infty} H_d^*(k+1)A_1(k)*,$$

and

$$Fz = \text{col}\left[\varepsilon G\left[Z(x_0(\cdot,c_r^0) + y(\cdot,\varepsilon),\cdot,\varepsilon)\right], \sum_{k=-\infty}^{+\infty} H_d^*(k+1)R(y(k,\varepsilon),k,\varepsilon), 0\right].$$

In view of the structure of the operator $L^{(1)}$ with zero blocks on the principal diagonal and above, the operator $(I_s - L^{(1)})^{-1}$ exists. System (7.62) can be transformed as follows:

$$z = Sz \quad (S = (I_s - L^{(1)})^{-1}, \;\; s = 2n + r), \tag{7.63}$$

where S is a contraction operator in a sufficiently small neighborhood of the point $x_0(n,c_r^0)$, $\varepsilon = 0$. Thus, to prove that the operator system (7.63) is solvable, one can use one of the existing versions of the fixed-point principle [87] applicable for sufficiently small $\varepsilon \in [0,\varepsilon_*]$. Using the method of simple iterations to find a solution of the operator system (7.63) and, hence, to find solutions of the original system (7.55) bounded on Z, we arrive at the following statement [31]:

Theorem 7.7 (sufficient condition). *Assume that the weakly nonlinear system (7.55) satisfies the conditions imposed above (and, thus, the homogeneous system (7.42) is dichotomous on the semiaxes Z_+ and Z_-) and the corresponding inhomogeneous linear system (7.43) possesses the r-parameter family of generating solutions $x_0(n, c_r)$ (7.53) bounded on Z. Also let*

$$P_{N(B_0^*)} = 0. \tag{7.64}$$

Then, for any vector $c_r = c_r^0 \in R^r$ satisfying the equation for generating constants (7.56), there exists at least one solution of system (7.55) bounded on Z. For $\varepsilon = 0$, this solution $x(n,\varepsilon)\colon x(n,\cdot) \in C[0,\varepsilon_0]$ turns into the generating solution $x(n,0) = x_0(n, c_r^0)$ (7.53). Moreover, it can be found by the method simple iterations convergent for $\varepsilon \in [0,\varepsilon_*] \subseteq [0,\varepsilon_0]$:

$$y_{s+1}^{(1)}(n,\varepsilon) = \varepsilon \left(G\left[Z(x_0(\cdot, c_r^0) + y_s, \cdot, \varepsilon)\right]\right)(n),$$

$$c_s = -B_0^+ \sum_{k=-\infty}^{+\infty} H_d^*(k+1)[A_1(k) y_s^{(1)}(k,\varepsilon) + R(y_s(k,\varepsilon), k, \varepsilon)],$$

$$y_{s+1}(n,\varepsilon) = X_r(n) c_s + y_{s+1}^{(1)}(n,\varepsilon),$$

$$x_s(n,\varepsilon) = x_0(n, c_r^0) + y_s(n,\varepsilon),$$

$$s = 0, 1, 2, \ldots,$$

$$y_0(n,\varepsilon) = y_0^{(1)}(n,\varepsilon) = 0,$$

$$x(n,\varepsilon) = \lim_{s\to\infty} x_s(n,\varepsilon).$$

In the case where the number $r = \operatorname{rank}[PP_{N(D)} = (I-Q)P_{N(D)}]$ of linear independent solutions of system (7.42) bounded on Z is equal to the number $d = \operatorname{rank}[P_{N(D^*)}(I-P) = P_{N(D^*)}Q]$ of linear independent bounded (on R) solutions of system (7.50) conjugate to (7.42), the condition $P_{N(B_0^*)} = 0$ implies that $P_{N(B_0)} = 0$ and, hence, $\det B_0 \neq 0$. In this case, Theorem 7.7 yields the following statement [22]:

Theorem 7.8 (sufficient condition). *Assume that the weakly nonlinear system (7.55) satisfies the conditions imposed above (and, thus, the homogeneous system (7.42) is dichotomous on the semiaxes Z_+ and Z_-) and the corresponding generating linear system (7.43) possesses an r-parameter family of generating solutions $x_0(n, c_r)$ (7.53) bounded on Z. Also let*

$$\det B_0 \neq 0 \quad (r = d). \tag{7.65}$$

Then, for any vector $c_r = c_r^0 \in R^r$ satisfying the equation for generating constants (7.56), there exists a unique solution of system (7.55) bounded on Z. For $\varepsilon = 0$, this solution $x(n,\varepsilon)$, $x(n,\cdot) \in C[0,\varepsilon_0]$, turns into the generating solution $x(n,0) = x_0(n,c_r^0)$ (7.53) and can be found by the method of simple iterations convergent for $\varepsilon \in [0,\varepsilon_*] \subseteq [0,\varepsilon_0]$:

$$x(n,\varepsilon) = \lim_{k\to\infty} x_k(n,\varepsilon).$$

Conclusions. The required estimates for the parameter ε_* and the error of approximation of the iterative process can be established in a standard way [87].

Condition (7.65) means that the constant $c_r^0 \in R^r$ is a simple root of equation (7.56) for generating constants of the problem of finding solutions of system (7.55) bounded on the entire real axis Z [38]. By using the procedure proposed in [38, p. 193] with some simplifying assumptions, we can generalize the proposed method to the case of multiple roots of equation (7.56).

Similar problems were studied for ordinary differential equations by Palmer [115, p. 248], Elaidy and Hajek [53], Gruendler [67], and Samoilenko, A. Boichuk, and An. Boichuk [136, 137]. For difference equations with $\operatorname{ind} L = 0$, these problems were studied by A. Boichuk [22]. It should be emphasized that the method used in the present book differs from the methods used in [53, 67, 115]. Actually, our method is similar to the method used in the theory of bounded solutions of ordinary differential equations. It was proposed in [21, 23].

Examples. **1.** Consider a system

$$x(n+1) = A(n)x(n) + f(n) + \varepsilon A_1(n)x(n), \tag{7.66}$$

where

$$A(n) = \operatorname{diag}\{2^{-\operatorname{sign}(n)}, 2^{-\operatorname{sign}(n)}, 2^{\operatorname{sign}(n)}\},$$

$$A_1(t) = \{a_{ij}(n)\}_{i,j=1}^{3} \in B(Z), \quad n \in Z.$$

It is easy to see that

$$X(n) = \begin{cases} \operatorname{diag}\{2^{-(n-1)}, 2^{-(n-1)}, 2^{n-1}\}, & n > 0, \\ I_3, & n = 0, \\ \operatorname{diag}\{2^{n},\ 2^{n},\ 2^{-n}\}, & n < 0. \end{cases}$$

The unperturbed homogeneous system $x(n+1) = A(n)x(n)$ is dichotomous on the semiaxes Z_+ and Z_- with the following projectors:

$$P = \operatorname{diag}\{1,1,0\} \quad \text{and} \quad Q = \operatorname{diag}\{0,0,1\}.$$

Then

$$D = 0, \quad D^+ = 0, \quad P_{N(D)} = P_{N(D^*)} = I_3,$$

$$r = \operatorname{rank} P P_{N(D)} = 2, \quad d = \operatorname{rank} P_{N(D^*)} Q = 1,$$

$$X_r(n) = X(n)[P P_{N(D)}]_{r=2} = \begin{pmatrix} \beta(n) & 0 \\ 0 & \beta(n) \\ 0 & 0 \end{pmatrix},$$

$$H_d(n) = X^{*-1}(n)[Q^* P_{N(D^*)}]_{d=1} = \begin{pmatrix} 0 \\ 0 \\ \beta(n) \end{pmatrix},$$

where

$$\beta(n) = \begin{cases} 2^{-(n-1)}, & n > 0, \\ 1, & n = 0, \\ 2^n, & n < 0. \end{cases}$$

The inhomogeneous system

$$x(n+1) = A(n)x(n) + f(n)$$

possesses the two-parameter family of solutions

$$x_0(n, c_r) = X_r(n)c_r + (G[f])(n), \quad c_r \in R^2,$$

bounded on R only in the case where the inhomogeneity

$$f(n) = \operatorname{col}\{f_1(n), f_2(n), f_3(n)\} \in B(Z)$$

satisfies following condition:

$$\sum_{k=-\infty}^{-1} f_3(k)2^{k+1} + f_3(0) + \sum_{k=1}^{+\infty} f_3(k)2^{-k} = 0$$

$$\forall f_1(n) \in B(Z) \quad \text{and} \quad \forall f_2(n) \in B(Z).$$

According to Theorems 7.6 and 7.7, the following assertion is true for system (7.66):

Let $P_{N(B_0^*)} = 0$. Then, for any vector $c_r = c_r^0 \in R^2$ satisfying the equation for generating constants (7.56), namely,

$$B_0 c_r^0 = - \sum_{k=-\infty}^{+\infty} H_d^*(k+1) A_1(k)(Gf)(k),$$

there exists a one-parameter set of solutions of system (7.66) bounded on Z ($\rho = \operatorname{rank} P_{N(B_0)} = r - \operatorname{rank} B_0 = r - d = 1$). For $\varepsilon = 0$, these solutions $x(n,\varepsilon)$: $x(n,\cdot) \in C[0,\varepsilon_0]$ turn into the generating solutions $x(n,0) = x_0(n, c_r^0)$, where

$$
\begin{aligned}
B_0 &= \sum_{k=-\infty}^{+\infty} H_d^*(k+1)A_1(k)X_r(k) \\
&= \Bigg[\sum_{k=-\infty}^{-1} a_{31}(k)2^{2k+1} + a_{31}(0) + \sum_{k=1}^{+\infty} a_{31}(k)2^{-2k+1}, \\
&\qquad \sum_{k=-\infty}^{-1} a_{32}(k)2^{2k+1} + a_{32}(0) + \sum_{k=1}^{+\infty} a_{32}(k)2^{-2k+1} \Bigg].
\end{aligned}
$$

If $a_{31}(n)$ or $a_{32}(n) \in B(Z)$ satisfies one of the conditions

$$
\sum_{k=-\infty}^{-1} a_{31}(k)2^{2k+1} + a_{31}(0) + \sum_{k=1}^{+\infty} a_{31}(k)2^{-2k+1} \neq 0
$$

and

$$
\sum_{k=-\infty}^{-1} a_{32}(k)2^{2k+1} + a_{32}(0) + \sum_{k=1}^{+\infty} a_{32}(k)2^{-2k+1} \neq 0,
$$

then $\operatorname{rank} B_0 = d = 1$. Thus, if $a_{31}(n) = \text{const} \neq 0$ or $a_{32}(n) = \text{const} \neq 0$, then one of these inequalities is true and, therefore, condition (7.17) is satisfied. In this case, the coefficients $a_{11}(n)$, $a_{12}(n)$, $a_{13}(n)$, $a_{21}(n)$, $a_{22}(n)$, $a_{23}(n)$, and $a_{33}(n)$ are arbitrary functions from the space $B(Z)$.

2. We now consider system (7.66) with

$$
A(n) = \operatorname{diag}\{2^{-\operatorname{sign}(n)}, 2^{\operatorname{sign}(n)}\} \quad \text{and} \quad A_1(n) = \{a_{ij}(n)\}_{i,j=1}^{2} \in B(Z).
$$

It is easy to see that

$$
X(n) = \begin{cases} \operatorname{diag}\{2^{-(n-1)}, 2^{n-1}\}, & n > 0, \\ I_2, & n = 0, \\ \operatorname{diag}\{2^n,\ 2^{-n}\}, & n < 0, \end{cases}
$$

and the homogeneous system $x(n+1) = A(n)x(n)$ is dichotomous on the semiaxes Z_+ and Z_- with the projectors $P = \operatorname{diag}\{1, 0\}$ and $Q = \operatorname{diag}\{0, 1\}$, respectively. Then

$$
D = 0, \quad D^+ = 0, \quad P_{N(D)} = P_{N(D^*)} = I_2,
$$

$$
r = \operatorname{rank} P P_{N(D)} = 1, \quad d = \operatorname{rank} P_{N(D^*)} Q = 1,
$$

$$X_r(n) = X(n)[PP_{N(D)}]_{r=1} = \begin{pmatrix} \beta(n) \\ 0 \end{pmatrix},$$

$$H_d(n) = X^{*-1}(n)[Q^* P_{N(D^*)}]_{d=1} = \begin{pmatrix} 0 \\ \beta(n) \end{pmatrix}.$$

The inhomogeneous system

$$x(n+1) = A(n)x(n) + f(n)$$

possesses the one-parameter family of solutions

$$x_0(n, c_r) = X_r(n)c_r + (G[f])(n), \quad c_r \in R,$$

bounded on Z only in the case where $f(n) = \operatorname{col}\{f_1(n), f_2(n)\} \in B(Z)$ satisfies the condition

$$\sum_{k=-\infty}^{-1} f_2(k)2^{k+1} + f_2(0) + \sum_{k=1}^{+\infty} f_2(k)2^{-k} = 0 \quad \forall f_1(n) \in B(Z).$$

According to Theorem 7.8, in this case, we get the following statement valid for system (7.66):

Assume that condition (7.18) is satisfied, i.e.,

$$\begin{aligned} B_0 &= \sum_{k=-\infty}^{+\infty} H_d^*(k+1)A_1(k)X_r(k) \\ &= \sum_{k=-\infty}^{-1} a_{21}(k)2^{2k+1} + a_{21}(0) + \sum_{k=1}^{+\infty} a_{21}(k)2^{-2k+1} \neq 0 \quad (r = d = 1). \end{aligned}$$

Then, for any constant $c_r = c_r^0 \in R$ *satisfying the equation for generating constants (7.56), namely,*

$$B_0 c_r^0 = -\sum_{k=-\infty}^{+\infty} H_d^*(k+1)A_1(k)(Gf)(k),$$

there exists a unique solution of system (7.66) bounded on Z. *For* $\varepsilon = 0$, *this solution* $x(k,\varepsilon)$, $x(k,\cdot) \in C[0,\varepsilon_0]$, *turns into the generating solution* $x(k,0) = x_0(k, c_r^0)$.

Thus, for $a_{21}(k) = \text{const} \neq 0$, the last inequality is true and condition (7.18) is satisfied. In this case, the coefficients $a_{11}(k)$, $a_{12}(k)$, and $a_{22}(k)$ are arbitrary functions from the space $B(Z)$.

Epilogue

In the development of constructive methods for the analysis of different classes of differential-operator equations, we tried, on the one hand, to get as general results as possible but, on the other hand, to explicitly specify the classes of systems for which the approaches proposed in the monograph are applicable. Both the first and second tasks were not realized completely. This seems to be quite natural because, in the process of accumulation of knowledge, the image of strictly marked boundaries permanently varies, enlarges, and escapes from the researcher. For a long period of time, we believed that the elimination of the Fredholm property (index zero) of the linear part of the analyzed operator leads to a class of problems so broad that no informative theory can be constructed for it [8, p. 29]. The theory of linear and nonlinear boundary-value problems for operator equations with a general Fredholm operator in the linear part presented in the monograph is also not the maximum possible generalization. The proposed approach can be used in the corresponding spaces (according to S. Krein classification [88]) and for boundary-value problems with normally solvable (reducibly invertible, d- and n-normal) operators. These problems have already been generalized in [159, 161, 163–165, 167–171, 173, 174, 176, 177, 189] and used in [162, 166, 171, 172, 182, 187], which enables us at least to keep the indicated image in the field of vision if it is impossible to reach it.

Bibliography

[1] N. I. Akhiezer and I. M. Glazman, *Theory of Operators in Hilbert Spaces*, Part 1, Vyshcha Shkola, Kharkov (1977) *(in Russian)*.

[2] A. A. Andronov, A. A. Vitt, and S. E. Khaikin, *Oscillation Theory*, Fizmatgiz, Moscow (1959) *(in Russian)*.

[3] A. V. Anokhin, "Functional differential equations with impulsive effect," *Dokl. Akad. Nauk SSSR*, **286**, No. 5, 1037–1040 (1988),

[4] A. B. Antonevich and Ya. V. Radyno, *Functional Analysis and Integral Equations*, Minsk University, Minsk (1984) *(in Russian)*.

[5] V. I. Arnol'd, *Mathematical Methods of Classical Mechanics*, Nauka, Moscow (1989) *(in Russian)*.

[6] F. V. Atkinson, "Normal solvability of linear equations in normed spaces," *Mat. Sb., New Series*, **28**, No. 1, 3–14 (1951).

[7] F. V. Atkinson, *Discrete and Continuous Boundary Problems*, Mir, Moscow (1968) *(Russian translation)*.

[8] N. V. Azbelev, V, V. Maksimov, and I. F. Rakhmatullina, *An Introduction to the Theory of Functional Differential Equations*, Nauka, Moscow (1991) *(in Russian)*.

[9] I. Babuška, M. Práger, and E. Vitásek, *Numerical Processes in Differential Equations*, Mir, Moscow (1969) *(Russian translation)*.

[10] D. Bainov and P. Simeonov, *Impulsive Differential Equations: Periodic Solutions and Applications*, Longman, Harlow (1993).

[11] Yu. M. Berezanskii, G. F. Us, and Z. G. Sheftel', *Functional Analysis*, Naukova Dumka, Kiev (1990) *(in Russian)*.

[12] C. D. Birkhoff, "Boundary value and expansion problems of ordinary linear differential systems," *Trans. Amer. Math. Soc.*, No. 9, 373–395 (1908).

[13] G. A. Bliss, "A boundary-value problem for a system of ordinary linear differential equations of the first order," *Trans. Amer. Math. Soc.*, No. 28, 561–584 (1926).

[14] N. N. Bogolyubov and Yu. A. Mitropol'skii, *Asymptotic Methods in the Theory of Nonlinear Oscillations*, Nauka, Moscow (1974) *(in Russian)*.

[15] N. N. Bogolyubov, Yu. A. Mitropol'skii, and A. M. Samoilenko, *Method of Accelerated Convergence in Nonlinear Mechanics*, Naukova Dumka, Kiev (1969) *(in Russian)*.

[16] A. A. Boichuk, "Construction of periodic solutions to nonlinear systems at resonance by using the method of simple iteration," in: *Abstracts of the 10th International Conference on Nonlinear Oscillations (Varna, September 12–17, 1984)*, Sofia (1984), pp. 36–37.

[17] A. A. Boichuk, *Boundary-Value Problems for Weakly Perturbed Systems in Critical Cases*, Preprint No. 39, Institute of Mathematics, Ukrainian Academy of Sciences, Kiev (1988) *(in Russian)*.

[18] A. A. Boichuk, "Construction of solutions of two-point boundary-value problems for weakly perturbed nonlinear systems in critical cases," *Ukr. Mat. Zh.*, **41**, No. 10, 1416–1420 (1989).

[19] A. A. Boichuk, *Constructive Methods for the Analysis of Boundary-Value Problems*, Naukova Dumka, Kiev (1990) *(in Russian)*.

[20] A. A. Boichuk, "Boundary-value problems for impulse differential systems," *Universitatis Iagellonicae Acta Mathematica*, **XXXVI**, 187–191 (1998).

[21] A. A. Boichuk, "Solutions of weakly nonlinear differential equations bounded on the whole line," *Nonlin. Oscillations*, **2**, No. 1, 3–10 (1999).

[22] A. A. Boichuk, "Solutions of linear and nonlinear difference equations bounded on the whole line," *Nonlin. Oscillations*, **4**, No. 1, 16–27 (2001).

[23] A. A. Boichuk, "Dichotomy, trichotomy, and solutions of nonlinear systems bounded on R," in: *Applications of Mathematics in Engineering'26 (Sozopol, Bulgaria, June 13–20, 2000)*, Heron, Sofia (2001), pp. 9–15.

[24] A. A. Boichuk, "A condition for the existence of a unique Green–Samoilenko function for the problem of invariant torus," *Ukr. Mat. Zh.*, **53**, No. 4, 556–559 (2001).

[25] A. A. Boichuk and S. M. Chuiko, *Autonomous Boundary-Value Problems in Critical Cases, Part 1. Autonomous Periodic Boundary-Value Problems in Critical Cases*, Preprint, Institute of Geophysics, Ukrainian Academy of Sciences, Kiev (1991); *Part 2*, Preprint, Institute of Geophysics, Ukrainian Academy of Sciences, Kiev (1992) *(in Russian)*.

[26] A. A. Boichuk and S. M. Chuiko, "Autonomous weakly nonlinear boundary-value problems," *Differents. Uravn.*, **28**, No. 10, 1668–1674 (1992).

[27] A. A. Boichuk and M. K. Grammatikopoulos, "Perturbed Fredholm BVP's for delay differential systems," *Technical Report, University of Ioannina, Department of Mathematics, June 9*, 1–19 (2003).

[28] A. A. Boichuk and M. K. Grammatikopoulos, "Perturbed Fredholm boundary-value problems for delay differential systems," *Abstract Appl. Analysis*, No. 15, 843–864 (2003).

[29] A. A. Boichuk and R. F. Khrashchevskaya, "Weakly nonlinear boundary-value problems for differential systems with impulsive effect," *Ukr. Mat. Zh.*, **45**, No. 2, 221–225 (1993).

[30] A. A. Boichuk, N. A. Perestyuk, and A. M. Samoilenko, "Periodic solutions of impulsive differential systems in critical cases," *Differents. Uravn.*, **27**, No. 9, 1516–1521 (1991).

[31] A. A. Boichuk and M. Ružičková, "Solutions of nonlinear difference equations bounded on the whole line," in: *Colloquium on Differential and Difference Equations, Brno, 2002. Folia FSN Universitatis Masarykianae Brunensis, Mathematica 13*, pp. 45–60.

[32] A. A. Boichuk and V. F. Zhuravlev, "Construction of solutions of boundary-value problems for delay differential systems in critical cases," *Dokl. Akad. Nauk Ukr. SSR, Ser. A*, No. 6, 3–6 (1990).

[33] A. A. Boichuk and V. F. Zhuravlev, "Construction of solutions of linear Noether operator equations in Hilbert spaces," *Dokl. Akad. Nauk Ukr. SSR, Ser. A*, No. 8, 3–6 (1990).

[34] A. A. Boichuk and V. F. Zhuravlev, "Generalized inverse operator for a Noether operator in a Banach space," *Dokl. Akad. Nauk Ukr. SSR, Ser. A*, No. 1, 5–8 (1991).

[35] A. A. Boichuk and V. F. Zhuravlev, "Construction of solutions of linear Noether operator equations in Banach spaces," *Ukr. Mat. Zh.*, **43**, No. 10, 1343–1350 (1991).

[36] A. A. Boichuk, V. F. Zhuravlev, and S. M. Chuiko, "Periodic solutions of nonlinear autonomous systems in critical cases," *Ukr. Mat. Zh.*, **42**, No. 9, 1180–1187 (1990).

[37] A. A. Boichuk, V. F. Zhuravlev, and A. M. Samoilenko, "Linear Noether boundary-value problems for impulsive delay differential systems," *Differents. Uravn.*, **30**, No. 10, 1677–1682 (1994).

[38] A. A. Boichuk, V. F. Zhuravlev, and A. M. Samoilenko, *Generalized Inverse Operators and Noether Boundary-Value Problems*, Institute of Mathematics, Ukrainian Academy of Sciences, Kiev (1995) *(in Russian)*.

[39] A. O. Boichuk, "A set of bounded solutions of a linear weakly perturbed system," *Nonlin. Oscillations*, **6**, No. 3, 309–318 (2003).

[40] Yn. F. Boyarintsev, *Methods for Solving Singular Systems of Ordinary Differential Equations*, Nauka, Novosibirsk (1988) *(in Russian)*.

[41] I. S. Bradley, "Generalized Green's matrices for compatible differential systems," *Mich. Math. J.*, **13**, No. 1, 97–108 (1966).

[42] E. L. Bunitsky, "Sur la fonction de Green des équations différentielles linéaires ordinaire," *J. Math. Pur. Appl., Ser. 6*, No. 5, 65–125 (1909).

[43] E. L. Bunitsky, *On the Theory of Green Function for Ordinary Differential Equations*, Sapozhnikov, Odessa (1913) *(in Russian)*.

[44] A. A. Bykov, "A stable numerical method for solving boundary-value problems for systems of ordinary differential equations," *Dokl. Akad. Nauk SSSR*, **251**, No. 5, 1040–1044 (1980).

[45] L. Cesari, *Functional Analysis and Periodic Solutions of Nonlinear Differential Equations, Contributions to Differential Equations*, Vol. 1, Interscience, New York (1962).

[46] S. N. Chow and J. K. Hale, *Methods of Bifurcation Theory*, Springer, New York–Berlin (1982).

[47] S. N. Chow, J. K. Hale, and J. Mallet-Paret, "An example of bifurcation to homoclinic orbits," *J. Different. Equat.*, **37**, 351–373 (1980).

[48] R. Conti, "Recent trends in the theory of boundary-value problems for ordinary differential equations," *Boll. Unione Math. ItaL*, **22**, No. 2, 35–178 (1967).

[49] W. A. Coppel, *Dichotomies in Stability Theory*, Springer, Berlin (1978).

[50] Yu. L. Daletskii and M. G. Krein, *Stability of Solutions of Differential Equations in Banach Spaces*, Nauka, Moscow (1970) *(in Russian)*.

[51] B. P. Demidovich, *Lectures on Mathematical Theory of Stability*, Nauka, Moscow (1967) *(in Russian)*.

[52] N. Dunford and J. T. Schwartz, *Linear operators. General Theory*, Izd. Inostr. Liter., Moscow (1962) *(Russian translation)*.

[53] S. Elaidy and O. Hajek, "Exponential trichotomy of differential systems," *J. Math. Anal. Appl.*, **123**, No. 2, 362–374 (1988).

[54] W. W. Elliot, "Generalized Green's functions for compatible differential systems," *Amer. J. Math.*, No. 50, 243–258 (1928).

[55] L. E. Él'sgol'ts and S. B. Norkin, *An Introduction to the Theory of Equations with Deviating Argument*, Nauka, Moscow (1971) *(in Russian)*.

[56] Fam Ki Ahn, "Approximate solution of nonlinear multipoint boundary-value problems in the case of resonance," *Ukr. Mat. Zh.*, **39**, No. 5, 619–621 (1987).

[57] E. I. Fredholm, "Sur une classe d'equations fonctionnelles," *Acta Math.*, **27**, 265–390 (1903).

[58] M. Furi and J. Mawhin, "Periodic solutions of some nonlinear differential equations of higher order," *Gas. Pest. Mat.*, **100**, 276–283 (1975).

[59] H. Gajewski, K. Gröger, and K. Zacharias, *Nichtlineare Operatorgleichungen und Operatordifferentialgleichungen*, Mir, Moscow (1978) *(Russian translation)*.

[60] F. R. Gantmakher, *Matrix Theory*, Nauka, Moscow (1988) *(in Russian)*.

[61] A. O. Gel'fond, *Calculus of Finite Differences*, Nauka, Moscow (1967) *(in Russian)*.

[62] S. K. Godunov, "On numerical solution of boundary-value problems for systems of ordinary differential equations," *Usp. Mat. Nauk*, **16**, No. 3, 171–174 (1961).

[63] I. Ts. Gokhberg and N. Ya. Krupnik, *Introduction to the Theory of One-Dimensional Singular Integral Operators*, Shtiintsa, Kishinev (1973) *(in Russian)*.

[64] V. V. Gorodetskii, N. I. Nagnibida, and P. N. Nastasiyev, *Methods for Solving Problems in Functional Analysis*, Vyshcha Shkola, Kiev (1990) *(in Russian)*.

[65] E. A. Grebenikov and Yu. A. Ryabov, *Constructive Methods for Analysis of Nonlinear Systems*, Nauka, Moscow (1979) *(in Russian)*.

[66] J. Gruendler, "The existence of homoclinic orbits and the method of Melnikov for systems in R^n," *SIAM J. Math. Anal.*, **16**, 907–931 (1985).

[67] J. Gruendler, "Homoclinic solutions for autonomous ordinary differential equations with nonautonomous perturbations," *J. Different. Equat.*, **122**, 1–26 (1995).

[68] J. Guckenheimer and P. Holmes, *Nonlinear Oscillations, Dynamical Systems and Bifurcations of Vector Fields.* Springer, New York (1983).

[69] A. Halanay and D. Wexler, *Qualitative Theory of Impulsive Systems*, Mir, Moscow (1971) *(Russian translation)*.

[70] J. K. Hale, *Oscillations in Nonlinear Systems*, Mir, Moscow (1966) *(Russian translation)*.

[71] J. K. Hale, *Theory of Functional Differential Equations*, Springer, New York (1975).

[72] P. R. Halmos, *Measure Theory*, van Nostrand, New York (1950).

[73] W. Hatson and J. Tim, *Application of Functional Analysis and Operator Theory*, Mir, Moscow (1983) *(Russian translation)*.

[74] C. Hayashi, *Nonlinear Oscillations in Physical Systems*, Mir, Moscow (1986) *(Russian translation)*.

[75] D. Henry, *Geometric Theory of Semilinear Parabolic Equations*, Springer, Berlin (1981).

[76] P. Holmes, "A nonlinear oscillator with a strange attractor." *Philos. Trans. Roy. Soc. London, Ser. A.*, **292**, 419–448.

[77] L. V. Kantorovich, "The principle of Lyapunov majorants and Newton method," *Dokl. Akad. Nauk SSSR*, **76**, No. 1, 17–20 (1951).

[78] L. V. Kantorovich and G. P. Akilov, *Functional Analysis*, Nauka, Moscow (1984) *(in Russian)*.

[79] L. V. Kantorovich, B. Z. Vulikh, and A. G. Pinsker, *Functional Analysis in Partially Ordered Spaces*, Gostekhizdat, Moscow–Leningrad (1950) *(in Russian)*.

[80] T. Kato, *Perturbation Theory for Linear Operators*, Springer, New York (1966).

[81] H. Kauderer, *Nichtlineare Mechanik*, Izd. Inostr. Liter., Moscow (1961) *(Russian translation)*.

[82] I. T. Kiguradze, "Boundary-value problems for systems of ordinary differential equations," in: *VINITI Series in Mathematics. Contemporary Results*,

Vol. 30, VINITI, Moscow (1987), pp. 3–103 *(in Russian)*. ***English translation:*** *J. Sov. Math.*, **43**, No. 2, 2259–2339 (1988).

[83] A. A. Kirillov and A. D. Gvishiani, *Theorems and Problems in Functional Analysis*, Nauka, Moscow (1988) *(in Russian)*.

[84] A. N. Kolmogorov and S. V. Fomin, *Elements of the Theory of Functions and Functional Analysis*, Nauka, Moscow (1968) *(in Russian)*.

[85] V. S. Korolyuk and A. F. Turbin, *Mathematical Foundations of Phase Lumping of Complex Systems*, Naukova Dumka, Kiev (1978) *(in Russian)*.

[86] M. A. Krasnosel'skii, V. Sh. Burd, and Yu. S. Kolesov, *Nonlinear Almost Periodic Oscillations*, Nauka, Moscow (1970) *(in Russian)*.

[87] M A. Krasnosel'skii, G. M. Vainikko, P. P. Zabreiko, Ya. B. Rutitskii, and V. Ya. Stetsenko, *Approximate Solution of Operator Equations*, Nauka, Moscow (1968) *(in Russian)*; ***English translation:*** Noordhoff, Groningen (1972).

[88] S. G. Krein, *Linear Equations in Banach Spaces*, Nauka, Moscow (1971) *(in Russian)*.

[89] A. N. Krylov, "Vibration of vessels," in: *Collected Works*, Vol. X, Academy of Sciences of the USSR, Moscow–Leningrad *(in Russian)*.

[90] N. M. Krylov and N. N. Bogolyubov, *Introduction to Nonlinear Mechanics*, Ukrainian Academy of Sciences, Kiev (1937) *(in Russian)*.

[91] V. I. Kublanovskaya, "On computation of the generalized inverse matrix and projectors," *Zh. Vychisl. Mat. Mat. Fiz.*, **6**, No. 2, 326–332 (1966).

[92] Yu. K. Lando, "On solvability of an integro-differential equation," *Differents. Uravn.*, **3**, No. 4, 695–697 (1967).

[93] D. K. Lika and Yu. A. Ryabov, *Iteration Methods and Majorizing Lyapunov Equations in the Theory of Nonlinear Oscillations*, Shtiintsa, Kishinev (1974) *(in Russian)*.

[94] J.-L. Lions, *Quelques Méthodes de Résolution des Problèmes aux Limites Nonlinéaires*, Mir, Moscow (1972) *(Russian translation)*.

[95] W. S. Loud, "Generalized inverses and generalized Green's functions," *SIAM J. Appl. Math.*, **14**, No. 2, 269–342 (1966).

[96] A. Yu. Luchka, *Projection Iterative Methods*, Naukova Dumka, Kiev (1993) *(in Russian)*.

[97] A. M. Lyapunov, *General Problem on Stability of Motion*, Gostekhizdat, Moscow (1950) *(in Russian)*.

[98] O. B. Lykova and A. A. Boichuk, "Periodic solutions and dichotomy for weakly perturbed linear systems," *Usp. Mat. Nauk*, **42**, No. 4, 135 (1987) *(in Russian)*.

[99] O. B. Lykova and A. A. Boichuk, "Construction of periodic solutions of nonlinear systems in critical cases," *Ukr. Mat. Zh.*, **40**, No. 1, 62–69 (1988) *(in Russian)*.

[100] V. P. Maksimov, "Noether property of a general boundary-value problem for linear functional differential equation," *Differents. Uravn.*, **10**, No. 12, 2288–2291 (1974) *(in Russian).*

[101] I. G. Malkin, *Some Problems in the Theory of Nonlinear Oscillations*, Gostekhizdat, Moscow (1956) *(in Russian).*

[102] D. I. Martynyuk, *Lectures on the Qualitative Theory of Difference Equations*, Naukova Dumka, Kiev (1972) *(in Russian).*

[103] J. Mary, *Nonlinear Differential Equations in Biology, Lectures on Models*, Mir, Moscow (1983) *(Russian translation).*

[104] J. Mawhin, *Topological Degree Methods in Nonlinear Boundary Value Problems*, American Mathematical Society, Providence, R.I. (1979).

[105] V. K. Melnikov, "On the stability of the center for time-periodic perturbations," *Trans. Moscow Math. Soc.*, **12**, 1–56 (1964).

[106] E. F. Mishchenko and N. Kh. Rozov, *Differential Equations with Small Parameter and Relaxation Oscillations*, Nauka, Moscow (1975) *(in Russian).*

[107] Yu. A. Mitropol'skii and D. I. Martynyuk, *Periodic and Quasiperiodic Oscillations in Delay Systems*, Vyshcha Shkola, Kiev (1979) *(in Russian).*

[108] Yu. A. Mitropol'skii, A. M. Samoilenko, and V. L. Kulik, *Investigation of Dichotomy for Linear Systems of Differential Equations Using Lyapunov Functions*, Naukova Dumka, Kiev (1990) *(in Russian).*

[109] E. H. Moore, "On reciprocal of the general algebraic matrix," *Abstract. Bull. Amer. Math. Soc.*, No. 26, 394–395 (1920).

[110] A. D. Myshkis, *Linear Delay Differential Equations*, Gostekhizdat, Moscow (1951) *(in Russian).*

[111] A. D. Myshkis and A. M. Samoilenko, "Systems with shocks at given moments of time," *Mat. Sb.*, **74**, No. 2, 202–208 (1967) *(in Russian).*

[112] M. Z. Nashed (ed.), *Generalized Inverses and Applications*, Academic Press, New York–San Francisco–London (1976).

[113] S. M. Nikol'skii, "Linear equations in linear normed spaces," *Izv. Akad. Nauk SSSR*, **7**, No. 3, 147–163.

[114] F. Noether, "Über eine Klasse singulärer Integralgleichungen," *Math. Ann.*, No. 82, 42–64 (1921).

[115] K. J. Palmer, "Exponential dichotomies and transversal homoclinic points," *J. Different. Equat.*, **55**, 225–256 (1984).

[116] K. J. Palmer, "Exponential dichotomies and Fredholm operators," *Proc. Amer. Math. Soc.*, **104**, 149–156 (1988).

[117] R. Penrose, "A generalized inverse for matrices," *Proc. Cambridge Philos. Soc.*, **51**, No. 3, 406–413 (1955).

[118] V. A. Pliss, *Nonlocal Problems in Oscillation Theory*, Nauka, Moscow (1964) *(in Russian).*

[119] V. A. Pliss, "Bounded solutions of inhomogeneous linear systems of differential equations," in: *Problems of the Asymptotic Theory of Nonlinear Oscillations*, Naukova Dumka, Kiev (1977), pp. 168–173 *(in Russian)*.

[120] H. Poincaré, *Sur les Courbes Définies par les Équations Différentielles*, OGIZ, Moscow–Leningrad (1947) *(Russian translation)*.

[121] B. van der Pol, *Nonlinear Theory of Electrical Oscillations*, Svyaz', Moscow (1935) *(Russian translation)*.

[122] A. P. Proskuryakov, *Poincaré Method in the Theory of Nonlinear Oscillations*, Nauka, Moscow (1977) *(in Russian)*.

[123] Yu. P. Pyt'yev, "Pseudoinverse operator. Properties and application," *Mat. Sb., New Series*, **118**, No. 1, 19–49 (1982).

[124] C. R. Rao and S. K. Mitra, *Generalized Inverse of Matrices and Its Applications*, Wiley, New York (1971).

[125] M. Reed and B. Simon, *Methods of Modern Mathematical Physics. 1. Functional Analysis*, Mir, Moscow (1977) *(Russian translation)*.

[126] W. T. Reid, "Generalized Green's matrices for compatible systems of differential equations," *Amer. J. Math.*, No. 53, 443–459 (1931).

[127] W. T. Reid, "Generalized Green's matrices for compatible systems for two-point boundary problems," *SIAM J. Appl. Math.*, **15**, No. 4, 856–870 (1967).

[128] W. T. Reid, *Riccati Differential Equations*, New York–London (1972).

[129] F. Riesz and B. Sz.-Nagy, *Leçons d'Analyse Fonctionnelle*, Mir, Moscow (1979) *(Russian translation)*.

[130] W. Rudin, *Functional Analysis*, Mir, Moscow (1975) *(Russian translation)*.

[131] R. J. Sacker, "Existence of dichotomies and invariant splittings for linear differential systems. IV," *J. Different. Equat.*, **27**, 106–137, (1978).

[132] R. J. Sacker, "The splitting index for linear differential systems," *J. Different. Equat.*, **33**, 368–405 (1979).

[133] R. J. Sacker and G. R. Sell, "Existence of dichotomies and invariant splittings for linear differential systems. I," *J. Different. Equat.*, **15**, 429–458 (1974).

[134] A. A. Samarskii and Yu. N. Karamzin, *Difference Equations*, Nauka, Moscow (1978) *(in Russian)*.

[135] A. M. Samoilenko and A. A. Boichuk, "Linear Noether boundary-value problems for differential systems with impulsive effect," *Ukr. Mat. Zh.*, **44**, No. 4, 564–568 (1992).

[136] A. M. Samoilenko, A. A. Boichuk, and An. A. Boichuk, "Solutions of weakly perturbed linear systems bounded on the entire axis," *Ukr. Mat. Zh.*, **54**, No. 11, 1517–1530 (2002).

[137] A. M. Samoilenko, A. A. Boichuk, and An. A. Boichuk, "Pseudoinverse matrices and solutions of linear and nonlinear systems bounded on *R*," in: *Proceedings of the 5th International Workshop on Computer Algebra in Scientific*

Computing (September 22–27, 2002, Yalta, Ukraine), Institut für Informatik, Technische Universität München, Germany (2002), pp. 269–278.

[138] A. M. Samoilenko, A. A. Boichuk, and S. A. Krivosheya, "Boundary-value problem for linear systems of integro-differential equations with degenerate kernel," *Ukr. Mat. Zh.*, **48**, No. 11, 1576–1579 (1996).

[139] A. M. Samoilenko and N. A. Perestyuk, *Impulsive Differential Equations*, Vyshcha Shkola, Kiev (1987) *(in Russian)*.

[140] A. M. Samoilenko and N. I. Ronto, *Numerical-Analytic Methods for the Investigation of Solutions of Boundary-Value Problems*, Naukova Dumka, Kiev (1986) *(in Russian)*.

[141] E. Schmidt, "Zur Theorie der linearen und nichtlinearen Integralgleichungen. Teil 3. Über die Auflösungen der nichtlinearen Integralgleichungen und die Verzweigung ihre Lösungen," *Math. Ann.*, No. 65 (1908).

[142] Š. Schwabik, M. Tvrdy, and O. Vejvoda, *Differential and Integral Equations, Boundary-Value Problems and Adjoints*, Academia, Prague (1979).

[143] S. N. Shimanov, "On the oscillation theory for quasilinear delay systems," *Prikl. Mat. Mekh.*, **23**, No. 5, 836–844 (1959) *(in Russian)*.

[144] A. N. Tikhonov, "On systems of differential equations with parameters." *Mat. Sb.*, No. 27, 147–156 (1950) *(in Russian)*.

[145] A. N. Tikhonov and V. Ya. Arsenin, *Methods for Solving Ill-Posed Problems*, Nauka, Moscow (1986) *(in Russian)*.

[146] V. A. Trenogin, *Functional Analysis*, Nauka, Moscow (1980) *(in Russian)*.

[147] A. F. Turbin, "Formulas for computing semiinverse and pseudoinverse matrices," *Zh. Vychisl. Mat. Mat. Fiz.*, **14**, No. 3, 772–776 (1974).

[148] M. M. Vainberg and V. A. Trenogin, *Theory of Branching of Solutions of Nonlinear Equations*, Nauka, Moscow (1969) *(in Russian)*.

[149] O. Vejvoda, "On perturbed nonlinear boundary-value problems," *Czech. Math. J.*, **11**, 323–364 (1961).

[150] M. I. Vishik and I. A. Lyusternik, "Solution of some perturbation problems for matrices, self-adjoint and nonself-adjoint differential equations," *Usp. Mat. Nauk*, **15**, No. 3, 3–80 (1960).

[151] V. V. Voevodin and Yu. A. Kuznetsov, *Matrices and Computations*, Nauka, Moscow (1984) *(in Russian)*.

[152] V. Volterra, *Una Teoria Matematica sulla Lotta per l'Esistenza*, Nauka, Moscow (1976) *(Russian translation)*.

[153] D. Wexler, "On boundary-value problems for ordinary linear differential systems," *Ann. Mat. Pura Appl.*, **80**, 123–136 (1968).

[154] V. A. Yakubovich and V. M. Starzhinskii, *Linear Differential Equations with Periodic Coefficients*, Nauka, Moscow (1972) *(in Russian)*.

[155] S. T. Zavalishin, A. N. Sesekin, and S. E. Drozdenko, *Dynamical Systems with Impulsive Structure*, Sredne-Ural'skoe Knizhnoe Izd., Sverdlovsk (1983) *(in Russian)*.

[156] W. Y. Zeng, "Exponential dichotomies of linear systems with small parameter," *Ann. Diff. Equat.*, **11**, No. 2, 249–253 (1995).

[157] V. F. Zhuravlev, *Noether Boundary-Value Problems for Delay Differential Systems*, Author's Abstract of the Candidate-Degree (Physics and Mathematics), Kiev (1991) *(in Russian)*.

[158] V. I. Zubov, "On the theory of the existence of solutions of boundary-value problems for systems of differential equations," *Differents. Uravn.*, **7**, No. 4, 632–633 (1970) *(in Russian)*.

Literature added to the second edition

[159] R. Agarwal, M. Bohner, A. Boichuk, and O. Strakh, "Fredholm boundary value problems for perturbed systems of dynamic equations on time scales," *Math. Methods Appl. Sci.*, (2014); DOI: 10.1002/mma.3356.

[160] A. Ben-Israel and T. N. E. Greville, *Generalized Inverses: Theory and Applications (2nd edition)*, Springer, New-York, 2003.

[161] B. A. Biletskyi, A. A. Boichuk, and A. A. Pokutnyi, "Periodic problems of difference equations and Ergodic Theory," *Abstr. Appl. Anal.*, **2011** (2011); Article ID 928587.

[162] A. M. Blokhin and A. S. Bushmanova, "Investigation of the stability of the equilibrium state for a gas dynamic model of charge transport in semiconductors," *Sib. Zh. Ind. Mat.*, **1**, No. 1, 41–56 (1998) *(in Russian)*.

[163] A. Boichuk, "Bounded solutions of differential equations in Banach space," in: *Colloquium on Differential and Difference Equations dedicated to Professor Jaroslav Kurzweil on the occasion of his 80-th birthday (Brno, Czech Republic, September 5–8, 2006)*, Abstracts, p. 35.

[164] A. A. Boichuk, J. Diblík, D. Ya. Khusainov, and M Růžičková, "Fredholm's boundary-value problems for differential systems with a single delay," *Nonlinear Anal.*, **72**, 2251–2258 (2010).

[165] O. A. Boichuk and I. A. Holovats'ka, "Boundary-Value Problems for Systems of Integrodifferential Equations," *Nelin. Kolyvannya*, **16**, No. 4, 460–474 (2013) *(in Ukrainian)*; ***English translation:*** *J. Math. Sci.*, **203**, No. 3, 306–321 (2014).

[166] A. A. Boichuk, I. A. Korostil, and M. Fečkan, "Conditions for the bifurcation of the solution of an abstract wave equation," *Differ. Uravn.*, **43**, No. 4, 481–487, 573 (2007) *(in Russian)*; ***English translation:*** *Differ. Equ.*, **43**, No. 4, 495–502 (2007).

[167] A. Boichuk, M. Langerová, M. Růžičková, and E. Voitushenko, "Systems of Singular Differential Equations with Pulse Action," *Adv. Difference Equ.*, **2013**, 2013:186, 2–8 (2013).

[168] A. Boichuk, M. Langerová, and J. Škoríková, "Solutions of linear impulsive differential equations bounded on the whole line," *Adv. Difference Equ.*, **2010**:494379 (2010); DOI: 10.1155/2010/494379.

[169] A. Boichuk, M. Medved', and V. Zhuravliov, "Fredholm boundary-value problems for linear delay systems defined by pairwise permutable matrices," *Electron. J. Qual. Theory Differ. Equ.*, No. 23, 1–9 (2015).

[170] A. A. Boichuk and A. A. Pokutnyi, "Bounded solutions of linear perturbed differential equations in a Banach space," *Tatra Mt. Math. Publ.*, **38**, No. 1, 29–40 (2007).

[171] A. Boichuk and O. Pokutnyi, "Solutions of the Schrödinger equation in a Hilbert space," *Bound. Value Probl.*, **2014**:4, 9 pp.

[172] A. A. Boichuk, A. A. Pokutnyi, and V. F. Chistyakov, "Application of perturbation theory to the solvability analysis of differential algebraic equations," *Comput. Math. Math. Phys.*, **53**, No. 6, 777–788 (2013).

[173] A. Boichuk and M. Růžičková, "Solutions bounded on the whole line for perturbed difference systems," *Proceedings of the 8th Conference on Difference Equations and Applications*, Chapmen & Hall/CR – Taylor & Francis Group, 2005, p. 51–60.

[174] A. A. Boichuk and A. M. Samoilenko, *Generalized inverse operators and Fredholm boundary-value problems*, VSP, Utrecht, 2004.

[175] O. A. Boichuk and L. M. Shehda, "Degenerate Fredholm boundary-value problems," *Nelin. Kolyvannya*, **10**, No. 3, 303–312 (2007); ***English translation:*** *Nonlinear Oscil.*, **10**, No. 3, 306–314 (2007).

[176] O. A. Boichuk and L. M. Shehda, "Degenerate nonlinear boundary-value problems," *Ukrain. Mat. Zh.*, **61**, No. 9, 1174–1188 (2009) *(in Ukrainian)*; ***English translation:*** *Ukrainian Math. J.*, **61**, No. 9, 1387–1403 (2009).

[177] A. A. Boichuk and L. M. Shegda, "Bifurcation of solutions of singular Fredholm boundary value problems," *Differ. Equ.*, **47**(4), 453–461 (2011).

[178] A. A. Boichuk, V. F. Zhuravlev, and A. A. Pokutnyi, "Normally solvable operator equations in a Banach space," *Ukrain. Mat. Zh.*, **65**, No. 2, 163–174 (2013) *(in Russian)*; ***English translation:*** *Ukrainian Math. J.*, **65**, No. 2, 179–192 (2013).

[179] Yu. E. Boyarintsev, V. A. Danilov, A. A. Loginov, and V. F. Chistyakov, *Numerical methods for solving singular systems*, Nauka, Novosibirsk, 1989 *(in Russian)*; ISBN: 5-02-028735-0.

[180] S. L. Campbell and L. R. Petzold, "Canonical forms and solvable singular systems of differential equations," *SIAM J. Alg. Discr. Meth.*, No. 4, 517–521 (1983).

[181] V. F. Chistyakov and A. A. Shcheglova, *Selected Chapters of the Theory of Algebraic-Differential Systems*, Nauka, Novosibirsk, 2003 *(in Russian)*.

[182] P. Drabek and M. Langerová, "First order periodic problem at resonance with nonlinear impulses," *Bound. Value Probl.*, **2014**:186, 9 pp.

[183] M. Fečkan, *Topological Degree Approach to Bifurcation Problems*, Springer, 2008.

[184] R. A. Horn and C. R. Johnson, *Matrix Analysis*, Cambridge University Press, Cambridge, 1985.

[185] D. Ya. Khusainov and G. V. Shuklin, "Relative controllability in systems with pure delay," *Prikl. Mekh.*, **41**, No. 2, 118–130 (2005); ***English translation:*** *Internat. Appl. Mech.*, **41**, No. 2, 210–221 (2005).

[186] J. Mallet-Paret, "The Fredholm alternative for functional-differential equations of mixed type," *J. Dynam. Differential Equations*, **11**, No. 1, 1–47 (1999).

[187] M. Medved' and M. Pospíšil, "Sufficient conditions for the asymptotic stability of nonlinear multidelay differential equations with linear parts defined by pairwise permutable matrices," *Nonlinear Anal.*, **75**, 3348–3363 (2012).

[188] J. R. Morris and P. J. Rabier, "Riccati inequality and functional properties of differential operators on the half line," *J. Differential Equations*, **225**, No. 2, 573–604 (2006).

[189] W. C. Rheinboldt, "Differential-algebraic systems as differential equations on manifolds," *Math. Comp.*, **43**, No. 168, 473–482 (1984).

[190] A. M. Samoilenko, A. A. Boichuk, and V. F. Zhuravlev, "Linear boundary value problems for normally solvable operator equations in a Banach space," *Differ. Uravn.*, **50**, No. 3, 317–326 (2014) *(in Russian)*; ***English translation:*** *Differ. Equ.*, **50**, No. 3, 312–322 (2014).

[191] A. M. Samoilenko, M. I. Shkil', and V. P. Yakovets', *Linear systems of differential equations with singularities*, Vyshcha Shkola, Kyiv. 2000 *(in Ukrainian)*.

[192] A. M. Samoilenko and V. P. Yakovets', "On the reducibility of a degenerate linear system to the central canonical form," *Dokl. Akad. Nauk Ukr.*, No. 4, 10–15 (1993).

[193] V. S. Vladimirov, *Equations of Mathematical Physics*, Nauka, Moscow, 1988 *(in Russian)*.

[194] V. P. Yakovets', "On some properties of degenerate linear systems," *Ukr. Mat. Zh.*, **49**, No. 9, 1278–1296 (1997); ***English translation:*** *Ukr. Math. J.*, **49**, No. 9, 1442–1463 (1997).

[195] A. Zettl, "Adjoint and self-adjoint BVP's with interface conditions," *SIAM J. Appl. Math.*, **16**, No. 4, 851–859 (1968); DOI: 10.1137/0116069.

Authors

Aleksandr Boichuk, Corresponding Member of the Ukrainian National Academy of Sciences, Prof. Dr., Winner of the State Prize of Ukraine in the Field of Science and Engineering, Winner of the Mitropol'skii Prize of the Ukrainian National Academy of Sciences. He was born in 1950 in Kirovograd (Ukraine). Graduated from the Department of Mechanics and Mathematics of the Shevchenko Kyiv National University. Head of the Laboratory of Boundary-Value Problems at the Institute of Mathematics and Professor of the Chair of Integral and Differential Equations at the Shevchenko Kyiv National University. The author of four monographs and more than 150 papers in the field of the theory of resonance boundary-value problems for a broad class of systems of ordinary differential and difference equations, equations with delayed argument and impulsive action, integrodifferential and singularly perturbed equations, including boundary-value problems with conditions imposed at infinity. The method of generalized inverse operators was proposed and extensively used for the investigation of these problems.

Anatoly Samoilenko, Academician of the Ukrainian National Academy of Sciences, Full Member of the European Academy Academy of Sciences Prof. Dr., Winner of the State Prize of Ukraine in the Field of Sience and Engineering, Winner of the State Prize of Ukraine in the Field of Education, Winner of the Ostrogradskii and Bogolyubov Prizes of the Ukraininan National Academy of Sciences. He was born in 1938 in Ukraine. Graduated from the Department of Mechanics and Mathematics of the Shevchenko Kyiv National University. Director of the Institute of Mathematics of the Ukrainian National Academy of Sciences. The author of more than 30 monographs and handbooks and numerous papers in the fields of multifrequency oscillations, theory of impulsive systems, and theory of boundary-value problems well-known both in Ukraine and abroad. These works promote the development of Ukrainian mathematics and develop the traditions of the famous Kiev Scientific School of Krylov, Bogolyubov, and Mitropol'skii.